A. V. Stepanov, V. V. Zaitsev,
and V. M. Nakariakov

Coronal Seismology

Related Titles

Stahler, S. W., Palla, F.

The Formation of Stars

2004
Softcover
ISBN: 978-3-527-40559-6

Rüdiger, G., Hollerbach, R.

The Magnetic Universe

Geophysical and Astrophysical Dynamo Theory

2004
Hardcover
ISBN: 978-3-527-40409-4

Foukal, P. V.

Solar Astrophysics

2004
Hardcover
ISBN: 978-3-527-40374-5

Spitzer, L.

Physical Processes in the Interstellar Medium

1998
Softcover
ISBN: 978-0-471-29335-4

A. V. Stepanov, V. V. Zaitsev, and V. M. Nakariakov

Coronal Seismology

Waves and Oscillations in Stellar Coronae

WILEY-VCH Verlag GmbH & Co. KGaA

The Authors

Prof. Alexander V. Stepanov
Russian Academy of Sciences
Pulkovo Observatory St. Petersburg
Russia Federation
stepanov@gao.spb.ru

Prof. Valery V. Zaitsev
Russian Academy of Sciences
Institute of Applied Physics
Nizhny Novgorod
Russia Federation

Prof. Valery M. Nakariakov
University of Warwick
Department of Physics
Coventry, United Kingdom

Cover picture
© NASA
The Transition Region and Coronal Explorer, TRACE, is a mission of the Stanford-Lockheed Institute for Space Research, and part of the NASA Small Explorer program.

Library of Congress Card No.: applied for

British Library Cataloguing-in-Publication Data
A catalogue record for this book is available from the British Library.

Bibliographic information published by the Deutsche Nationalbibliothek
The Deutsche Nationalbibliothek lists this publication in the Deutsche Nationalbibliografie; detailed bibliographic data are available on the Internet at <http://dnb.d-nb.de>.

Typesetting Laserwords Private Limited, Chennai, India
Printing and Binding Markono Print Media Pte Ltd, Singapore
Cover Design Grafik-Design Schulz, Fußgönheim

Printed in Singapore
Printed on acid-free paper

Print ISBN: 978-3-527-40994-5
ePDF ISBN: 978-3-527-64601-2
oBook ISBN: 978-3-527-64598-5
ePub ISBN: 978-3-527-64600-5
mobi ISBN: 978-3-527-64599-2

Contents

Preface

Wave and oscillatory phenomena, which are intrinsically inherent in the activity of solar and stellar coronae, present the subject of *coronal seismology* – a new rapidly developing branch of astrophysics. The philosophy of this novel method of remote diagnostics of plasma parameters is analogous to the study of the Earth's interior by surface and body waves, *geoseismology*. In solar and stellar physics, a similar technique for probing of the interiors, *helio-* and *asteroseismology*, is successfully applied. Moreover, the observational detection of waves and oscillations in accretion disks recently gave rise to the topic of *diskoseismology*. Likewise, magnetoseismology, the diagnostics of the Earth's (and, potentially, planetary) magnetospheres with magnetohydrodynamic waves and oscillations, is another successful manifestation of this technique.

Hundreds of papers and reviews as well as national and international meetings are devoted now to the current problems in the field of the coronal seismology. Flares, charged particle acceleration and emission, plasma heating, prominence dynamics, and coronal mass ejections are related to the magnetic structures of active regions of the stars. Numerous ground- and space-borne observations provide evidence that magnetic flux tubes and loops, being a typical structure of solar and stellar coronae, act as waveguides and resonators for magnetohydrodynamical (MHD) waves and oscillations. Moreover, modern view suggests that coronae of accretion disks and neutron stars also consist of magnetic loops. Waves and oscillations in open and closed magnetic structures modulate solar and stellar emission in various wave bands. Hence, coronal seismology can provide a good diagnostic tool for flare plasma as well as for energy transfer and energy transformation in solar and stellar atmospheres. More opportunities in the coronal seismology were made more than a decade ago with the use of space missions: the Solar and Heliospheric Observatory and Transition Region and Coronal Explorer. New good chances for coronal plasma diagnostics open now with the Solar Dynamic Observatory and Solar Terrestrial Relations Observatory.

This book is devoted to the successive exposition of the main problems of the coronal seismology. Two approaches are mainly used for the interpretation of the wave and oscillatory phenomena in solar and stellar coronae. The first approach represents the coronal magnetic loops and flux tubes as resonators and waveguides for MHD oscillations and waves. The second one describes the coronal loop in terms of an equivalent electric circuit with the effective resistance, inductance, and capacitance. Both these approaches complement one another effectively in the process of diagnostics of coronal plasma. Because waves and oscillations cause quasi-periodic modulations of plasma parameters of coronal magnetic structures, the emission mechanisms from coronal structures of both thermal and nonthermal origins should be studied in this context. Most pronounced phase of quasi-periodic pulsations is in the course of flaring event. Hence, the flare plasma heating and charged particle acceleration are also the subjects of this book.

It should be noted that the author's views are biased by their own preferences and interests. However, they tried to address wider problems of the coronal seismology. Sometimes the main formulas and notations are repeated in various chapters for the reader's convenience.

December 2011

A. V. Stepanov
V. V. Zaitsev
V. M. Nakariakov

1
Introduction

Uchida [1], who suggested the idea of plasma and magnetic field diagnostics in the solar corona on the basis of waves and oscillations in 1970, and Rosenberg [2], who first explained the pulsations in type IV solar radio bursts in terms of the loop magnetohydrodynamic (MHD) oscillations, can be considered to be pioneers of coronal seismology.

Various approaches have been used to describe physical processes in stellar coronal structures: kinetic, MHD, and electric circuit models are among them. Two main models are presently very popular in coronal seismology. The first considers magnetic flux tubes and loops as wave guides and resonators for MHD waves and oscillations, whereas the second describes a loop in terms of an equivalent electric (RLC) circuit. Several detailed reviews are devoted to problems of coronal seismology (see, i.e., [3–7]). Recent achievements in the solar coronal seismology are also referred in *Space Sci. Rev.* vol. 149, No. 1–4 (2009). Nevertheless, some important issues related to diagnostics of physical processes and plasma parameters in solar and stellar flares are still insufficiently presented in the literature. The main goal of the book is the successive description and analysis of the main achievements and problems of coronal seismology.

There is much in common between flares on the Sun and on late-type stars, especially red dwarfs [8]. Indeed, the timescales, the Neupert effect, the fine structure of the optical, radio, and X-ray emission, and the pulsations are similar for both solar and stellar flares. Studies of many hundreds of stellar flares have indicated that the latter display a power–law radiation energy distribution, similar to that found for solar flares. Thus, we can use the solar–stellar analogy to study flaring stars.

The next goal of this book is to illustrate the efficiency of coronal seismology as a diagnostic tool for the analysis of stellar flares.

1.1
Magnetic Loops and Open Flux Tubes as Basic Structural Elements in Solar and Stellar Coronae

Magnetic loops constitute the basic structural element in the coronae of the Sun and late-type stars [9–11]. They play an important role in solar activity. Observations

Coronal Seismology: Waves and Oscillations in Stellar Coronae, First Edition.
A. V. Stepanov, V. V. Zaitsev, and V. M. Nakariakov.

made with *Skylab*, Solar and Heliospheric Observatory (SOHO), *Yohkoh*, Reuven Ramaty High-Energy Solar Spectroscopic Imager (RHESSI), Transition Region and Coronal Explorer (TRACE), Complex Orbital Near-Earth Observations of Activity of the Sun – Photon (CORONAS–F), *Hinode*, Solar Dynamic Observatory (SDO) space missions, as well as with large optical telescopes Vacuum Tower Telescope (VTT), and radio telescopes (Very Large Array (VLA), Siberian Solar Radio Telescope (SSRT), NoRH (Nobeyama Radioheliograph)) have shown that solar flares originate in coronal loops [3, 12]. Eruptive prominences and coronal transients result in giant coronal mass ejections (CMEs) and also frequently display the loop shape [13]. The flaring activity of dMe-stars and close binaries is also spawned by the energy release in magnetic loops [9, 14, 15]. In some late-type stars, magnetic spots cover up to 70–80% of the surface, whereas solar spots occupy $\sim$0.04% of the photosphere. This implies that, in fact, loops form the magnetic structure of stellar coronae. In addition, loops are typical for the magnetic structure of the atmospheres of accretion disks, young stellar objects, and neutron stars [16–19]. Owing to space-borne observations and advances in the physics of the loops, some progress has been recently made in finding a solution for the problem of the origin of coronal loops. Alfvén and Carlqvist [20] have suggested that a flaring loop can be considered to be an equivalent electric circuit. This phenomenological approach was nevertheless very productive in understanding the energy pattern of flare processes. The description of the loops in terms of resonators and wave guides for MHD waves explains various kinds of modulations of stellar flare emissions and serves as a diagnostic tool for the flare plasma. The concept of a coronal loop as a magnetic mirror trap for energetic particles makes it possible to efficiently describe particle dynamics and peculiarities of emission generated by energetic particles.

Open magnetic structures – the flux tubes – are wave guides for MHD waves, which make them important channels of energy transfer from one part of the stellar atmosphere to another, from the photosphere and chromosphere to the corona, and further to the solar and stellar wind. Similar to magnetic loops, the flux tubes provide a necessary link in the mechanism of coronal heating. Flux tubes are exemplified by solar spicules, which are energy/mass bridges between the dense and dynamic photosphere and the tenuous hot solar corona [21].

Prominence dynamics and oscillations also present important subjects for coronal seismology [22, 23] since prominences play a crucial role both in triggering flares [24, 25] and in CME's origin [26]. Therefore, the study of ballooning instability presents a very important point in the context of flaring and CME's activity.

1.2
Data of Observations and Types of Coronal Loops

The corona of the Sun (a main-sequence G2 star) in its active phase consists predominantly of magnetic loops filled with comparatively dense and hot plasma, which is observed in soft X rays and constitutes an essential part of the total mass of the corona. The presence of magnetic loops indicates the complexity of the subphotospheric magnetic field, which is most likely linked to convective motions of the photosphere matter. Observations indicate that there are at least five morphologically distinct types of loops present in the solar atmosphere (see, i.e., [27, 28]):

1) **Loops connecting different active regions**. Their lengths reach 700 000 km, the plasma temperature in such loops is $(2-3) \times 10^6$ K, and the density is about 10^9 cm^{-3}. The loop footpoints are located in islands of the strong magnetic field on the periphery of active regions. The characteristic lifetime of such loops is about one day.
2) **Loops in quiescent regions**. They do not connect active regions; their lengths are the same as those of the previous types of loops. Their temperature, however, is somewhat lower, within the interval $(1.5-2.1) \times 10^6$ K, while the density is in the range $(2-10) \times 10^8$ cm^{-3}.
3) **Loops in active regions**. Their lengths span from 10 000 to 100 000 km and temperature and density are within the intervals $10^4-2.5 \times 10^6$ K and $(0.5-5.0) \times 10^9$ cm^{-3}, respectively.
4) **Post-flare loops**. They commonly connect footpoints of two-ribbon flares, and display lengths of 10 000–100 000 km, temperature $10^4-4 \times 10^6$ K, and density up to 10^{11} cm^{-3}.
5) **Single-flare loops**. These are separate loops, in which flare energy is released. Hard X-ray bursts last for about a minute and are the most pertinent feature of such flares. In soft X rays, these loops are characterized by small volumes and low heights. The loops are 5000–50 000 km in length, their temperature is less than 4×10^7 K, and their plasma density reaches 10^{12} cm^{-3} [12].

Closed magnetic structures resembling coronal loops also exist in stars of other types. Data obtained with *Einstein*, *ROSAT* (Röntgensatellit), and *XMM-Newton* (XMM stands for X-ray multimirror mission) space missions indicate that virtually all stars on the Hertzsprung–Russel diagram possess hot coronae with temperatures ranging between 10^7 and10^8 K [29–31]. They are not confined gravitationally, which implies the presence of magnetic fields. Of special interest are late-type stars, particularly dMe red dwarfs, which display high flare activity and represent nearly 80% of the total number of stars in the Galaxy and in its close neighborhood. Although red dwarfs are morphologically similar to the Sun, especially in their radio-wavelength radiation (the slowly varying component, rapidly drifting bursts, sudden reductions, spike bursts, and quasi-periodical pulsations [15, 32]), these objects present some peculiarities, which stem from the high activity of their coronae.

First, note the high brightness temperature of the "quiescent" radio emission of dMe-stars (up to 10^{10} K), which cannot be described in terms of thermal coronal plasma with the temperature 10^7-10^8 K. This radiation is commonly interpreted as gyrosynchrotron emission of nonthermal electrons abundant in the coronae of red dwarfs. The coexistence of the hot plasma and subrelativistic particles is also suggested by the correlation between the radio and soft X-ray emission over six orders of magnitude in intensity [33]. This phenomenon is not observed in the solar corona. In the quiescent state, the corona of the Sun does not host a sufficient

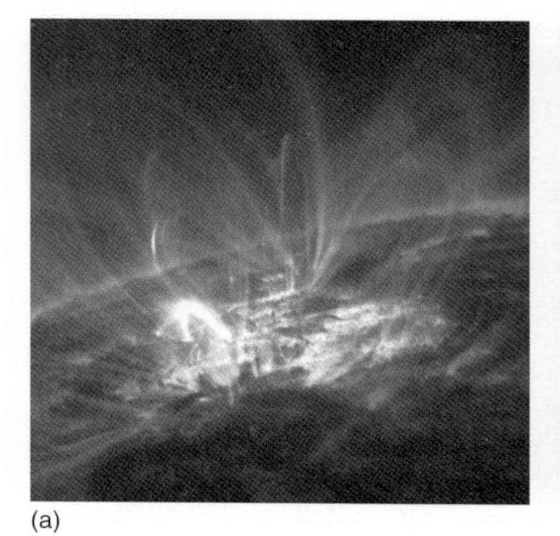
(a)

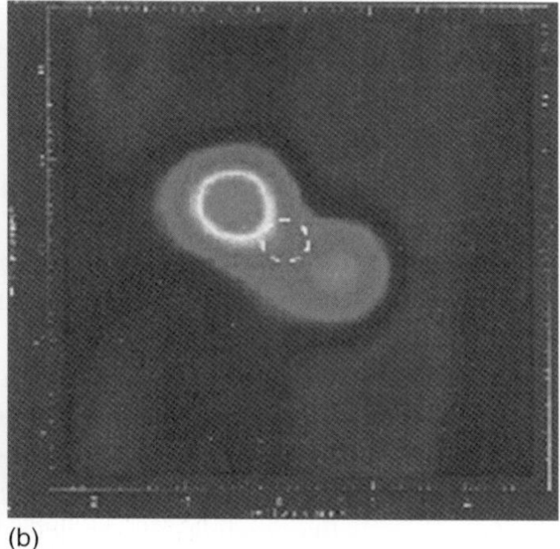
(b)

Figure 1.1 (a) Coronal magnetic loops of an active region on the Sun observed with the space laboratory TRACE in the ultraviolet range, $\lambda = 171$ Å [11]. A hot flare loop stands out sharply. (b) Radio map of UV Cet B on the wavelength of 3.6 cm obtained with VLBA (very long baseline array) and VLA (very large array). The dashed line marks the optical disk of the star [10].

Table 1.1 Parameters of coronal flare loops.

Parameter	The Sun	Red draft UV Ceti	Close binary stars	
			RS CVn	Algol
Length (cm)	$(1-10) \times 10^9$	$2 \times 10^9 - 3 \times 10^{11}$	$(5-10) \times 10^{10}$	$(2-6) \times 10^{11}$
Lateral size (cm)	$(1-5) \times 10^8$	$10^8 - 3 \times 10^9$	–	–
Plasma density n (cm^{-3})	$10^9 - 10^{12}$	$10^{10} - 10^{12}$	$10^8 - 10^{10}$	$10^9 - 10^{12}$
Plasma temperature T (K)	$10^6 - 10^7$	$3 \times 10^6 - 10^8$	$(3-9) \times 10^7$	$(3-7) \times 10^7$
Magnetic field induction B (G)	$10^2 - 10^3$	$3 \times 10^2 - 10^3$	$(0.3-6) \times 10^2$	$(1-5) \times 10^2$
Emission measure (EM) (cm^{-3})	$10^{47} - 10^{50}$	$10^{50} - 10^{53}$	$10^{53} - 10^{55}$	$10^{52} - 10^{54}$

number of high-energy particles, and the brightness temperature of solar radio emission does not exceed 10^6 K.

Second, the difference is manifested in the extremely high brightness temperature of flare radio emission of red dwarves, sometimes exceeding 10^{16} K, which is three to four orders of magnitude higher than that of the most powerful radio bursts in the Sun. This suggests the presence of an efficient coherent (maser) mechanism of emission from stellar coronae. Moreover, in red dwarfs, radio flares can frequently occur independently of optical flares.

Third, the sizes of loops on the Sun and red dwarves are different [9, 10]. As a rule, the length of the solar loops, with the exception of trans-equatorial ones, is substantially smaller than that of the solar radius (Figure 1.1a). On red dwarves, the size of magnetic loops can be comparable to the stellar radii or can exceed them by a few factors (Figure 1.1b). The magnetic field in stellar loops can exceed that in solar loops by an order of magnitude.

Table 1.1 presents parameters of flare loops on the Sun and late-type stars derived from multiwavelength (optical, radio, and X-ray) observations and various diagnostic methods.

1.3 The MHD Approach for Coronal Plasma

In our study of magnetic structures in coronal plasma and their evolution, we frequently use MHD approximation. It includes Maxwell equations, generalized Ohm's law, and also equations of motion and continuity of mass, the energy equation, and the gas equation of state. In MHD approximation, it is common to take the simplified Maxwell equations, without taking into account the bias current, and also the simplified form of the Ohm's law in the approximation of isotropic conductivity of plasma. For a number of problems of coronal seismology, such approximations appear to be insufficient, and more general equations are used. In addition to that, when the dynamics and radiation of fast particles in coronal magnetic loops are analyzed, the kinetic equation for the velocity distribution of particles should be used (see, for example, Chapters 6 and 7). Below, we present the basic equations of magnetic hydrodynamics in their generalized form, which is applied further on. For the sake of convenience, we repeatedly reproduce these equations along with the explanation for the previously used notation in different sections.

The *Maxwell equations* in the Gaussian system of units (cgs) may be written in the form

$$\mathrm{rot}\vec{B} = \frac{4\pi}{c}\mu\vec{j} + \frac{1}{c}\mu\varepsilon\frac{\partial\vec{E}}{\partial t} \tag{1.1}$$

$$\mathrm{rot}\vec{E} = -\frac{1}{c}\frac{\partial\vec{B}}{\partial t} \tag{1.2}$$

$$\mathrm{div}\vec{B} = 0 \tag{1.3}$$

$$\mathrm{div}\vec{E} = \frac{1}{\varepsilon}4\pi\rho_e \tag{1.4}$$

Here, the magnetic induction $\vec{B}$ and magnetic field $\vec{H}$ are related as follows: $\vec{B} = \mu\vec{H}$, while the electric field $\vec{E}$ may be expressed through electric induction: $\vec{E} = \vec{D}/\varepsilon$, where μ and ε are magnetic and dielectric permeabilities. For coronal plasma, μ and ε are usually taken to be equal to their values in vacuum; in the Gaussian system of units (cgs), we further assume $\mu = \varepsilon = 1$. In Eqs. (1.1) and (1.4), $\vec{j}$ and ρ_e are the density of the electric current and electric charge, respectively.

The *generalized Ohm's law* follows the "three-fluid" model for electrons, ions, and neutral atoms; it connects the electric current with the velocity of the center of mass, and also with the electric and magnetic fields. When the inertia of electrons is neglected, the generalized Ohm's law is written as follows:

$$\vec{E} + \frac{1}{c}\vec{V} \times \vec{B} = \frac{\vec{j}}{\sigma} + \frac{\vec{j} \times \vec{B}}{enc} + \frac{\vec{f}_e}{en} + F\frac{\left[\vec{f}_a \times \vec{B}\right]}{cnm_i\nu'_{ia}} - \frac{F^2\rho}{cnm_i\nu'_{ia}}\left[\frac{d\vec{V}}{dt} \times \vec{B}\right] \tag{1.5}$$

Here, $\rho = n_a m_a + n_e m_e + n_i m_i$ is the plasma density, $p = p_a + p_e + p_i$ is its pressure, $\vec{V} = \left(\sum_k n_k m_k \vec{V}_k\right) / \left(\sum_k n_k m_k\right)$ is the average velocity of plasma motion, $k =$ a, i, e (a, atoms; i, ions; and e, electrons), $\sigma = \frac{ne^2}{m_e(\nu'_{ei}+\nu'_{ea})}$ is the Coulomb (Spitzer) conductivity, $F = \rho_a/\rho$ is the relative density of neutrals, ν_{kl} is the frequency of collisions between particles of k and l types, $\nu'_{kl} = \left[m_l/(m_k + m_l)\right]\nu_{kl}$ is the effective frequency of collisions, $n_i = n_e = n, \vec{f}_e = -\text{grad}\, p_e, \vec{f}_a = -\text{grad}\, p_a + n_a m_a \vec{g}$, and $\vec{g}$ is the gravitational acceleration. Two last summands in the right-hand part of Eq. (1.5) take into account dissipation related to ion–atom collisions. Under the conditions in the chromospheres and even of the corona, this dissipation frequently appears to be more efficient than that caused by Spitzer conductivity σ.

The induction equation. Excluding the electric field from Eq. (1.5) with the use of Eq. (1.2), we obtain the following equation for magnetic induction:

$$\begin{aligned}\frac{\partial \vec{B}}{\partial t} &= \eta\nabla^2\vec{B} + \text{rot}\,[\vec{V} \times \vec{B}] + \frac{F^2\rho}{n_i m_i \nu'_{ia}}\text{rot}\left[\frac{d\vec{V}}{dt} \times \vec{B}\right] \\ &- \frac{F}{n_i m_i \nu'_{ia}}\text{rot}\,[\vec{f}_a \times \vec{B}] - \frac{1}{en_i}\text{rot}[\vec{j} \times \vec{B}]\end{aligned} \tag{1.6}$$

where $\eta = c^2/4\pi\sigma$ is the magnetic viscosity. Frequently, only the first two terms in the right-hand part of the induction Eq. (1.6) are taken into account, which correspond to the simplest form of the Ohm's law

$$\vec{j} = \sigma\left(\vec{E} + \frac{1}{c}\vec{V} \times \vec{B}\right) \tag{1.5a}$$

This form of Ohm's law may be applied, for instance, in the case of totally ionized plasma ($\vec{f}_a = 0$), provided the Ampere force becomes zero ($\vec{j} \times \vec{B} = 0$, the so-called force-free approximation). In a more general case, for example, in the lower chromosphere or in the prominence, plasma is partly ionized, and the anisotropy of conductivity becomes essential. Therefore, the more general induction Eq. (1.6) should be used.

Equations of plasma. The induction Eq. (1.6) indicates that the behavior of the magnetic field is related to the motion of plasma, since in this equation some summands contain the matter velocity. In turn, the motion of plasma is specified by the continuity equation, the equation of motion, and the energy equation:

$$\frac{\partial \rho}{\partial t} + \operatorname{div}(\rho \vec{V}) = 0 \tag{1.7}$$

$$\rho \frac{\mathrm{D}\vec{V}}{\mathrm{D}t} = -\nabla p + \frac{1}{c}\vec{j} \times \vec{B} + \rho \vec{g} \tag{1.8}$$

$$\frac{\rho^{\gamma}}{\gamma - 1} \frac{\mathrm{D}}{\mathrm{D}t}\left(\frac{p}{\rho^{\gamma}}\right) = -Q \tag{1.9}$$

In Eqs. (1.8) and (1.9), $D/Dt = \partial/\partial t + (\vec{v}\nabla)$ means the convective derivative that specifies time variations of parameters related to plasma motion. In Eq. (1.9), $\gamma = C_P/C_V$ is the ratio between specific heat capacities and Q is the function equal to the difference between the energy loss and energy inflow rates per unit volume of plasma. The energy loss commonly consists of the thermal conductivity and radiative losses. In coronal magnetic fields, the thermal conductivity across the magnetic field may commonly be neglected. In addition to that, in the optically thin part of the atmosphere ($T \geq 2{\times}10^4$ K in the chromosphere and in the corona), the radiative losses do not depend on the radiation intensity, which results in simplification of the radiative loss function. If along with that the energy inflow rate is specified by dissipation of electric currents, the function Q may be presented as follows:

$$Q = -\frac{1}{A}\frac{\mathrm{d}}{\mathrm{d}s}\left(\kappa_{\mathrm{e}}\frac{\mathrm{d}T}{\mathrm{d}s}A\right) + n_{\mathrm{e}}n_{\mathrm{H}}\chi(T) - \left(\vec{E} + \frac{1}{c}\vec{V} \times \vec{B}\right)\vec{j} \tag{1.10}$$

Here, $A(s)$ is the cross section of the force tube related to the magnetic field, s is the coordinate along the tube, and $\kappa_{\mathrm{e}} = 0.92 \times 10^{-6}$ erg cm^{-1}K$^{-7/2}$ is the coefficient of thermal conductivity along the magnetic field,

$$\begin{aligned} \chi(T) &\approx 10^{-21.85} \quad (10^{4.3} \leq T \leq 10^{4.6}\mathrm{K}) \\ &\approx 10^{-31}T^2 \quad (10^{4.6} \leq T \leq 10^{4.9}\mathrm{K}), \\ &\approx 10^{21.2} \quad (10^{4.9} \leq T \leq 10^{5.4}\mathrm{K}) \\ &\approx 10^{-10.4}T^{-2} \quad (10^{5.4} \leq T \leq 10^{5.75}\mathrm{K}) \\ &\approx 10^{-21.94} \quad (10^{5.75} \leq T \leq 10^{6.3}\mathrm{K}) \\ &\approx 10^{-17.73}T^{-2/3} \quad (10^{6.3} \leq T \leq 10^{7}\mathrm{K}) \end{aligned} \tag{1.11}$$

is the temperature function, the values of which are presented, for example, in Ref. [34]; n_{H} is the number of hydrogen atoms in the unit volume (for totally ionized plasma, $n_{\mathrm{H}} = n_{\mathrm{e}}$). The last summand in Eq. (1.10) is determined by taking Ohm's law into account (Eq. (1.5)). Within the temperature interval 10^5 K $< T < 10^7$ K, a possible approximation for the radiative loss function is $\chi(T) = \chi_0 T^{-1/2}$ erg cm^{3}s^{-1}, where $\chi_0 = 10^{-19}$ [27].

The connection between the pressure and the density is specified by the equation of state; for ideal gas,

$$p = nk_{\mathrm{B}}T \tag{1.12}$$

where $n = n_{\mathrm{e}} + n_{\mathrm{i}} + n_{\mathrm{a}}$ is the total number of particles in unit volume and k_{B} is the Boltzmann's constant. Different versions of the equations used in MHD approximation are considered in more detail in the monograph by Priest [27].

References

1. Uchida, Y. (1970) *Publ. Astron. Soc. Jpn.*, **22**, 341.
2. Rosenberg, H. (1970) *Astron. Astrophys.*, **9**, 159.
3. Aschwanden, M.J., Poland, A.I., and Rabin, D. (2001) *Ann. Rev. Astron. Astrophys.*, **39**, 175.
4. Aschwanden, M.J. (2003) NATO Advances Research Workshop, NATO Science Series II, p. 22.
5. Nakariakov, V.M. and Verwichte, E. (2005) *Living Rev. Sol. Phys.* Coronal waves and oscillations, **2**, No 3, pp. 3–65
6. Nakariakov, V.M. and Stepanov, A.V. (2007) *Lect. Notes Phys.*, **725**, 221.
7. Zaitsev, V.V. and Stepanov, A.V. (2008) *Phys. Usp.*, **51**, 1123.
8. Gershberg, R.E. (2005) *Solar-Type Activity in Main-Sequence Stars*, Springer, Berlin, Heidelberg.
9. Bray, R.J., Cram, L.E., Durrant, C.J., and Loughhead, R.E. (1991) *Plasma Loops in the Solar Corona*, Cambridge University Press.
10. Benz, A., Conway, J., and Güdel, M. (1998) *Astron. Astrophys.*, **331**, 596.
11. Schrijver, C.J., Title, A.M., Berger, T.E., Fletcher, L. *et al.* (1999) *Sol. Phys.*, **187**, 261.
12. Sakai, J.-I. and de Jager, C. (1996) *Space Sci. Rev.*, **77**, 1.
13. Plunkett, S.P., Vourlidas, A., Šimberová, S., Karlický, M. *et al.* (2000) *Sol. Phys.*, **194**, 371.
14. Lestrade, J.F. (1988) *Astrophys. J.*, **328**, 232.
15. Bastian, T.S., Bookbinder, J.A., Dulk, G.A., and Davis, M. (1990) *Astrophys. J.*, **353**, 265.
16. Galeev, A.A., Rosner, R., Serio, S., and Vaiana, G.S. (1981) *Astrophys. J.*, **243**, 301.
17. Kuijpers, J. (1995) *Lect. Notes Phys.*, **444**, 135.
18. Feigelson, E.D. and Montmerle, T. (1999) *Ann. Rev. Astron. Astrophys.*, **37**, 363.
19. Beloborodov, A.M. and Thompson, C. (2007) *Astrophys. J.*, **657**, 967.
20. Alfvén, H. and Carlqvist, P. (1967) *Sol. Phys.*, **1**, 220.
21. Zaqarashvili, T.V. and Erdelyi, R. (2009) *Space Sci. Rev.*, **149**, 355.
22. Oliver, R. (2009) *Space Sci. Rev.*, **149**, 175.
23. Tripathi, D., Isobe, H., and Jain, R. (2009) *Space Sci. Rev.*, **149**, 283.
24. Pustyl'nik, L.A. (1974) *Sov. Astron.*, **17**, 763.
25. Zaitsev, V.V., Urpo, S., and Stepanov, A.V. (2000) *Astron. Astrophys.*, **357**, 1105.
26. Gopalswamy, N. (2006) *Space Sci. Rev.*, **124**, 145.
27. Priest, E.R. (1982) *Solar Magnetohydrodynamics*, D. Reidel Publishing Company, Dordrecht.
28. Aschwanden, M.J. (2005) *Physics of the Solar Corona. An Introduction with Problems and Solutions*, Springer.
29. Haisch, B.M. (1983) in *Activity in Red-Dwarf Stars* X-ray Oscillations of Stellar Flares (eds P.B. Birne and M. Rodono), Reidel, p. 255–268.
30. Schmitt, J.H.M.M., Collura, A., Sciortino, S., Vaiana, G.S. *et al.* (1990) *Astrophys. J.*, **365**, 704.
31. Mullan, D.J., Mathioudakis, M., Bloomfield, D.S., and Christian, D.J. (2006) *Astrophys. J. Suppl.*, **164**, 173.
32. Stepanov, A.V., Kliem, B., Zaitsev, V.V. *et al.* (2001) *Astron. Astrophys.*, **374**, 1072.
33. Güdel, M. and Benz, A.O. (1993) *Astrophys. J.*, **405**, L63.
34. Rosner, R., Tucker, G.S., and Vaiana, R. (1978) *Astrophys. J.*, **220**, 643.

2
Coronal Magnetic Loop as an Equivalent Electric Circuit

In Chapter 1, we presented numerous evidences for the fact that the solar and stellar coronae are structured and consist of magnetic loops and open flux tubes filled with plasma. Therewith the parameters of the loops vary within a broad range [1, 2]. For example, hot X-ray loops with a temperature of up to 10 MK, observed with the *Yohkoh* mission, may be located at quite a large distance from the spots. In contrast, "warm" loops with a temperature of (1.0–1.5) MK, observed with the Transition Region and Coronal Explorer (*TRACE*) mission, are, as a rule, situated in the vicinity of the spots; the footpoints of these loops are located in the penumbral regions. This fact, as well as the difference in the density and size of these types of loops, may provide evidence of different mechanisms of formation and heating of hot X-ray and "warm" loops [3].

Essentially, two different types of magnetic flux tubes are possible. The first type originates in the course of "raking" the background magnetic field up by convective flows of photospheric plasma. Footpoints of these tubes are commonly located either in the nodes of several supergranulation cells, where horizontal convective flows converge, or close to the boundary of two adjacent supergranules. In the latter case, arcades of coronal magnetic loops may be formed along the boundary of the supergranules. Such loops may be located in the distance from sunspots; inside them, a large (up to 10^{12}A) electric current may occur due to the interaction between the convective plasma flow and the intrinsic magnetic field of the tubes. Magnetic flux tubes with parallel currents may contain large nonpotential energy and may therefore be a source of powerful flares.

The second type is represented by numerous magnetic flux tubes originating in the vicinity of sunspots as a result of filling thin filamentary volumes extended along magnetic field lines of a spot up by chromospheric plasma. Some parameters of these tubes were studied with Solar and Heliospheric Observatory (*SOHO*) and *TRACE* missions. In particular, it was found that plasma inside the loops displays a temperature of 1.0–1.5 MK and density exceeding that of the surrounding plasma by more than an order of magnitude. The magnetic field in the vicinity of sunspots may become filled up with plasma of increased density (which results in the formation of magnetic loops with increased plasma density inside), owing to the development of the ballooning mode of the flute instability in chromospheric footpoints of magnetic loops. In this region, the magnetic field extends and develops

Coronal Seismology: Waves and Oscillations in Stellar Coronae, First Edition.
A. V. Stepanov, V. V. Zaitsev, and V. M. Nakariakov.

the curvature directed from the surrounding chromosphere inside the loops. The radius of curvature of the magnetic lines is of the order of the height of the inhomogeneous atmosphere. The curvature of the magnetic field results in the centrifugal force being directed inward of the loops; under certain conditions, this leads to flute instability and to filling of some part of magnetic lines of the loops by denser outer plasma, which surrounds the loops.

2.1
A Physical Model of an Isolated Loop

In their footpoints, isolated magnetic loops commonly display the shape of compact ($r_0 \sim 100–500$ km) magnetic flux tubes with the typical magnetic field $B \sim 1000–2000$ G, roughly vertical to the solar surface. Traditionally, the formation of these tubes was explained by the effect of amplification of the frozen-in magnetic field by converging plasma flows [4–6] in the approximation of the one-fluid magnetic hydrodynamics for totally ionized plasma. It appears, however, that this approximation does not always adequately describe the process of formation of magnetic tubes by photospheric flows. In weakly ionized ($n/n_a \sim 10^{-4}$), relatively cold and dense plasma of the solar photosphere on characteristic scales typical for intense magnetic flux tubes, the magnetic field cannot be considered to be frozen into the plasma [7–9]. In this medium, the magnetic field related to ionized fraction may slide through moving neutral gas. This process is called *ambipolar diffusion*.

Interacting with the magnetic field in the footpoint of a coronal magnetic loop, convective flows of photospheric plasma generate electric currents, which provide

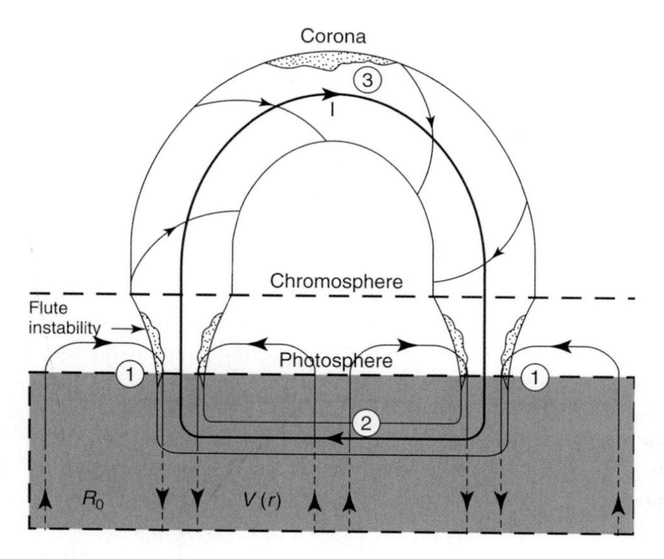

Figure 2.1 A sketch of a coronal magnetic loop formed by converging convective flows of photospheric plasma [8].

free energy to heat plasma and to accelerate charged particles. Figure 2.1 illustrates this idea, which initially dates back to studies of Alfven and Carlqvist [10] and Sen and White [7]. Figure 2.1 presents a magnetic loop with the footpoints submerged into the photosphere and formed by converging horizontal flows of photospheric matter. Such a situation may occur, for example, when the footpoints of the loop are in the nodes of several supergranulation cells. Indirectly, the existence of strong electric currents in coronal loops is confirmed by the virtually invariable cross section of a loop along its total length, which is hardly possible for a potential magnetic field.

In the presented magnetic structure, three important domains may be delineated. In region 1 (so-called dynamo region), located in the upper photosphere and lower chromosphere, the magnetic field and the related electric current are generated. In this region, $\omega_e/\nu'_{ea} \gg 1$, $\omega_i/\nu'_{ia} \ll 1$, where ω_e and ω_i are gyrofrequencies of electrons and ions, while ν'_{ea} and ν'_{ia} are effective frequencies for electron–atom and ion–atom collisions. Consequently, the neutral atoms in the convective flow are able to carry the positive ions better, and the electrons are bound to the magnetic field lines. Therefore, a charge separation develops, which results in the origination of the radial electric field of E_r [7]. The electric field E_r, along with the initial magnetic field B_z, generates the Hall's current j_φ, which amplifies the initial magnetic field B_z [8, 9]. This amplification continues until the "raking up" of the background magnetic field has been compensated by the diffusion of the magnetic field due to anisotropic conductivity of the plasma. As a result, a stationary magnetic flux tube is formed, the magnetic field in which is specified by the total energy contribution from the plasma convective flow for the time of the formation of the flux tubes (of the order of R_0/V_r, where $R_0 \sim 30\,000$ km is the scale of the supergranulation cell and $V_r \sim 0.1$–0.5 km s^{-1} is the horizontal velocity of convective motion). The density of the magnetic field energy inside flux tubes may substantially exceed that of the kinetic energy of convective motion. In the steady state, the ampere force $(1/c)\left[\vec{V} \times \vec{B}\right]$ inside the tubes counterbalances the gas-kinetic pressure gradient, while the kinetic energy of the convective flow goes toward the maintenance of the radial electric field of charge separation E_r and the Hall's current j_φ.

Region 3 is the coronal part of the loop. Here, the gas-kinetic pressure is lower than that of the magnetic field (the plasma parameter $\beta \ll 1$), and the magnetic loop structure is force free, that is, the electric current lines are directed almost along the magnetic lines.

Region 2 is situated in the lower photosphere, or directly under the photosphere. It is suggested that the closure of electric current I that flows through the magnetic loop occurs in this region. The distribution of electric currents in the photosphere, obtained from measurements of magnetic fields [11, 12], provides evidence of noncompensated electric currents [13–15]. These data suggest that the electric current in a magnetic flux tube flows through the coronal part of the loop from one footpoint to the other, and no inverse current is detected. The closure of the current occurs in the subphotospheric region, where the plasma conductivity becomes isotropic and the current flows along the shortest way from one footpoint of the loop to the other. For the magnetic field in the flux tube

$B = 1000$ G and the temperature $T = 5 \times 10^3$ K, the conductivity is isotropic for the plasma density $n = 5 \times 10^{16}$ cm^{-3}, which approximately corresponds to the level $\tau_{5000} = 1$. Estimates [8, 9] indicate that, for the velocity of the convective motion $V_r = 0.1$ km s^{-1}, the radius of a formed tube at an altitude of 500 km above the level $\tau_{5000} = 1$ is $r \approx 3.3 \times 10^7$ cm, while the parallel current is $I_z \approx 3 \cdot \times 10^{11}$A for the magnetic field on the axis $B = 1000$ G.

Note that the electromagnetic energy of the current in a flaring loop may come from the kinetic energy of the rotation of a sunspot with the velocity $\sim$0.1 km s^{-1} [16]. This situation may generally occur if the footpoints of the flare loops are located in the regions of spots. For loops formed beyond the sunspots, the electric current is generated by photospheric convection.

The above magnetic loop with the current is an equivalent electric circuit. This idea was first formulated by Alfven and Carlquist [10] within the framework of their circuit model of a flare. The model [10] was based on measurements made by Severny [17], who detected electric currents $I \geq 10^{11}$A in the vicinity of sunspots, and also on the analogy with a circuit containing a mercury-vapor rectifier, which may produce a sharp transition from high conductivity to the state with large resistance. A sudden interruption of the current in the circuit results in explosive release of energy. Mechanisms for this disruption may be different. One of the above may be due to flute instability in regions close to the top or footpoints of the loop, and also due to the actuation of Cowling conductivity [18] related to ion–atom collisions, which increases the electric resistance of the magnetic loop by eight to nine orders of magnitude [19, 20]. This is accompanied by efficient heating of plasma and acceleration of particles in the coronal loop, which is discussed in detail below.

2.2
The Formation of Magnetic Tubes by Photospheric Convection

Let us consider the simplest examples of the formation of magnetic tubes by converging flows of photospheric plasma. The effect of the concentration of the magnetic field may occur in different geometries of converging flows. For example, the interfaces between two adjacent supergranules display opposing convective flows; along the interface, a thin extended layer of the magnetic field with relatively large strength may originate (Figure 2.2a).

Under some conditions, the extended layer may manifest the interchanging instability of Rayleigh–Taylor type or the flute instability. As a result, the layer is subdivided into a system of magnetic tubes with a radii of the order of the thickness of the layer (Figure 2.2b), which may form an arcade of coronal magnetic loops. In the case when a magnetic flux tube is formed at the point of contact between more than two cells rather than on the boundary of two supergranulation cells, the axially symmetrical (cylindrical) approximation for the converging plasma flow is more adequate. In this case, the converging flow of photospheric plasma transforms

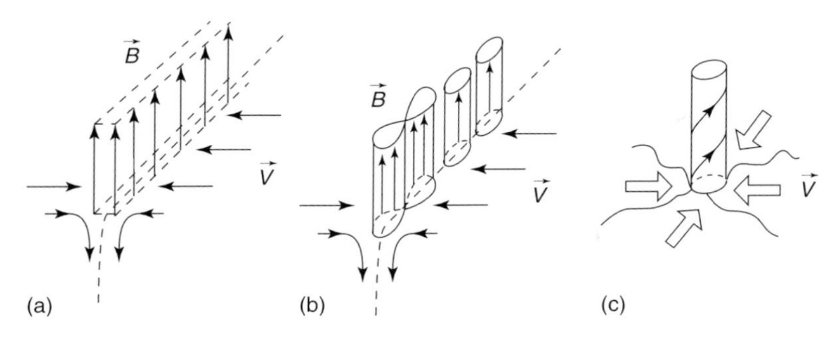

Figure 2.2 The formation of intense magnetic flux tubes on the boundaries of supergranulation cells [9]: (a) the formation of an extended thin magnetic layer; (b) the fragmentation of the layer to a system of magnetic tubes due to interchanging instability; and (c) the formation of an intense magnetic tube in a node of several supergranulation cells.

the initial background magnetic field into a compact cylindrical magnetic tube (Figure 2.2c). This case was analyzed in more detail in Ref. [21].

For the sake of simplicity, we consider the formation of an axially symmetrical magnetic tube $\boldsymbol{B}(0, B_\varphi, B_z)$ with current $\boldsymbol{j}(0, j_\varphi, j_z)$ for the case of a stationary axially symmetrical flow of photospheric matter with the velocity $\mathbf{V}(V_r, V_\varphi, V_z)$, $V_r < 0$. Here, r, φ, and z are cylindrical coordinates with the vertical axis z. We assume that the magnetic tube is located vertically in the node of several supergranules and that the velocity of the converging convective flow is much lower than the velocity of sound, the Alfven velocity, and the free-fall velocity. Then the evolution of the magnetic field in the tube is quasi-static, and the following system of equations may be used:

$$\rho\vec{g} - \nabla p + \frac{1}{c}\vec{j} \times \vec{B} = 0 \tag{2.1}$$

$$\text{div}\rho\vec{V} = 0 \tag{2.2}$$

$$\vec{\boldsymbol{E}} + \frac{1}{c}\vec{V} \times \vec{B} = \frac{\vec{\boldsymbol{j}}}{\sigma} + \frac{\vec{\boldsymbol{j}} \times \vec{\mathbf{B}}}{enc} - \frac{F^2}{(2-F)c^2 nm_i\nu'_{ia}}\left[\vec{j} \times \vec{\mathbf{B}}\right] \times \vec{B} \tag{2.3}$$

$$\text{rot}\vec{E} = -\frac{1}{c}\frac{\partial\vec{B}}{\partial t} \tag{2.4}$$

Here, $\rho = n_a m_a + n_e m_e + n_i m_i$ is the density of partially ionized photospheric plasma, $p = p_a + p_e + p_i$ is the pressure, $\vec{V} = (\sum\limits_k n_k m_k \vec{V}_k)/(\sum\limits_k n_k m_k)$ is the mean velocity of the plasma motion, $k = a,i,e$ (a – atoms, i – ions, and e – electrons), $\sigma = \frac{ne^2}{m_e(v'_{ei}+v'_{ea})}$ is the Coulomb conductivity, $F = \rho_a/\rho$ is the relative density of neutrals, ν_{kl} is the frequency of collisions between particles of k and l types, and $\nu'_{kl} = \left[m_l/(m_k + m_l)\right]\nu_{kl}$ is the effective frequency of collisions, $n_i = n_e = n$. The generalized Ohm's law (Eq. (2.3)) is written by taking into account the quasi-stationary type of process in the approximation of three-fluid MHD. The last summand in the right-hand part of Eq. (2.3) takes into account the dissipation related to ion–atom collisions. Under the conditions of the chromosphere and even the corona, this dissipation frequently appears to be more efficient than classical Spitzer dissipation. There exist two reasons for this. First, the ampere force that acts on the plasma under certain conditions may impart to it the velocity substantially exceeding the relative velocity of electrons and ions, which specifies the parallel electric current. Second, in collisions with neutral atoms, ions accelerated by the ampere force transfer a substantial part of the energy of the directed motion to them even in a single collision, whereas, due to the large difference in their mass, electrons only gradually transfer the energy to ions in electron–ion collisions, which specify the classical conductivity.

Since the degree of ionization in the region l_1 (with the height $h = 0$–500 km above the level $\tau_{5000} = 1$) is small, we can further take $F \approx 1$, $p \approx p_a$. Applying the "rot" operation to both parts of Eq. (2.3), taking into account Eq. (2.4) and relations

$$j_r = 0, \; j_\varphi = -\frac{c}{4\pi}\frac{\partial B_z}{\partial r}, \; j_z = \frac{c}{4\pi}\frac{1}{r}\frac{\partial (rB_\varphi)}{\partial r} \tag{2.5}$$

we obtain the equation that describes the slow evolution of the components of the magnetic field of the vertical magnetic flux tube [21]:

$$\frac{\partial B_r}{\partial t} = 0 \tag{2.6}$$

$$\frac{\partial B_z}{\partial t} = \frac{1}{r}\frac{\partial}{\partial r}\left\{r\left[\frac{c^2}{4\pi\sigma}\left(1+\alpha B_z{}^2\right)\frac{\partial B_z}{\partial r} + \frac{c^2}{4\pi\sigma}\alpha B_\varphi B_z \frac{1}{r}\frac{\partial (rB_\varphi)}{\partial r} - V_r B_z\right]\right\} \tag{2.7}$$

$$\frac{\partial B_\varphi}{\partial t} = \frac{\partial}{\partial r}\left[\frac{c^2}{4\pi\sigma}\left(1+\alpha B_\varphi{}^2\right)\frac{1}{r}\frac{\partial (rB_\varphi)}{\partial r} + \frac{c^2}{4\pi\sigma}\alpha B_\varphi B_z \frac{\partial (B_z)}{\partial r} - V_r B_\varphi\right] \tag{2.8}$$

Here $\alpha = \sigma F^2 \left[c^2 n m_i \nu'_{ia}(2-F)\right]^{-1}$. Note that Eqs. (2.7) and (2.8) take into account the anisotropy of conductivity.

A stationary convective flow of plasma should follow Eq. (2.2); hence, the components of the flow should be determined by the following relations:

$$\rho V_r = -\frac{1}{r}\frac{\partial \psi(r,z)}{\partial z}, \; \rho V_z = \frac{1}{r}\frac{\partial \psi(r,z)}{\partial r}, \; \rho V_\varphi = \rho V_{\varphi 0}(r,z) \tag{2.9}$$

where $\psi(r,z)$ and $V_{\varphi 0}(r,z)$ are some functions of r and z. If we specify the exponential distribution of density with the height $\rho = \rho_0 \exp(-z/H)$ and a certain dependence for $V_r(r)$, we can determine the velocity V_z from Eq. (2.9). For example, if the radial component of the velocity is specified in the form

$$V_r(r) = \frac{V_0 r}{R_0}, \; r \le R_0, \; V_r(R) = \frac{V_0 R_0^2}{r^2}, \; r \ge R_0, \tag{2.10}$$

where $V_0 < 0$, then from Eq. (2.9), we obtain $V_z = 2HV_0/R_0$ for $r < R_0$ and $V_z = \text{const}/r^3$ for $r > R_0$. The selection of the azimuthal velocity $V_{\varphi 0}(r,z)$ is to some extent arbitrary. Further, we select this velocity so that $\left[\vec{V} \times \vec{B}\right]_r = V_\varphi B_z - V_z B_\varphi = 0$ to remove any contribution from convective motion to the radial component of the electric field (see Ohm's law in the form of Eq. (2.3)), which will simplify the subsequent analysis.

At the initial stage of the evolution of a magnetic flux tube, the diffusion of the magnetic field due to conductivity may be neglected compared to its amplification by the convective motion of the plasma. Assuming $V_r = V_0 r/R_0$, where $V_0 < 0$ (a converging flow) and R_0 is the scale of the variation of the horizontal velocity of convection, comparable to the radius of a supergranulation cell, from Eqs. (2.7) and (2.8), we obtain the following solution for the components of the magnetic field:

$$B_\varphi(r,t) = B_{\varphi 0}\left(r \exp\left(-\frac{V_0}{R_0}t\right)\right) \times \exp\left(-\frac{V_0}{R_0}t\right)$$
$$B_z(r,t) = B_{z0}\left(r \exp\left(-\frac{2V_0}{R_0}t\right)\right) \times \exp\left(-\frac{V_0}{R_0}t\right) \tag{2.11}$$

where $B_{\varphi 0}(r)$ and $B_{z0}(r)$ are the components of the field for $t = 0$. It follows from Eq. (2.11) that the magnetic field increases in the vicinity of the magnetic tube ($r = 0$) with the characteristic time $t_0 = R_0/|V_0|$ comparable to the lifetime of the supergranule. The strengthening of the magnetic field in the tube results in an increase in the role of the field diffusion. In a stationary flux tube, the amplification of the field due to the "raking up" by the convective flow is compensated by the diffusion of the magnetic field caused by conductivity. Projecting Eqs. (2.1–2.4) onto the coordinate axes and taking into account the relations (Eq. (2.5)), one may ascertain that the stationary state corresponds to the following equations for the components of the electric and magnetic fields:

$$E_\varphi = 0 \tag{2.12}$$
$$E_r = \frac{1}{(2-F)en}\frac{\partial p}{\partial r} \tag{2.13}$$
$$-E_z B_\varphi = \frac{c}{\sigma}\left(1 + \alpha B^2\right)\frac{\partial p}{\partial r} + \frac{V_r}{c}B^2 \tag{2.14}$$
$$\frac{\partial B_z}{\partial r} = \frac{4\pi\sigma}{c}\frac{V_r}{c}B_z + 4\pi\alpha B_z\frac{\partial p}{\partial r} \tag{2.15}$$
$$\frac{1}{r}\frac{\partial(rB_\varphi)}{\partial r} = \frac{4\pi\sigma}{c}\left(\frac{V_r}{c}B_\varphi + E_z\right) + 4\pi\alpha B_\varphi\frac{\partial p}{\partial r} \tag{2.16}$$

where $B^2 = B_\varphi^2 + B_z^2$. Excluding the component of the electrical field E_z from Eqs. (2.15) and (2.16), we obtain the equations for the gas pressure and the components of the magnetic field [22]:

$$B_z\frac{\partial B_z}{\partial r} = \frac{4\pi\sigma V_r}{c^2}B_z^2 + 4\pi\alpha B_z^2\frac{\partial p}{\partial r} \tag{2.17}$$

$$B_\varphi\frac{1}{r}\frac{\partial(rB_\varphi)}{\partial r} = -\frac{4\pi\sigma V_r}{c^2}B_z^2 - 4\pi(1 + \alpha B_z^2)\frac{\partial p}{\partial r} \tag{2.18}$$

From Eqs. (2.17) and (2.18), the well-known equilibrium condition for a magnetic flux tube follows:

$$\frac{\partial p}{\partial r} + \frac{1}{8\pi}\frac{\partial}{\partial r}\left(B_\varphi^2 + B_z^2\right) + \frac{1}{4\pi}\frac{B_\varphi^2}{r} = 0 \tag{2.19}$$

Equations (2.17) and (2.18) determine the magnetic structure for a specified distribution of the pressure of plasma in the tube, or otherwise the pressure for a specified distribution of the electric current flowing along the flux tube. This is due to the fact that the number of the unknown functions in the system of Eqs. (2.1), (2.3), and (2.4) exceeds the number of the equation. The exception might be the photospheric footpoints of the flux tube, where, as a matter of fact, the coronal magnetic loop is formed.

If the footpoints of the magnetic loop are situated in the nodes of several supergranulation cells, then, with the availability of converging convective flows of photospheric plasma, the interaction of the plasma flow, and the magnetic field of the loop occurs, as a result of which the electromotive force originates and electric current is generated, which specifies the structure of the magnetic flux tube in the vicinity of the footpoints. If in this case $E_z = 0$, Eqs. (2.14), (2.17), and (2.18) self-consistently determine the pressure and magnetic field in the tube, taking the form [21]

$$\frac{\partial p}{\partial r} = -\frac{\sigma V_r B^2}{c^2 \left(1 + \alpha B^2\right)} \tag{2.20}$$

$$\frac{\partial B_z}{\partial r} = \frac{4\pi \sigma V_r B_z}{c^2 \left(1 + \alpha B^2\right)} \tag{2.21}$$

$$\frac{1}{r}\frac{\partial (r B_\varphi)}{\partial r} = \frac{4\pi \sigma V_r B_\varphi}{c^2 \left(1 + \alpha B^2\right)} \tag{2.22}$$

Assume for the sake of definiteness that the radial component of the velocity of the convective motion of the plasma in the vicinity of the magnetic flux tube is specified in the form (Eq. (2.10)), where the radius of the tube r_1 occurs instead of R_0. Provided the magnetic field of the tube is sufficiently large, so that $\alpha \left(B_\varphi^2 + B_z^2\right) \gg 1$, the solution for Eq. (2.20) is in the form

$$p(r) = p(0) - \frac{\sigma V_0 r^2}{2c^2 \alpha r_1}, \quad r \le r_1 \tag{2.23}$$

$$p(r) = p_\infty + \frac{\sigma V_0 r_1^2}{c^2 \alpha r}, \quad r \ge r_1 \tag{2.24}$$

For $r = r_1$, both solutions should coincide, whence we obtain the radius of the tube

$$r_1 = \frac{p_\infty - p_0}{\frac{3}{2}\frac{\sigma\,|V_0|}{c^2 \alpha}} = \frac{2}{3}\frac{F^2}{2 - F}\frac{p_\infty - p(0)}{n m_i \nu'_{ia}\,|V_0|} \tag{2.25}$$

Since $V_0 < 0$, it follows from Eq. (2.23) that the gas pressure on the axis of the tube is smaller than that at its periphery. The converging flow of plasma enhances the magnetic field inside the tube, partially supplanting the plasma, so that the increased magnetic pressure inside the tube is counterbalanced by the outer gas pressure.

Let us now consider the structure of the magnetic field in the footpoint of the flux tube, that is, in the region of the photosphere and lower chromosphere where the structure of the magnetic field is described by Eqs. (2.21) and (2.22). These equations have the integral $rB_\varphi(r) = \text{const} \times B_z(r)$. In addition, the form of the system of equations does not vary if we substitute in the right parts of the equations

$$B_\varphi \longrightarrow B'_\varphi + \frac{\text{const}_1}{r}, \quad B_z \longrightarrow B'_z + B_0 \tag{2.26}$$

since the left parts of the equations are also invariantly relative to such substitution. According to Eq. (2.5), the substitution does not alter the components of the electric current that specify the structure of the magnetic field of the flux tube. In other words, this implies that the system of Eqs. (2.21) and (2.22) specifies the magnetic field of the flux tube with an accuracy to a certain potential field. Taking the above into account, we may write the general expression for the azimuthal component in the form

$$B_\varphi = C_1 \frac{B_z(r)}{r} + C_2 \frac{1}{r} \tag{2.27}$$

We select the constant C_2 so that for $r \to 0$ the component $B_\varphi(0)$ is finite. This is possible under the condition $C_2 = -C_1 B_z(0)$. The constant C_1 is selected from the condition that, for $r = r_1$, the components of the magnetic field are equal to $B_\varphi(r_1)$ and $B_z(r_1)$. As a result, we obtain

$$B_\varphi(r) = b\frac{r_1}{r}\left[B_z(r) - B_z(0)\right], \quad b = \frac{B_\varphi(r_1)}{B_z(r_1) - B_z(0)} \tag{2.28}$$

The expression for $B_z(r)$ can be easily found from Eq. (2.21) in the approximation of $\alpha\left(B_z^2 + B_\varphi^2\right) \gg 1$, $B_z^2 \gg B_\varphi^2$:

$$B_z^2(r) = B_z^2(0) - \frac{4\pi\sigma\,|V_0|\,r^2}{c^2\alpha r_1}, \quad r \le r_1 \tag{2.29}$$

$$B_z^2(r) = B_z^2(\infty) + \frac{8\pi\sigma\,|V_0|\,r_1^2}{c^2\alpha r}, \quad r \ge r_1 \tag{2.30}$$

The radius of the flux tube is expressed in terms of the boundary values of the magnetic field:

$$r_1 = \frac{\left[B_z^2(0) - B_z^2(\infty)\right]c^2\alpha}{12\pi\sigma\,|V_0|} = \frac{F^2}{2-F}\,\frac{B_z^2(0) - B_z^2(\infty)}{12\pi n m_i \nu'_{ia}\,|V_0|} \tag{2.31}$$

It can be easily seen that Eq. (2.31) for the radius of the tube exactly coincides with Eq. (2.25), since in our approximation $\left(B_z^2 \gg B_\varphi^2\right)$, the condition of the energy balance is fulfilled

$$p_\infty - p(0) = \frac{B_z^2(0)}{8\pi} - \frac{B_z^2(\infty)}{8\pi} \tag{2.32}$$

For height $h \approx 500$ km above the level $\tau_{5000} = 1$, where $n \approx 10^{11}\ \text{cm}^{-3}$, $n_a \approx 10^{15}\ \text{cm}^{-3}$, and $T \approx 10^4$ K, we obtain $r_1 \approx 2.7 \times 10^7$ cm, if we suggest that the magnetic field $B_z(0) = 10^3$ G and the rate of convection $V_0 = 0.1\ \text{km s}^{-1}$.

The component of the magnetic field $B_z(r)$ reaches its maximum on the axis of the tube and decreases with the distance from the axis. The component $B_\varphi(r)$

increases linearly with r if $r \ll r_1$, and decreases roughly as $1/r$ for $r > r_1$. In other words, the dependence $B_\varphi(r)$ is similar to that of the azimuthal component of the magnetic field of the cylindrical straight current with radius r_1.

Equations. (2.5) and (2.28) make it possible to determine the total electric current through the cross section of the magnetic flux tube:

$$I_z = \int_0^\infty 2\pi j_z r \mathrm{d}r = \frac{bcr_1}{2}\left[B_z(\infty) - B_z(0)\right] \tag{2.33}$$

which depends on the magnetic field $B_z(0)$ on the axis of the tube, the radius of the tube, and the degree of twist of the magnetic field b. This current may be expressed through parameters of plasma and the convection rate if in Eq. (2.33) the expression (2.31) for the tube radius is substituted:

$$I_z = -\frac{F^2 cb\left[B_z(0) - B_z(\infty)\right]^2\left[B_z(0) + B_z(\infty)\right]}{24\pi(2-F)nm_i\nu'_{\mathrm{ia}}\,|V_0|} \tag{2.34}$$

Assuming for the sake of definiteness $B_\varphi(r_1) = B_z(r_1)$, we estimate the lateral currents $I_z \approx 10^{11}$–10^{12}A for the rate of convection $|V_0| = 0.1$–$1.0\ \mathrm{km\,s^{-1}}$ and typical parameters of the photosphere at the height of 500 km above the level $\tau_{5000} = 1$ if we take the magnetic field on the axis of the tube $B_z(0) = 2\times10^3$ G. The obtained estimates apparently characterize the maximum values for electric currents in coronal magnetic loops, since they correspond to the maximum magnetic fields observed in active regions.

2.3 The Structure of the Coronal Part of a Flux Tube

Let us consider the peculiarities of the structure of a magnetic flux tube in the corona, where convection is absent and it may be adopted that $V_r = 0$. In this case, from Eqs. (2.17) and (2.18), we obtain the following equations for the components of the magnetic field:

$$B_z\frac{\partial B_z}{\partial r} = 4\pi\alpha B_z^2\frac{\partial p}{\partial r},\quad \frac{B_\varphi}{r}\frac{\partial(rB_\varphi)}{\partial r} = -4\pi\left(1+\alpha B_z^2\right)\frac{\partial p}{\partial r} \tag{2.35}$$

Adding up Eq. (2.35), we again obtain condition (Eq. (2.19)) for the magnetic tube balance. If we assume that neutral atoms in the corona are totally absent ($\alpha = 0$), then the component of the magnetic field B_z inside the tube does not vary, while the φ-component of the electric current is absent. The B_φ component of the field and the z-component of the electric current display the same scale of variation across the section of the flux tube as the pressure. In reality, however, the presence of even a small number of neutrals in the corona (of the order of 10^{-6} to 10^{-7} of the mass) may substantially alter the distribution of currents in the tube, since, as seen from further estimates, in Eq. (2.35) the parameters $4\pi\alpha p$ may substantially exceed unity.

Let us assume for definiteness that the plasma pressure inside the tube is specified and distributed as follows:

$$p(r) = p_0 + p_1\left(1 - \frac{r^2}{r_2^2}\right), \; r \le r_2, \; p(r) = p_\infty, \; r \ge r_2 \tag{2.36}$$

With a specified pressure, Eq. (2.35) can be easily integrated, and we can find the distribution for the components of the magnetic field across the tube section:

$$B_z^2(r) = B_{z0}^2 \exp\left(-8\pi\alpha p_1 \frac{r^2}{r_2^2}\right), \quad r \le r_2 \tag{2.37}$$

$$B_z^2(r) = B^2 z\infty, \quad r > r_2 \tag{2.38}$$

$$(rB_\varphi)^2 = 4\pi p_1 \frac{r^4}{r_2^2} + r_2^2 \frac{B_{z0}^2}{8\pi\alpha p_1}\left[1 - \left(1 + 8\pi\alpha p_1 \frac{r^2}{r_2^2}\right)\exp\left(-8\pi\alpha p_1 \frac{r^2}{r_2^2}\right)\right],$$
$$r \le r_2 \tag{2.39}$$

$$(rB_\varphi)^2 = 4\pi p_1 r_2^2 + r_2^2 \frac{B_{z0}^2}{8\pi\alpha p_1}\left[1 - (1 + 8\pi\alpha p_1)\exp(-8\pi\alpha p_1)\right], \quad r > r_2 \tag{2.40}$$

Note some peculiarities of the obtained solution. First, it follows from Eq. (2.39) that the solution exists only if $p_1 > 0$, that is, when the plasma pressure decreases with the distance from the axis of the flux tube. This coincides with the data obtained with the use of *SOHO* and *TRACE* missions [26], which indicate that the plasma inside the loops is denser than the outer coronal plasma. The B_z component of the magnetic field inside the tube also decreases from the center to the periphery.

The azimuthally component of the magnetic field B_φ increases from the zero value on the axis of the tube to maximum

$$B_\varphi(r_2) = \sqrt{4\pi p_1\left(1 + \alpha B_{z0}^2\right)}, \quad 8\pi\alpha p_1 \ll 1 \tag{2.41}$$

$$B_\varphi(r_2) = \sqrt{4\pi p_1}\left[1 + \frac{\alpha B_{z0}^2}{2(4\pi\alpha p_1)^2}\right]^{1/2}, \quad 8\pi\alpha p_1 \gg 1 \tag{2.42}$$

and later decreases as r^{-1} for $r > r_2$. The electric current that flows through the cross section of the tube is specified by the formula [22]

$$\begin{aligned} I_z &= \frac{cr_{20} B_\varphi(r_2)}{2} \\ &= \frac{cr_2}{2}\left\{4\pi p_1 + \frac{B_{z0}^2}{8\pi\alpha p_1}\left[1 - (1 + 8\pi\alpha p_1)\exp(-8\pi\alpha p_1)\right]\right\}^{1/2} \end{aligned} \tag{2.43}$$

If the condition $B_{z0}^2/8\pi\alpha p_1 \ge 4\pi p_1$ is fulfilled, then ~35% of the total current I_z is concentrated in a thin paraxial region with the radius

$$r_* = \frac{r_2}{\sqrt{8\pi\alpha p_1}}, 8\pi\alpha p_1 \gg 1 \tag{2.44}$$

This is related to the rapid growth of the B_ϕ component of the magnetic field in the vicinity of the axis of the magnetic flux tube caused by the substantial impact of the

Cowling conductivity, which increases the "diffusion" of the magnetic field from the peripheral part of the tube toward its central region. The azimuthal current

$$j_\varphi = -\frac{2cB_{z0}\alpha p_1 r}{r_2^2}\exp\left(-4\pi\alpha p_1 \frac{r^2}{r_2^2}\right) \tag{2.45}$$

is essentially totally concentrated in the vicinity of the tube axis inside a radius $r \le r_*$. In the opposite limiting case of $8\pi\alpha p_1 \ll 1$, the electric current I_z displays uniform distribution across the section of the flux tube.

Let us estimate the value of $\alpha = \sigma F^2 \left[c^2 n m_i \nu'_{ia}(2 - F)\right]^{-1}$ for characteristic conditions in coronal plasma. Assuming that the effective cross section of ion–atom collisions for coronal temperature is determined by the process of recharging, we obtain the effective frequency $\nu'_{ia} \approx 1.6\times 10^{-11} nFT^{1/2}$ for the temperature interval $10^5 < T < 10^7$ K. In addition to that, note that for $T > 10^5$ K the relative abundance of neutrals in the corona is specified by the formula (see in this connection [23, 24])

$$F = 0,32 \times 10^{-3}\frac{1 + \frac{T}{6T_{\mathrm{H}}}}{\left(\sqrt{\frac{T}{T_1}}\right)^{2-b}\left(1 + \sqrt{\frac{T}{T_1}}\right)^{1+b}\sqrt{T}} e^{\frac{T_{\mathrm{H}}}{T}} \tag{2.46}$$

$$T_1 = 7.036 \times 10^5\ \mathrm{K},\quad T_{\mathrm{H}} = 1.58 \times 10^5\ \mathrm{K},\quad b = 0.748$$

where it is taken into account that the helium concentration in the corona is roughly 10%. As a result, within the temperature interval $10^6 < T < 10^7$ K, α only weakly depends on temperature and is equal to

$$\alpha = \frac{1,5 \times 10^{19}}{n^2} \tag{2.47}$$

Assuming, for example, for loops filled with evaporating chromospheric plasma, $n \approx 10^{10}\mathrm{cm}^{-3}$, $T \approx 10^7$ K, $p_1 \approx p$, we obtain the estimate $8\pi\alpha p_1 \approx 10^2 \gg 1$. Then it follows from Eq. (2.44) that in the paraxial part of the magnetic tube with the radius $r_* \approx 10^{-1} r_2$, rather large electric current density occurs, which may result in larger heating of the plasma in the paraxial region of the magnetic tube.

We have outlined the situation with model problems in the case of infinite magnetic flux tubes. The question arises on how the current closes in a real coronal magnetic loop. Two basic versions are essentially possible here. In the first case, the closure occurs due to the flow of the surface current in the inverse direction. The inverse current flows inside the skin layer, the thickness of which

$$\delta r \approx \left(\frac{c^2\tau}{4\pi\sigma}\left(1 + \alpha B_z^2\right)\right)^{1/2} \tag{2.48}$$

depends on the effective conductivity and the characteristic timescale of the process. If the characteristic time of the formation of an isolated loop is taken of the order of the size of a supergranulation cell divided into the rate of photospheric convection, we obtain the thickness of the skin layer for quasi-stationary loops in the solar corona of the order of 10^4 km. Such a scale, first, allows experimental verification and, second, makes it possible for closely located magnetic loops to interact inductively. The second possible way for the closure of the current that flows along loops is the closure through the photosphere or, more exactly, through layers where conductivity becomes isotropic with an increase in the density. In this case, the current flows from one footpoint of the loop through its coronal part to the other footpoint, and then is closured through the photosphere. The distribution of the vertical electric current in the vicinity of a footpoint of a coronal magnetic loop was obtained in the study [11] with the use of the vector magnetograph of Marshall Space Flight Center, with a spatial resolution of around 2.5 arcseconds (about 1800 km). Figure 2.3 presents the distribution of the vertical electric current in the vicinity of a footpoint of a coronal magnetic loop.

The measurements indicated the presence of a vertical electric current outflowing from the photosphere at one of the footpoints of the loop and inflowing into the photosphere at the other footpoint, without any signs of skinning current. This may imply that the current is closured through the photosphere rather than through the tube surface. In this case, the current was 1.5×10^{11} A, while the magnetic field in the footpoints of the loop was about 1000 G. The thickness of the skin layer was $\sim$5 arcseconds; therefore, the spatial resolution of the vector magnetograph would have been sufficient to detect the inverse current if it had existed.

2.4 Diagnostics of Electric Currents

2.4.1 "Warm" Loops

Let us consider a possibility for the diagnostics of electric currents in warm coronal magnetic loops with the temperature of the order of 1–2 MK, a number of which were observed in UV lines with *SOHO* and *TRACE* missions. The enhanced plasma density in such flux tubes, in comparison with the surrounding corona, may be related to the injection of dense chromospheric plasma into magnetic tubes, as a result of flute instability that develops in the vicinity of the footpoints of magnetic loops [25]. The data obtained with *SOHO* and *TRACE* missions make it possible to derive the thickness of a loop, $w = 2r_2$, and the average pressure in it, which appears to be substantially larger than that in the surrounding corona, and also to estimate the magnetic field, which subsequently yields the total current through the cross section of the magnetic flux tube calculated with Eq. (2.43).

Table 2.1 presents the observed values for the above parameters and the results of the diagnostics of electric currents made for 30 coronal magnetic loops observed with the *SOHO* UV telescope in the 171 Å line, as well as with the *SOHO*/Michelson–Doppler Imager magnetograph [26]. It follows from the table that, in the case of warm loops, the current displays an average value of about 2.5×10^{9} A.

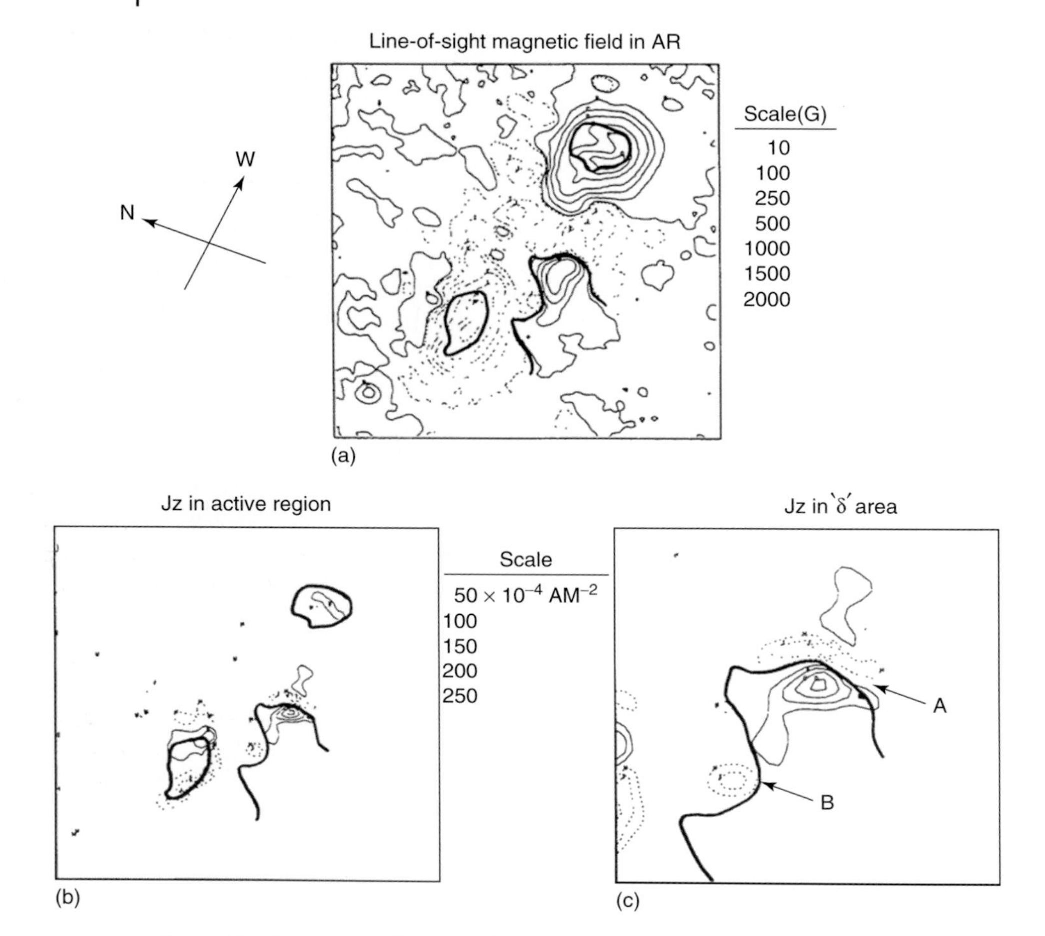

Figure 2.3 Distribution of vertical electric current density in AR 2372 on 6 April 1980 at 21 : 10 UT [11]. (a) The observed line-of-sight component of the magnetic field. Solid (dashed) contours represent positive (negative) polarities. (b) Distribution of the vertical electric current density of the active region. Solid (dashed) contours represent currents flowing out of (into) the photosphere over the 4×4 arc min field of view. (c) An enlargement of (b) in the 1.6×1.6 arc min area. The dark curve is the magnetic inversion line in δ region shown in (a) and (b). The spatial resolution is around 2.5 arcseconds (roughly 1800 km). No signs of skinning current are seen.

2.4.2
"Hot X-RAY" Loops

Similar diagnostic may be used to determine electric currents in hot X-ray loops. For example, for 16 hot X-ray loops observed by the *Yohkoh* mission [27], the average temperature of plasma was $T = 5$ MK, the plasma number density $n_e = (0.6-3) \times 10^{10}\ \text{cm}^{-3}$, the thickness of the loops $w = 9 \times 10^8$ cm, and the average

Table 2.1 Parameters of "warm" magnetic loops, according to the study [26], and the results of the diagnostics of electric currents in the loops.

Number	W (Mm)	n_e (10^9 cm^{-3})	T (MK)	B (G)	I_z (10^9 A)
1	6.8	2.5	1.08	−413	2.89
2	6.1	2.3	1.09	−413	2.49
3	7.4	2.1	1.25	−285	2.88
4	6.9	2.6	1.26	−285	3.00
5	6.3	2.5	1.22	−270	2.64
6	7.0	3.7	0.93	−298	3.29
7	7.1	1.7	1.36	−261	2.57
8	7.1	2.2	1.26	−114	2.76
9	8.1	2.1	1.32	−333	3.26
10	7.9	1.4	1.16	−148	2.37
11	6.8	1.6	1.44	7	2.40
12	7.8	2.0	1.31	−208	2.98
13	6.4	1.8	1.31	−252	2.34
14	6.4	1.7	0.73	−157	1.75
15	6.8	2.0	1.04	−294	2.43
16	6.7	1.4	1.15	−269	2.06
17	7.4	1.7	1.03	−178	2.33
18	5.6	1.1	2.03	−159	1.95
19	7.9	1.3	1.43	−129	2.51
20	7.7	1.8	1.22	−140	2.68
21	7.6	2.1	0.98	−97	2.55
22	7.2	1.0	1.27	−159	1.91
23	8.0	2.3	1.31	54	3.23
24	8.7	1.3	1.41	951	3.42
25	8.1	1.8	0.93	76	2.45
26	5.1	1.4	1.30	60	1.60
27	7.1	1.5	1.26	61	2.27
28	7.5	2.6	1.24	128	3.15
29	6.0	2.4	1.31	128	2.49
30	7.4	1.7	0.84	142	2.11

magnetic field $B = 100$ G. For these parameters, Eq. (2.43) yields the electric current through the cross section of a loop $I_z \approx 3 \times 10^{10}$ A, which exceeds that in "warm" magnetic loops by an order of magnitude.

2.4.2.1 Flare Magnetic Loops

We estimate electric currents in flaring magnetic loops using the example of the well-studied flare on 14 July 2000, known as the *Bastille Day Flare*. In the course of the flare, an arcade of coronal magnetic loops was observed. Figure 2.4, taken from [28], presents the photo of the arcade, obtained by *TRACE* in 171 Å (1 MK).

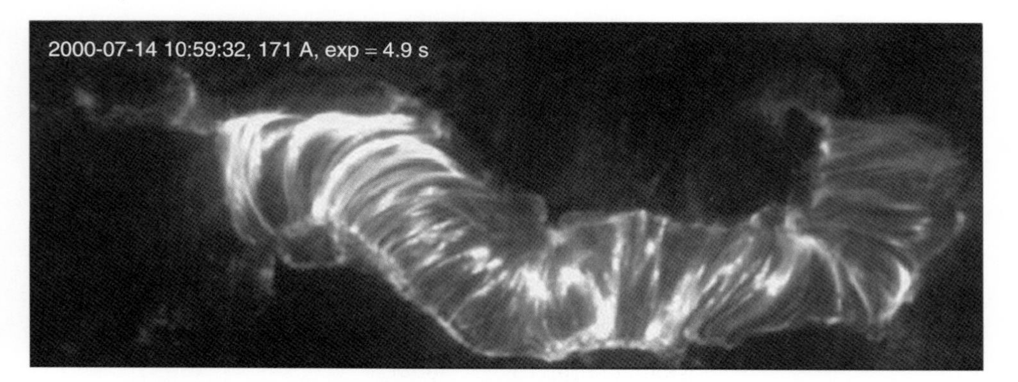

Figure 2.4 The arcade of coronal magnetic loops observed by *TRACE* mission in 171 Å line 40 min after the peak of the solar flare on 14 July 2000. (From Aschwanden and Alexander [28].)

In total, the arcade contained about 100 loops; the active phase of the flare was observed in $N_{\text{fl}} = 23$ of the loops. According to estimates [29], free magnetic energy involved in the flare process was $E_{\text{magn}} \approx 1.6 \times 10^{32}$ erg, that is, each flare loop accounted for free energy of the order of $E_{\text{magn}}^{\text{loop}} \approx (E_{\text{magn}}/N_{\text{fl}}) \approx 7 \times 10^{30}$erg. On the other hand, free or nonpotential energy of a coronal magnetic loop is the energy of electric current that flows through the loop, that is,

$$E_{\text{magn}}^{\text{loop}} = \frac{LI_z^2}{2c^2} \tag{2.49}$$

where L is the inductance of an individual loop, which for a thin magnetic loop, in which the major radius R is large compared to the minor radius r_2, that is, $R \gg r_2$, may be written in the form [30]

$$L = 2l\left(\ln\frac{4l}{\pi r_2} - \frac{7}{4}\right) \tag{2.50}$$

where $l \approx \pi R$ is the length of the coronal part of the loop. In the event on 14 July 2000, all the loops of the magnetic arcade were approximately the same, with the average size being $R \approx 1.75 \times 10^9$ cm, $r_2 \approx 0.9 \times 10^8$ cm, and the distance between the loops $d \approx 1.8 \times 10^8$ cm. Then from Eqs. (2.49) and (2.50), we obtain an estimate for auto-inductance of a single magnetic loop $L \approx 3 \times 10^{10}$ cm ≈ 30 H, and of the electric current in the flare loop $I_z \approx 2.6 \times 10^{11}$ A. This is close to the current value measured in Ref. [11] with the use of the vector magnetograph for the flare loop observed on 6 April 1980, as mentioned in Section 2.3.

Thereby, the considered data indicate that the transition from coronal magnetic loops with moderate temperature of the order of 1–2 MK to hot X-ray loops with the temperature of the order of 5–7 MK, and subsequently to flare magnetic loops is accompanied with a systematic increase in the electric current in the loop from 2.5×10^9A for "warm" loops to 3×10^{10}A for "hot" loops and to $(1-3) \times 10^{11}$A for flare loops. This may mean that the electric current that flows through the cross section of a coronal magnetic loop plays an important role in the heating of plasma and in the origin of flares in the considered magnetic structures.

2.5
The Equivalent Electric Circuit

It was shown in the previous sections that convective motions of the photosphere matter result in generation of thin magnetic flux tubes with radii ranging 10^7–10^8 cm and with the electric current inside the tubes probably exceeding 10^{11} A. The current flows from one footpoint of a loop through its coronal part to the other and is closured in subphotospheric layers. The depth of the current closure should be specified from the conditions of isotropy of the conductivity. The electric current in the plasma with a magnetic field may be presented in the form

$$\vec{j} = \sigma \vec{E}_{||} + \sigma_{\rm p} \vec{E}_{\perp} + \frac{\sigma_{\rm H}}{B} \vec{B} \times \vec{E}_{\perp} \tag{2.51}$$

Here, σ is the conductivity in the absence of the magnetic field. The Pedersen conductivity is

$$\sigma_{\rm p} = \sigma \frac{1 + F\omega_{\rm e}\omega_{\rm i}\tau_{\rm e}\tau_{\rm ia}}{1 + \omega_{\rm e}^2\tau_{\rm e}^2 + 2F^2\omega_{\rm e}\omega_{\rm i}\tau_{\rm e}\tau_{\rm ia} + (F^2\omega_{\rm e}\omega_{\rm i}\tau_{\rm e}\tau_{\rm ia})^2} \tag{2.52}$$

while the Hall conductivity is

$$\sigma_{\rm H} = \sigma_{\rm p} \frac{\omega_{\rm e}\tau_{\rm e}}{1 + F^2\omega_{\rm e}\omega_{\rm i}\tau_{\rm e}\tau_{\rm ia}} \tag{2.53}$$

where $\tau_{\rm e} = (\nu'_{\rm ea} + \nu'_{\rm ei})^{-1}$, $\tau_{\rm ia} = (\nu'_{\rm ia})^{-1}$, $\omega_e = eB/m_e c, \omega_{\rm i} = eB/m_{\rm i}c$. In the coronal part of the loop, where $\omega_{\rm e}\tau_{\rm e} \gg 1$, $F \approx 0$, the condition $\sigma \gg \sigma_{\rm H} \gg \sigma_{\rm p}$ is fulfilled, that is, the conductivity is strongly anisotropic. The relation between the current and the electric field has the form $\vec{j} = \sigma \vec{E}_{||}$, and the basic part of the current flows along the magnetic field lines. Under the photosphere, the situation is inversed. Here, $\omega_{\rm e}\tau_{\rm e} \ll 1$, $F^2\omega_{\rm e}\omega_{\rm i}\tau_{\rm e}\tau_{\rm ia} \ll 1$, and the conditions $\sigma \approx \sigma_{\rm p}$, $\sigma_{\rm H} \ll \sigma$ are fulfilled. Therefore, Ohm's law takes the same form as that in the absence of magnetic field $\vec{j} = \sigma \vec{E}$, and the current starts to flow from one footpoint of the arch to the other along the line of least electric resistance, rather than along the magnetic tube, as in the corona and in the chromosphere. For the magnetic field in the loop in the photosphere $B \sim 10^3$ G, the isotropic conductivity volume and the region of the current closure are at a depth of around 75 km under the photosphere if we use the well-known solar atmosphere model [31]. Thereby, supergranulation convection generates magnetic loops whose footpoints are not deeply submerged into subphotospheric layers. A loop, together with the subphotospheric current channel, is similar to a turn with electric current for which the equation of equivalent electric circuit may be written.

A phenomenological approach, based on the analogy between magnetic loops in the solar and stellar atmospheres and an electric or RLC-circuit, appeared to be fruitful not only in describing the flare processes in the Sun and stars [10, 20] but also in the problem of heating of stellar coronae, in electrodynamics of hot stars, and in accretion disks of magnetic neutron stars [32, 33]. Considering the electric analogy for a coronal magnetic loop, Ionson [34] analyzed the case when a loop magnetic field is potential and is formed by a generator of static field located deep under the photosphere. Here, we consider a current-carrying magnetic loop with the generator of the current related to photospheric convection. In this case, the loop magnetic field is essentially nonpotential. As a result, the effective resistance and capacitance of the circuit become dependent on the current value through self-consistent values for ω_e and ω_i.

The slow variation of the current for a time substantially exceeding the period of free oscillations of the circuit is described by the equation [35]:

$$\frac{1}{c}\frac{\mathrm{d}(LI)}{\mathrm{d}t} + RI = \Xi \tag{2.54}$$

where I is the total current that flows through the cross section of a loop parallel to its axis,

$$R(I) = \frac{l_1}{\pi r_1^2 \sigma_1} + \frac{l_2}{\pi r_2^2 \sigma_2} + \frac{l_3}{\pi r_3^2 \sigma_3} + \frac{3\xi l_1 F^2 I^2}{\pi r_1^4 (2-F) c^4 n m_i \nu'_{ia}}\left[1 + b^{-2}\right] \tag{2.55}$$

l_1, l_2, and l_3 are the lengths of the portions of electric circuit in the region of action of photospheric *emf*, in the corona, and in the region of the current closure under the photosphere, respectively (Figure 2.1), r_1, r_2, and r_3 are the radii of the current channel in these regions, σ_1, σ_2, and σ_3 are the corresponding isotropic conductivities, and $\xi \approx 0,5$ is the form factor that arises in the integration over the volume of the arch; b is specified by Eq. (2.28). The right-hand side of Eq. (2.54) contains the *emf* caused by photospheric convection in the footpoint of the loop:

$$\Xi = \frac{l_1}{\pi c r_1^2} \int_0^{r_1} V_r B_\varphi 2\pi r dr \approx \frac{\left|\overline{V}_r\right| I l_1}{c^2 r_1} \tag{2.56}$$

where $\left|\overline{V}_r\right|$ is the average radial component of the velocity of the convective flow inside the tube in the dynamo region of the photosphere. The basic contribution to the circuit resistance is introduced by the last summand in Eq. (2.55) related to the dynamo region of the loop. In this region, the density of electric current is virtually perpendicular to the magnetic field, while the plasma is partially ionized, as follows from Eqs. (2.51) and (2.53). Therefore, the Cowling conductivity [18] plays a crucial role

$$\sigma_K = \frac{\sigma_1}{1 + F^2 \dfrac{\omega_e \omega_i}{(2-F)\nu'_{ei}\nu'_{ia}}} \tag{2.57}$$

and for $F^2\omega_e\omega_i/(2-F)\nu'_{ei}\nu'_{ia} \gg 1$, it is substantially smaller than the isotropic conductivity σ_1. Since ω_e and ω_i depend on the self-consistent magnetic field of the loop, the circuit resistance appears to be dependent on the current I. If we approximate a current-carrying loop with a thin turn with length l and thickness $2r$, its inductance may be estimated with Eq. (2.50), in which the values for r and l may be taken for the coronal part of the loop.

The characteristic time of the increase in the current in the loop to its maximum is specified by the smaller timescales of the two:

$$t_R = \frac{L}{c^2 R}, \quad t_L = \left(\frac{1}{L}\frac{\mathrm{d}L}{\mathrm{d}t}\right)^{-1} \tag{2.58}$$

The value of t_R specifies the time of increase in the current in a stationary magnetic loop. If the inductance of a loop varies with time, for example, the size of the magnetic loop increases, and if in this case $t_L < t_R$, then the time of increase in the current is specified by t_L. By an order of magnitude, t_L is equal to several hours, that is, it is comparable with the lifetime of a supergranulation cell. The time t_R displays the same order of magnitude if we estimate it using the characteristic size of a single flare loop in the Sun ($l \approx 5 \cdot \times 10^9$ cm) and the parameters of the dynamo region in the photosphere.

An established current value is determined according to Eq. (2.54) from the relation

$$R(I)I = \Xi(I) \tag{2.59}$$

As can be easily verified, this value virtually coincides with the current Eq. (2.34) for infinite uniform magnetic flux tube if we neglect the insignificant contribution from the coronal part of the loop and the region of the current closure into the total resistance of the circuit.

To obtain an equation for eigenoscillations of a loop as an equivalent RLC-circuit, we should exclude velocity variations from the equation of motion of the bulk plasma

$$\rho\frac{\mathrm{d}\boldsymbol{V}}{\mathrm{d}t} = \rho\boldsymbol{g} - \nabla p + \frac{1}{c}\boldsymbol{j} \times \boldsymbol{B} \tag{2.60}$$

and from the generalized Ohm's law in the nonstationary form

$$\begin{aligned} \boldsymbol{E} + \frac{1}{c}V \times B = \frac{j}{\sigma} + \frac{j \times \boldsymbol{B}}{enc} - \frac{\nabla p_e}{en} - \frac{F^2}{(2-F)cnm_i\nu'_{ia}}[j \times \boldsymbol{B}] \times B \\ + \frac{(1-F)F^2}{(2-F)cnm_i\nu'_{ia}}\rho\frac{dv}{dt} \times \boldsymbol{B} \end{aligned} \tag{2.61}$$

expressing the electric field through variations of electric current. Then, integration over the volume of the magnetic loop should be made, taking into account the equation

$$\oint \frac{\partial E_z}{\partial t} dl = -\frac{L}{c^2}\frac{\partial^2 I}{\partial t^2} \tag{2.62}$$

As a result, for current oscillations with a minor amplitude $|I_\sim| \ll I$, the following equation is obtained [35]:

$$\frac{1}{c^2}L\frac{\partial^2 I_\sim}{\partial t^2} + \left[R(I) - \frac{|\overline{V}_r|\, l_1}{c^2 r_1}\right]\frac{\partial I_\sim}{\partial t} + \frac{1}{C(I)}I_\sim = 0 \tag{2.63}$$

The effective capacitance of the circuit is determined from the relation

$$\frac{1}{C(I)} = \frac{I^2}{\pi c^4}\left(\frac{l_1}{\rho_1 r_1^4} + \frac{l_2}{\rho_2 r_2^4}\right)(1 + b^{-2}) \tag{2.64}$$

Since $l_1/\rho_1 \ll l_2/\rho_2$, the basic contribution to the capacity of the circuit is made by the coronal part of the loop. It follows from Eq. (2.63) that circuit oscillations increase if $R(I) \lessdot |V_r| l_1/c^2 r_1$, that is, if the current in the circuit is smaller than the established value, and attenuate if the photospheric *emf* ceases its action, for example, due to a decrease in the convection rate when plasma is heated during a flare. This attenuation is slow, since the circuit's Q-factor $Q = \left[1/cR(I)\right]^{-1}\sqrt{L/C(I)}$ reaches $10^3 - 10^4$ for typical parameters of flare loops. The frequency of oscillations of the RLC-circuit is specified by the formula

$$\nu_{\mathrm{RLC}} \approx \frac{1}{2\pi\sqrt{2\pi\Lambda}}\left(1 + b^{-2}\right)^{1/2}\frac{I}{c r_2^2\sqrt{n_2 m_{\mathrm{i}}}} \tag{2.65}$$

where $\Lambda = \ln\frac{4l}{\pi r_2} - \frac{7}{4}$ and r_2 and n_2 refer to the coronal part of the loop.

The dependence of frequency on current in Eq. (2.65) is related to the fact that the equivalent capacity of a loop is specified by Alfven velocity, which not only depends on z-component but also on the φ-component of the magnetic field in the loop. The parameter b is specified by the magnetic field components and the pressure on the flux tube axis and beyond it:

$$b = \frac{B_{\varphi 0}(r_2)}{B_{z0}(r_2) - B_{z0}(0)} \approx 6\frac{B_{\varphi 0}(r_2)B_{z0}(0)}{8\pi\left[p(\infty) - p(0)\right]} \tag{2.66}$$

In the coronal part of a loop, $b^2 \gg 1$ may be assumed; therefore, for the frequency of LRC-oscillations of an equivalent electric circuit, we obtain the following formula:

$$\nu_{\mathrm{RLC}} = \frac{c}{2\pi\sqrt{LC(I_0)}} \approx \frac{1}{(2\pi)^{3/2}\sqrt{\Lambda}}\frac{I}{c r_2^2\sqrt{n_2 m_{\mathrm{i}}}} \tag{2.67}$$

The dependence of this frequency on the electric current may be used for diagnostics of electric currents in coronal magnetic loops based on temporal fluctuations of radio- and X-ray radiation of flares in the Sun and stars. Microwave emission of coronal magnetic loops is commonly interpreted as gyrosynchrotron radiation of some population of fast electrons at gyrofrequency harmonics in the magnetic field of a loop. In the case of power–law energy distribution of electrons $f(E) \propto E^{-\delta}$, the intensity of gyrosynchrotron radiation of an optically thin source is specified by the relation [36] $I_\nu \propto B^{-0.22+0.9\delta}(\sin\vartheta)^{0.43+0.65\delta}$. For typical values of the index of the energy spectrum of electrons ($2 \lessdot \delta < 7$), we obtain relatively strong dependence of the intensity on the magnetic field. Thereby, oscillations of the magnetic field in a coronal magnetic loop caused by current oscillations may be reduced to a modulation in the intensity of the detected radio emission.

Equation (2.63) suggests that oscillations of electric current are cophased in all points of a loop as an equivalent electric circuit. On the other hand, variations of the current propagate along the loop with the Alfven velocity. Therefore, for the condition of phase coincidence, the Alfven time $\tau_{\mathrm{A}} = l_2/C_{\mathrm{Ai}}$ should be smaller than the period of RLC oscillations $P_{\mathrm{RLC}} = 1/\nu_{\mathrm{RLC}}$. Since $I_0 \approx c r_2 B_{\varphi 0}(r_2)/2$, the in-phase condition takes the form

$$\frac{B_{\varphi 0}(r_2)}{B_{z0}(0)} < \pi\sqrt{2\Lambda}\frac{r_2}{l_2} \tag{2.68}$$

that is, the shear of the magnetic field in a coronal magnetic loop should be sufficiently small. Note that the magnetic loops observed in the solar corona, as a rule, display a very small twist of the magnetic field; therefore, Eq. (2.67) is applicable at least for magnetic loops that are not very long.

2.6 Inductive Interaction of Magnetic Loops

Equations (2.54) and (2.63) for slow and rapid variations of the electric current in a coronal magnetic loop are valid for a loop, which is magnetically isolated from surrounding loops, that is, these equations do not include effects of mutual induction related to variations of the outer magnetic flux through the circuit of the loop. These effects may be included into the integration of the generalized Ohm's law by adding the electromotive force of mutual induction $\frac{1}{c^2}\frac{\partial^2}{\partial t^2}\left[\sum_j^N M_j I_j\right]$ to $\frac{\partial}{\partial t}\oint E_z dz$, where I_j is the current in the jth loop, M_j is the factor of mutual induction between the jth and considered loops, and the summation is made over all loops that surround the selected loop. In the case of slow variations of the currents in the surrounding loops (for the time substantially exceeding the period of RLC-oscillation, $1/\nu_{\mathrm{RLC}} \ll \tau_{\mathrm{curr}}$, where τ_{curr} is the characteristic timescale of a current variation), the mutual induction affects only slow variations of the equilibrium current I_{z0} in the given loop. Therefore, in the studies of relatively rapid RLC oscillations of the electric current in a loop, the impact of the surrounding loops may be ignored. In particular, this impact does not alter the type of functional dependence of frequency ν_{RLC} on the current I_{z0} that flows in the loop. Along with this, slow variations I_{z0} caused by induction interaction with surrounding loops result in a definite drift in the frequency ν_{RLC} of RLC – oscillations – which, in turn, modulate microwave radiation, and may be detected in low-frequency spectra. In this connection, we distinguish two different types of the dynamics of electric current in magnetic loops: "rapid processes," that is, RLC oscillations of electric current around the equilibrium value I_{z0}, and "slow processes," that is, variations in equilibrium current, which result in drift of the frequency ν_{LCR} of RLC oscillations. One of the reasons for slow variation I_{z0}, which was considered in Section 2.5, is the increase in the current driven by the photospheric electromotive force, or its dissipation in the course of the development of the flare process in a coronal magnetic loop. Another reason may be the inductive *emf*, which arises in the interaction of a magnetic loop with other loops in the course of their floating upward, or their relative motion [8, 37, 38]. In the latter case, the system of equations for slow variations of the current in two inductively interacting magnetic loops is written as follows [37]:

$$\frac{1}{c^2}\frac{\partial}{\partial t}(L_1 I_1 + M_{12} I_2) + I_1 R_1(I_1) = \Xi_1 \tag{2.69}$$

$$\frac{1}{c^2}\frac{\partial}{\partial t}(L_2 I_2 + M_{21} I_1) + I_2 R_2(I_2) = \Xi_2 \tag{2.70}$$

Here, $L_{1,2}$ are inductances of the loops, specified by Eq. (2.50), with corresponding values for parameters for each loop, $R_{1,2}(I_{1,2})$ is the total resistance of each loop, determined from Eq. (2.55), and $\Xi_{1,2}$ are electromotive forces in photospheric footpoints of the loops (Eq. (2.56)). In the case of the loops approximated by thin torus, when the main radii $R_{\text{loop}}^{(i)} \gg r_i$ (r_i are minor radii of the torus), the factors of mutual induction may be approximated by the formula [39, 40]

$$M_{12} = M_{21} = 8(L_1 L_2)^{1/2}\left[\frac{R_{\text{loop}}^{(1)} R_{\text{loop}}^{(2)}}{\left(R_{\text{loop}}^{(1)} + R_{\text{loop}}^{(2)}\right)^2 + d_{1,2}^2}\right]\cos\varphi_{12} \tag{2.71}$$

In Eq. (2.71), $R_{\text{loop}}^{(1,2)}$ are major radii of the loops, $d_{1,2}$ is the distance between the centers of the loops' tori, and φ_{12} is the angle normal to the planes of the loops.

Equations (2.69) and (2.70) were used in the study [37] for modeling of double tracks, which sometimes emerge in the spectra of low-frequency modulation of the intensity of microwave radiation of flares.

Figure 2.5 presents time profiles of microwave radiation of three flares at a frequency of 37 GHz, dynamical spectra of low-frequency modulation of this radiation, obtained using the Wigner–Ville method, and results of numerical modeling of these dynamical spectra as RLC oscillations of two inductively interacting coronal magnetic loops. Table 2.2 presents the parameters of the loops, the velocities of their floating upward, and their mutual orientation, for which the best consistency is attained between the results of the simulations and the experimental data.

2.7 Waves of Electric Current in Arcades of Coronal Magnetic Loops

As another example of inductive interaction of coronal magnetic loops, we consider waves of electric current, which may propagate in an arcade of magnetic loops. For the sake of simplicity, we suggest that all loops in the arcade are of the same geometric size (major radii R_{loop} and minor radii r), their planes are perpendicular to z-axis, and they are situated at distance $d \ll R_{\text{loop}}$ from each other. In addition, we suggest that $d \leq \delta r$ are the thicknesses of the skin layer, specified by Eq. (2.48), so that inductive interaction is possible. Presenting each loop as an electric circuit with the same inductance L, capacity C, resistance R, and electromotive force, we write the equation for the current $I(z)$ in the magnetic loop, the center of which displays the coordinate z on the oz-axis:

$$\frac{L}{c^2}\frac{\partial^2 I(z)}{\partial t^2} + (R - \varepsilon)\frac{\partial I(z)}{\partial t} + \frac{1}{C}I(z) = -\frac{M}{c^2}\left[\frac{\partial^2 I(z-d)}{\partial t^2} + \frac{\partial^2 I(z+d)}{\partial t^2}\right] \tag{2.72}$$

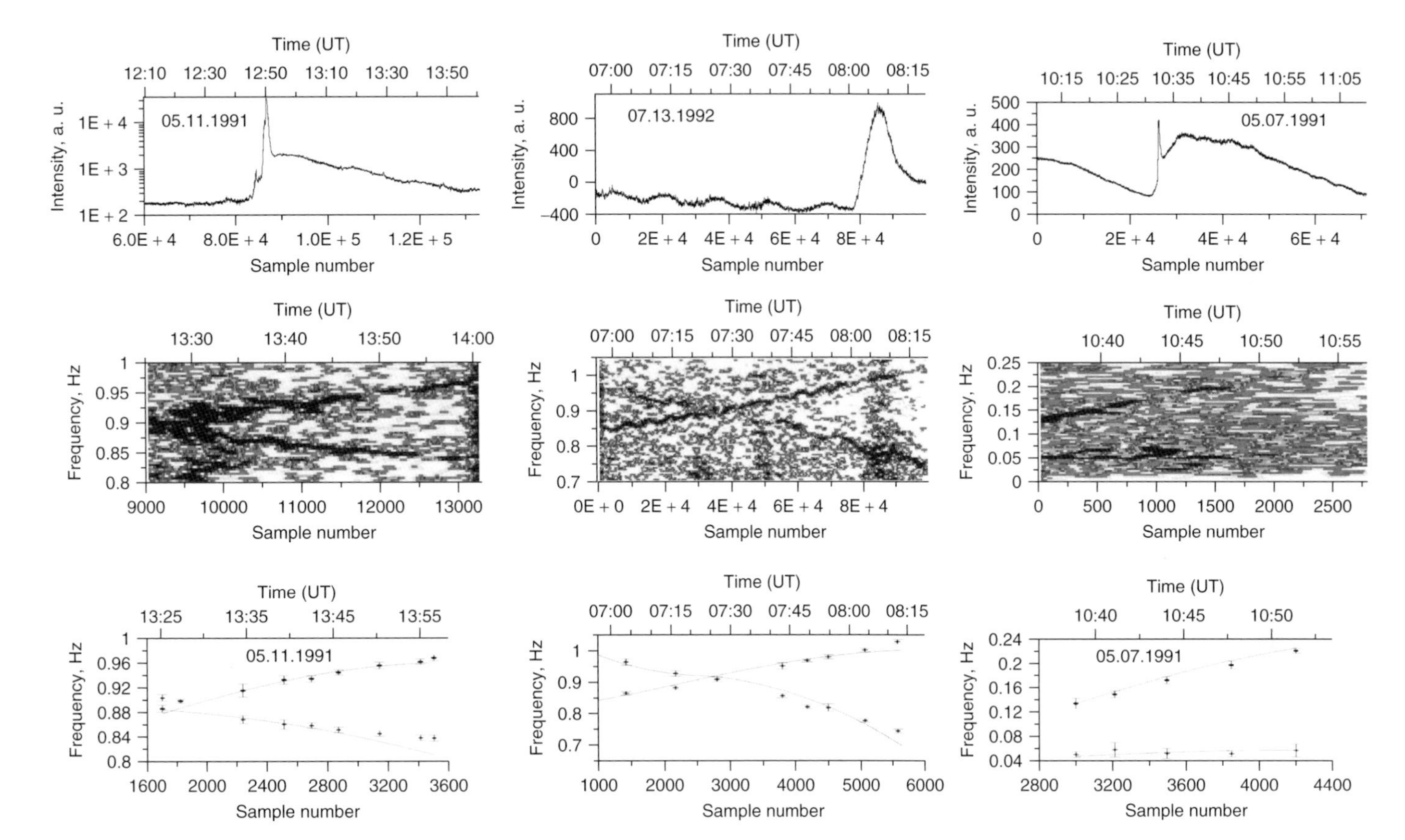

Figure 2.5 Inductive interaction of two current-carrying loops in the course of their floating upward or relative motion. Slow variations of the current due to their inductive interaction with the adjacent loop result in a drift in the frequency of modulation of microwave radiation [37].

Table 2.2 Parameters of coronal loops for three flares presented in Figure 2.5.

Event	**7 May 1991**	**11 May 1991**	**13 July 1992**
$r_0^{(1)}$	600 km	600 km	590 km
$r_0^{(2)}$	550 km	700 km	510 km
$R_{\text{loop0}}^{(1)}$	20 000 km	20 000 km	20 000 km
$R_{\text{loop0}}^{(2)}$	1000 km	1 000 km	3000 km
v_1	$0.5\ \text{km s}^{-1}$	$1.2\ \text{km s}^{-1}$	$0.5\ \text{km s}^{-1}$
v_2	$4.5\ \text{km s}^{-1}$	$4.9\ \text{km s}^{-1}$	$2.5\ \text{km s}^{-1}$
T_{01}	10^7 K	10^7 K	2×10^6 K
T_{02}	10^7 K	5.0×10^6 K	1.0×10^6 K
n_1	$1.3 \times 10^9\ \text{cm}^{-3}$	$2.0 \times 10^9\ \text{cm}^{-3}$	$3.5 \times 10^9\ \text{cm}^{-3}$
n_2	$2.4 \times 10^9\ \text{cm}^{-3}$	$4.5 \times 10^9\ \text{cm}^{-3}$	$5.0 \times 10^9\ \text{cm}^{-3}$
I_{01}	1.7×10^8 A	-1.58×10^{10} A	-1.8×10^{10} A
I_{02}	1.7×10^8 A	1.58×10^{10} A	2.3×10^{10} A
B_1	85 Gs	120 Gs	170 Gs
B_2	85 Gs	120 Gs	100 Gs
d_{12}	2500 km	2500 km	2500 km
φ_{12}	$\pi/20$	$\pi/4$	$\pi/4$

Major R_{loop} and minor r_0 radii of the arches, velocities of floating upward v of the loops, temperature T, magnetic field B, and plasma density n in the loops are indicated.

Here, we have taken the inductive interaction of the considered loop with two adjacent loops into account, on the right and on the left, assuming the interaction with other loops to be weak. For example, for the arcade of coronal magnetic loops in Figure 2.4, the characteristic time of propagation of the flare process to the total arcade was of the order of 1 h, and the skin-layer thickness estimated from Eq. (2.48) was $\delta r \approx (0.3 - 1.0) \times 10^8$ cm, which is of the order of the thickness of an individual magnetic flux tube. In Eq. (2.72), ε is the negative resistance related to the presence of an effective electromotive force in the footpoints of the loop, caused by the interaction of the convective flow of photospheric plasma with the magnetic field of the loop, and M is the factor of mutual induction, specified by Eq. (2.71). Presenting the right-hand part of Eq. (2.72) in the form of a Taylor series expansion in the vicinity of the z-coordinate, we obtain the following equation for $I(z,t)$:

$$\frac{(L+2M)}{c^2}\frac{\partial^2 I}{\partial t^2} + (R-\varepsilon)\frac{\partial I}{\partial t} + \frac{1}{C}I + \frac{M}{c^2}\frac{\partial^4 I}{\partial t^2 \partial z^2}d^2 = 0 \tag{2.73}$$

Hence, for perturbations of the type $I \propto \exp(i\omega t - ikr)$, we obtain a dispersion equation for waves of electric current, propagating along the arcade of coronal magnetic loops [41]:

$$\omega^2\left(\frac{L+2M}{c^2} - \frac{M}{c^2}k^2 d^2\right) - i\omega(R-\varepsilon) - \frac{1}{C} = 0 \tag{2.74}$$

The real and imaginary parts of the frequency are equal, respectively, to

$$\omega_r^2 = \omega_0^2 - \frac{1}{4}(R-\varepsilon)^2 C^2 \omega_0^4, \quad \gamma = \frac{1}{2}(R-\varepsilon)C\omega_0^2, \tag{2.75}$$

where

$$\omega_0^2 = C^{-1}\left(\frac{L+2M}{c^2} - \frac{M}{c^2}k^2 d^2\right)^{-1} \tag{2.76}$$

It follows from Eq. (2.75) that, for $R < \varepsilon$, current perturbations in the system of loops may increase. If $\gamma^2 \ll \omega_0^2$ and $k^2 d^2 \ll 1$, the perturbations propagate along the arcade with the group velocity

$$V_g = \frac{\partial \omega_r}{\partial k} \approx \frac{c}{\sqrt{(L+2M)C}} \frac{M}{(L+2M)} k d^2 \tag{2.77}$$

As an example, we estimate the group velocity of propagation of perturbations of electric current in the known arcade of coronal magnetic loops observed during the solar flare on 14 July 2000, presented in Figure 2.4. For typical values of the parameters of coronal magnetic loops $\omega_r \approx 1\ \text{s}^{-1}, d \approx 10^8$ cm, $kd \approx 0,1$ and $M \approx 2\ L$, Eq. (2.77) yields the group velocity $V_g \approx 3 \times 10^6\ \text{cm s}^{-1}$, which is close to the observed values of $(3-5) \times 10^6\ \text{cm s}^{-1}$ for the velocity of propagation of the flare process along the arcade of magnetic loops in the *Bastille Day event* on 14 July 2000.

Thereby, in coronal magnetic loops, large electric currents may exist, which are generated as a result of either interaction of convective flows of photospheric plasma with the magnetic field of the loops or injection of evaporating chromospheric plasma into a loop. This current flows from one footpoint of a loop through the coronal part to the other footpoint, and closures through the photosphere. No experimental evidence of the existence of the inverse current in the outer shell of a loop has been found. The data considered here indicate that the transition from coronal magnetic loops with a moderate temperature of the order of 1–2 MK to hot X-ray loops with the temperature of the order of 5–7 MK, and to flare magnetic loops is accompanied by a systematic increase in the electric current in the loop from 2.5×10^9A for "warm" loops to 3×10^{10}A for "hot" loops and to $(1-3) \times 10^{11}$A for flare loops. This may mean that the electric current flowing through the cross section of a coronal magnetic loop plays an important role in the heating of plasma and in origination of flares in the considered magnetic structures. A coronal magnetic loop is an equivalent electric circuit, the intrinsic frequency of which is proportional to the electric current in the circuit. This makes it possible to diagnose electric currents from low-frequency modulation of microwave radiation generated in coronal magnetic loops.

2.8 Magnetic Loops above Spots

The process of formation of coronal magnetic loops above spots apparently consists of two phases, substantially different in their physical nature. First, a subphotospheric magnetic flux tube floats upward into the corona. As a result, a dipole

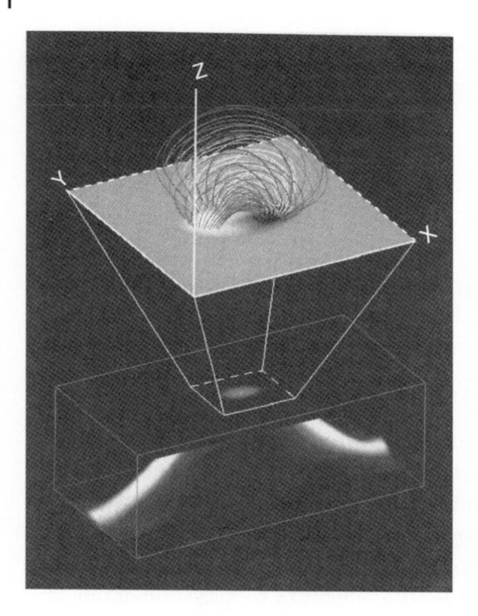

Figure 2.6 Modeling of floating of a magnetic tube upward from the convective zone (lower box) into the corona. The photosphere is indicated by the X–Y plane [43].

magnetic structure originates in the form of two spots of opposite polarity at the photosphere level and numerous thin loops in the corona. The other important process is filling of the coronal magnetic loops with hot dense plasma; as a result, the emission measure of the loops increases. It cannot be excluded that both the processes occur simultaneously. Three-dimensional MHD modeling [42] indicates that, in the course of floating of a magnetic flux tube upward, instabilities develop, in particular, the interchanging instability, which results in the formation of a filament structure of the magnetic field at coronal levels; this structure manifests itself as an aggregation of coronal magnetic loops above the spots. Figure 2.6 presents a model example of floating a magnetic tube upward, with the formation of coronal magnetic loops [43].

Filling of the potential magnetic field in the vicinity of sunspots with plasma, as a result of which magnetic loops with enhanced plasma density are formed, may occur due to the development of the ballooning mode of flute instability in the chromospheric footpoints of magnetic spots [44]. This instability is developed on the boundary between the spot penumbra, where $\beta \ll 1$, and the surrounding chromosphere, where $\beta \leq 1$. In this region, the magnetic field expands and gains curvature directed from the surrounding chromosphere into the spot. The radius of the curvature of the magnetic field lines is of the order of the scale of the height of nonuniform atmosphere,

$$R_c \approx \frac{k_B T}{\mu m_i g} \tag{2.78}$$

Owing to the curvature of the magnetic field, the centrifugal force appears

$$\vec{f}_{\mathrm{c}} = \frac{2nk_{\mathrm{B}}T}{R_{\mathrm{c}}^2}\vec{R}_{\mathrm{c}} \tag{2.79}$$

which acts on a cubic centimeter of plasma with the density $\rho \approx (n + n_{\mathrm{a}})m_{\mathrm{i}}$. Therefore, the effective centrifugal acceleration acting on the chromosphere plasma, which surrounds the spot, is equal to

$$\vec{g}_{\mathrm{c}} = \left(\frac{\vec{f}_{\mathrm{c}}}{\rho}\right) = \frac{2k_{\mathrm{B}}T}{m_{\mathrm{i}}R_{\mathrm{c}}^2}\frac{n}{n + n_{\mathrm{a}}}\vec{R}_{\mathrm{c}} \tag{2.80}$$

Substituting Eqs. (2.78–2.80), we obtain

$$g_{\mathrm{c}} = 2g\frac{n}{n + n_{\mathrm{a}}} \tag{2.81}$$

The condition of the ballooning instability is of the form

$$g_{\mathrm{c}} - g\ \cos\theta\ > 0 \tag{2.82}$$

where θ is the angle between the direction of the radius of the curvature $\boldsymbol{R}_{\mathrm{c}}$ and the vertical line. In the case when $\boldsymbol{R}_{\mathrm{c}}$ is approximately perpendicular to $\boldsymbol{g}$, the time of the increase in the ballooning mode of the flute instability is equal to

$$\tau_b = \frac{1}{2}\left(\frac{\lambda}{\pi g}\right)^{1/2}\left(\frac{n}{n + n_{\mathrm{a}}}\right)^{-1/2} \tag{2.83}$$

where λ is the wavelength of the perturbation. Equation (2.83) corresponds to the most unstable case, when the wave vector of the perturbations is perpendicular to the magnetic field [45]. In the case when $n_{\mathrm{a}} \ll n$

$$\tau_b \approx 2 \times 10^{-3}\ (\lambda cm)^{1/2}\ s \tag{2.84}$$

and perturbations with the wavelength $\lambda \approx \Lambda \approx 3 \times 10^8$ cm for $T = 10^5$ K increase for the time of the order of 35 s. The perturbation wavelength $\lambda \approx \Lambda$ may be taken for the maximum wavelength for the ballooning mode of the flute instability. This may be associated with magnetic loops with the thickness of the order of $(3\text{–}6) \times 10^8$ cm, observed with *SOHO*/Extreme ultraviolet Imaging Telescope (*EIT*) and *TRACE*. However, it is seen from Eq. (2.84) that short-wavelength perturbations increase more rapidly. This implies that the initial long-wavelength perturbation should be fractioned to smaller scales.

The minimum scale of the instability may be related to the heating of the plasma inside the flute due to the dissipation of diamagnetic currents with the scale λ_{min}. If the time of plasma heating t_{H} in a flute with the thickness λ_{min} is substantially smaller than the time of development of the instability for a flute with then wavelength λ_{min}, the plasma with increased pressure will have time to leave the flute along the magnetic field lines under the action of the pressure gradient that originated due to heating. As a result, the instability for this scale disappears.

The assumption of the importance of filamentation of magnetic structures for the heating of the solar and stellar coronae was made in the study [46]. In addition, it was shown in Ref. [47] that almost uniform temperature distributions along coronal magnetic loops obtained with *TRACE* may be explained by the presence of subarcsecond structures inside loops, extended along magnetic field. In the study [48], 41 loops observed with *TRACE* were analyzed, and the observed isothermality of the loops was found to be consistent with the inner threadlike structure of the loops.

As a possible reason for the filamentation of the currents in the magnetic flux tube, in Ref. [49] the thermal instability was also considered, which requires currents up to 10^{12}A in a flux tube with a diameter of 10^8 cm, along with the presence of turbulent conductivity, and displays the time of development of filamentation of the order of 2×10^3 s for the above parameters. Such currents occur only in flare loops, whereas numerous magnetic loops, which surround sunspots, show no signs of azimuthal component of the magnetic field, which provides evidence of the small value of parallel electric currents in such loops. Another possibility for filamentation is related to tearing instability [50]; this case, however, has been studied in application to a plane current sheet.

References

1. Vaiana, G.S., Krieger, A.S., and Timothy, A.F. (1973) *Sol. Phys.*, **32**, 81.
2. Rosner, R., Tucker, W.H., and Vaiana, G.S. (1978) *Astrophys. J.*, **220**, 643.
3. Schmieder, B., Rust, D.M., Georgoulis, M.K., Demoulin, P., and Bernasconi, P.N. (2004) *Astrophys. J.*, **601**, 530.
4. Parker, E.N. (1963) *Astrophys. J.*, **138**, 552.
5. Weiss, N.O. (1966) *Proc. R. Soc.*, **A293**, 310.
6. Galloway, D.J. and More, D.R. (1979) *Geophys. Astrophys. Fluid Dyn.*, **12**, 73.
7. Sen, N.K. and White, M.L. (1972) *Sol. Phys.*, **23**, 146.
8. Zaitsev, V.V. and Khodachenko, M.L. (1997) *Radiophys. Quantum Electron.*, **40**, 114.
9. Khodachenko, M.L. and Zaitsev, V.V. (2002) *Astrophys. Space Sci.*, **279**, 389.
10. Alfven, H. and Carlquist, P. (1967) *Sol. Phys.*, **1**, 220.
11. Hagyard, M.J. (1988) *Sol. Phys.*, **115**, 107.
12. Leka, K.D., Canfield, R.C., McClymont, A.N. *et al.* (1993) *Astrophys. J.*, **411**, 370.
13. Hudson, H.S. (1987) *Sol. Phys.*, **113**, 315.
14. Melrose, D.B. (1991) *Astrophys. J.*, **381**, 306.
15. Melrose, D.B. (1995) *Astrophys. J.*, **451**, 391.
16. Stenflo, J.O. (1969) *Sol. Phys.*, **8**, 115.
17. Severny, A. (1964) *Space Sci. Rev.*, **3**, 451.
18. Cowling, O. (1957) *Magnetohydrodynamics*, Interscience Publication, London.
19. Zaitsev, V.V. and Stepanov, A.V. (1991) *Sov. Astron.*, **35**, 189.
20. Zaitsev, V.V. and Stepanov, A.V. (1992) *Sol. Phys.*, **199**, 343.
21. Khodachenko, M.L. and Zaitsev, V.V. (2000) *Proceedings of the 3rd Internatinal Workshop 'The Solar Wind – Magnetosphere System 3', Graz, September 23–27, 1998*, Verlag der Österreichischen Academie der Wissenschaften, Wien, p. 33.
22. Zaitsev, V.V. and Kruglov, A.A. (2009) *Astron. Rep.*, **52**, 925.
23. McWhirter, R.P. (1965) *Plasma Diagnostics Techniques*, Academic Press, New York.
24. Verner, D.A. and Ferland, G.J. (1996,) *Astrophys. J. Suppl.*, **103**, 467.
25. Zaitsev, V.V. and Shibasaki, K. (2005) *Astron. Rep.*, **49**, 1009.
26. Aschwanden, M.J., Newmark, J.S., Delaboudinière, J.P. *et al.* (1999) *Astrophys. J.*, **515**, 842.
27. Kano, R. and Tsuneta, S. (1995) *Astrophys. J.*, **454**, 934.

28. Aschwanden, M.J. and Alexander, D. (2001) *Sol. Phys.*, **204**, 93.
29. Yan, Y., Deng, Y., Karlický, M. *et al.* (2001) *Astrophys. J.*, **551**, L115.
30. Landau, L.D. and Lifshits, E.M. (1984) *Electrodynamics of Continouos Media*, Pergamon Press, Oxford.
31. Vernazza, J.E., Avrett, E.H., and Loeser, R. (2000) *Astrophys. J. Suppl.*, **45**, 635.
32. Conti, P.S. and Underhill, A.B. (1998) O Stars and Wolf-Rayet Stars, NASA SP-497.
33. Miller, M.C., Lamb, F.K., and Hamilton, R.J. (1994) *Astrophys. J. Suppl.*, **90**, 833.
34. Ionson, J. (1982) *Astrophys. J.*, **254**, 318.
35. Zaitsev, V.V., Stepanov, A.V., Urpo, S., and Pohjolainen, S. (1998) *Astron. Astrophys.*, **337**, 887.
36. Dulk, G.A. (1985) *Ann. Rev. Astron. Astrophys.*, **23**, 169.
37. Khodachenko, M.L., Zaitsev, V.V., Kislyakov, A.G. *et al.* (2005) *Astron. Astrophys.*, **433**, 691.
38. Khodachenko, M.L., Zaitsev, V.V., Kislyakov, A.G., and Stepanov, A.V. (2009) *Space Sci. Rev.*, **143**, 112.
39. Melrose, D.B. (1997) *Astrophys. J.*, **486**, 521.
40. Aschwanden, M.J., Fletcher, L., Schrijver, C.J., and Alexander, D. (1999) *Astrophys. J.*, **520**, 880.
41. Zaitsev, V.V. and Kruglov, A.A. (2009) *Radiophys. Quantum Electron.*, **52**, 323.
42. Matsumoto, R., Tajima, T., Shibata, K. *et al.* (1993) *Astrophys. J.*, **414**, 357.
43. Abbett, W.P. and Fisher, G.H. (2003) *Astrophys. J.*, **582**, 475.
44. Shibasaki, K. (2001) *Astrophys. J.*, **557**, 326.
45. Priest, E. (1982) *Solar Magnetohydrodynamics*, D. Reidel Publishing Company, Dordrecht.
46. Litwin, C. and Rosner, R. (1993) *Astrophys. J.*, **412**, 375.
47. Reale, F. and Peres, G. (2000) *Astrophys. J.*, **528**, L45.
48. Aschwanden, M.J., Alexander, D., Hurlburt, N. *et al.* (2000) *Astrophys. J.*, **531**, 1129.
49. Heyvaertz, J. (1974) *Astron. Astrophys.*, **37**, 65.
50. Galeev, A.A., Rosner, R., Serio, S., and Vaiana, G.S. (1981) *Astrophys. J.*, **243**, 301.

3
Resonators for MHD Oscillations in Stellar Coronae

An important event in the astrophysics of the 1960s was the discovery of 5 min oscillations on the Sun [1], which gave birth to a new branch of solar physics – *helioseismology*. Five minute oscillations are acoustic-type vibrations (p-modes), and their spectrum contains extremely valuable information on the solar interior. The basic results of helioseismology, which owes its progress to the Solar and Heliospheric Observatory (SOHO), Complex Orbital Near-Earth Observations of Activity of the Sun – Photon (CORONAS–F), Solar Dynamic Observatory (SDO), and Picard space missions, are related to probing the solar interior and to the verification of the "standard" model of the Sun. The sound speed was measured in the range of depths from 0.2 to $0.98R_\odot$; the position and steepness of the tachocline were determined, which is important for the theory of generation of the solar magnetic field (solar dynamo). The position of the base of the convective zone at $0.29R_\odot$ was clarified. Later on, the methods of helioseismology were applied to the study of oscillations of local structures, such as sunspots and prominences. The advances in helioseismology and the studies of oscillations on the surfaces of other stars gave rise to the new, rapidly developing field of *asteroseismology* [2].

Observations of extreme ultraviolet (EUV) solar radiation by *transition region and coronal explorer* (*TRACE*) with high spatial resolution revealed oscillations of coronal loops [3, 4], inspiring a new promising direction in astrophysics, *coronal seismology*. As similar oscillations are seen in the light curves of flares observed in other stars, coronal seismology can be used for the diagnostics of coronae of other stars.

There is much in common between flares on the Sun and on late-type stars, especially red dwarfs [5]. Indeed, the timescales, the Neupert effect, the fine structure of the optical, radio, and X-ray emission, and the pulsations are quite similar both in solar and stellar flares. Studies of many hundreds of stellar flares indicated that they display the power-law distribution for the radiation energy; a similar distribution was found for solar flares. Thus, the solar–stellar analogy in studies of flare stars is justified.

Methods of coronal seismology are, in particular, attractive for the interpretation of pulsations in red dwarfs at various wavelengths [6–8] and for the explanation of the high-frequency quasi-periodic oscillations (QPOs) (20–2400 Hz) in SGRs – soft gamma-ray repeaters (magnetars) [9–11]. Interest in waves and oscillations in stellar coronae is related not only to the problem of the origin of and mechanisms

Coronal Seismology: Waves and Oscillations in Stellar Coronae, First Edition.
A. V. Stepanov, V. V. Zaitsev, and V. M. Nakariakov.

for solar and stellar coronal heating and wind acceleration but also to the necessity for finding an advanced diagnostic tool in the studies of the parameters and physical processes in stellar coronae, particularly in flaring loops.

3.1
Eigenmodes of Coronal Loops: The Plasma Cylinder Approach and the Dispersion Equation

Large-scale (with scales substantially exceeding ion gyro-radii) and slow (with the characteristic time much longer than the ion gyro-period) dynamics of a magnetized plasma is adequately described by magnetohydrodynamics (MHD). MHD is essential in the interpretation of wave and oscillatory phenomena in solar and stellar coronae. The standard model for linear MHD modes of coronal structures is based on the theory of interaction of MHD waves with a plasma cylinder, which reproduces the main properties of loops, jets, filaments, solar spicules, and other approximately axially symmetric plasma structures. Specific impedances for MHD waves undergo a jump at the interface of such structures and the surrounding medium; therefore, these structures can be considered as resonators and wave guides.

In the first approximation, the oscillations of coronal magnetic loops can be studied considering a column of homogeneous plasma with the radius a and length l, the foot points of which are "frozen" in superconducting dense plasma of the photosphere. The plasma inside the cylinder has the density ρ_i, temperature T_i, and the magnetic field induction B_i, aligned with the axis of the cylinder. Outside the cylinder, the corresponding parameters are ρ_e, T_e, and B_e. The cylinder approach in the incompressible plasma limit was used by Ershkovich and Nusinov [12] for the study of comet tails and the tail of the Earth magnetosphere. Zaitsev and Stepanov [13] took into account plasma compression and showed that fast radial magnetosonic (sausage) oscillations of the plasma column should undergo substantial damping. The physical mechanism of such damping, which is called *acoustic damping*, was described by Lord Rayleigh: loop oscillations excite waves in the ambient medium and the energy of oscillations decreases. There is no "acoustic damping" in a sufficiently dense and thick loop, which presents an ideal resonator for sausage oscillations. This situation is essentially analogous to the total internal reflection. In this case, MHD modes are called *trapped modes*, and the main role in their damping is associated with dissipative processes inside the loop.

In one-fluid ideal MHD, the equation of motion, induction equation, mass continuity equation, and equation of state can be written as

$$\rho\left(\frac{\partial \vec{v}}{\partial t} + (\vec{v}\nabla)\vec{v}\right) = -\nabla p + \frac{\left(\nabla \times \vec{B}\right) \times \vec{B}}{4\pi} \tag{3.1}$$

$$\frac{\partial \vec{B}}{\partial t} = \nabla \times \left(\vec{v} \times \vec{B}\right) \tag{3.2}$$

$$\frac{\partial \rho}{\partial t} + \nabla(\rho \vec{v}) = 0 \tag{3.3}$$

$$\frac{\partial S}{\partial t} + (\vec{v}\nabla)S = 0 \tag{3.4}$$

where $S = p\rho^{-\gamma}$ is the entropy and $\gamma = C_{\mathrm{p}}/C_{\mathrm{V}} = 5/3$ is the polytropic index.

Assume that an axially symmetrical magnetic flux tube $\boldsymbol{B}= (0,0,B(r))$ is out of equilibrium state due to some external impact, that is, $f_{\sim} = f + \delta f$, where f and δf are equilibrium and disturbed values, respectively. Then, taking $\delta f \sim f(r)\,\exp(-i\omega t + im\varphi + ik_{||}z)$, Eqs. (3.1)–(3.4) give us the set of linearized equations:

$$-i\omega\rho\delta v_{\mathrm{r}} = -\frac{\partial}{\partial r}\left(\delta p + \frac{\delta B_z B}{4\pi}\right) + \frac{ikB}{4\pi}\delta B_{\mathrm{r}} \tag{3.5}$$

$$-i\omega\rho\delta v_{\varphi} = -\frac{im}{r}\left(\delta p + \frac{\delta B_z B}{4\pi}\right) + \frac{ikB}{4\pi}\delta B_{\varphi} \tag{3.6}$$

$$-i\omega\rho\delta v_z = -ik\delta p - \frac{kB}{4\pi\omega}\delta v_{\mathrm{r}}\frac{\partial B}{\partial r} \tag{3.7}$$

$$\delta B_{\mathrm{r}} = -\frac{kB}{\omega}\delta v_{\mathrm{r}} \tag{3.8}$$

$$\delta B_{\varphi} = -\frac{kB}{\omega}\delta v_{\varphi} \tag{3.9}$$

$$-i\omega\delta B_z = -\frac{1}{r}\frac{\partial}{\partial r}rB_z\delta v_{\mathrm{r}} - \frac{imB}{r}\delta v_{\varphi} \tag{3.10}$$

$$-i\omega\delta\rho + \frac{1}{r}\frac{\partial}{\partial r}r\rho\delta v_{\mathrm{r}} + \frac{im\rho}{r}\delta v_{\varphi} + ik\rho\delta v_z = 0 \tag{3.11}$$

$$-i\omega(\delta p - c_{\mathrm{s}}{}^2\delta\rho) + \delta v_{\mathrm{r}}\left(\frac{\partial p}{\partial r} - c_{\mathrm{s}}{}^2\frac{\partial\rho}{\partial r}\right) = 0 \tag{3.12}$$

Here, $c_{\mathrm{s}} = (\gamma p/\rho)^{1/2}$ is the sound velocity, and $k = k_{||}$, directed along the field. Taking into account the equilibrium condition

$$\frac{\partial}{\partial r}\left(p(r) + \frac{B^2(r)}{8\pi}\right) = 0 \tag{3.13}$$

after some transformations, Eqs. (3.5)–(3.12) lead to the equations first obtained by Hain and Lüst [14] (see also [15]):

$$i\rho\left(\omega^2 - k_{||}{}^2c_{\mathrm{A}}{}^2\right)\frac{1}{r}\frac{\partial}{\partial r}(r\delta v_{\mathrm{r}}) = \omega\left(\mu^2 - \frac{m^2}{r^2}\right)\delta P \tag{3.14}$$

$$\omega\frac{\partial\delta P}{\partial r} = i\rho\left(\omega^2 - k_{||}{}^2c_{\mathrm{A}}{}^2\right)\delta v_{\mathrm{r}} \tag{3.15}$$

where $c_{\mathrm{A}} = B/(4\pi\rho)^{1/2}$ is the Alfvén speed, $\delta P = \delta p + \delta B_{\mathrm{z}}B_{\mathrm{z}}/4\pi$ is a variation of the total pressure, and

$$\mu^2 = \frac{\left(k_{||}^2c_{\mathrm{s}}{}^2 - \omega^2\right)\left(\omega^2 - k_{||}^2c_{\mathrm{A}}{}^2\right)}{\left(c_{\mathrm{A}}{}^2 + c_{\mathrm{s}}{}^2\right)\left(k_{||}^2c_{\mathrm{T}}{}^2 - \omega^2\right)} \tag{3.16}$$

Here, $c_T{}^2 = c_A{}^2 c_s{}^2/(c_A{}^2 + c_s{}^2)$ is the "tube" (or "cusp") speed. If there are a fixed number of half-wavelengths in perturbation of the magnetic flux tube with line-tied footpoints, which corresponds to a coronal magnetic loop with footpoints "frozen" in the photosphere, then a standing wave can be excited. Thereby, presenting the solution as

$$\delta f = \delta f(r)\exp(-i\omega t)\left[C_1\exp\left(ik_{||}z + im\varphi\right) + C_2\exp\left(-ik_{||}z + im\varphi\right)\right] \quad (3.17)$$

we conclude that Eqs. (3.14) and (3.15) describe also the standing waves.

Excluding the variation of the plasma radial velocity v_r from Eqs. (3.14) and (3.15) and assuming that the plasma both inside and outside the magnetic flux tube is homogenous, we obtain the Bessel equation

$$\frac{1}{r}\frac{\partial}{\partial r}\left(r\frac{\partial \delta P}{\partial r}\right) - \left(\frac{m^2}{r^2} - \mu^2\right)\delta P = 0 \quad (3.18)$$

describing the transverse structure of the perturbations. In the general case, μ is complex. Then, taking into account the properties of cylindrical functions $Z_m(\mu r)$ at $r \to 0$ and $r \to \infty$, we obtain the solutions inside (i) and outside (e) the tube in the following form:

$$\delta P_i = C_1 J_m(\mu_i r), \quad \delta P_e = C_2 H_m^{(1)}(\mu_e r) \quad (3.19)$$

Here, C_1 and C_2 are constants, $J_m(\mu_i r)$ and $H_m{}^{(1)}(\mu_e r)$ are the Bessel and Hankel functions of the order m, respectively, and the prime denotes the derivative with respect to the argument μr. Functions μ_i and μ_e may be considered as radial wave numbers "$k_\perp$" of the perturbations inside and outside the cylinder, respectively. Indeed, from Eq. (3.16) with $\mu = k_\perp$, one can obtain the frequencies of fast and slow magnetoacoustic waves:

$$\omega_\pm = \frac{1}{2}\left[\sqrt{\chi^2\left(c_A{}^2 + c_s{}^2\right) + 2\chi k_{||}c_A c_s} \pm \sqrt{\chi^2\left(c_A{}^2 + c_s{}^2\right) - 2\chi k_{||}c_A c_s}\right],$$
$$\chi^2 = k_\perp{}^2 + k_{||}{}^2 \quad (3.20)$$

where $k_\perp$ and $k_{||}$ are "quantized," since the plasma volume is finite.

To connect the solutions inside and outside the flux tube, we use the conditions of the continuity of the total pressure and the radial displacement at the tube boundary $r = a$. This means that the following equalities are fulfilled:

$$\delta P_i(a) = \delta P_e(a), v_{r_i}(a) = v_{r_e}(a) \quad (3.21)$$

The first condition follows from equation of motion (Eq. (3.1)), while the second is due to the integral of Eq. (3.14) from $a - \xi$ to $a + \xi$ with $\xi \to 0$, subsequently.

From Eqs. (3.14) and (3.21), we obtain the dispersion relation for MHD eigenmodes of the plasma cylinder [12]:

$$\rho_e\left(\omega^2 - k_{||}{}^2 c_{Ae}{}^2\right)\mu_i\frac{J_m'(\mu_i a)}{J_m(\mu_i a)} = \rho_i\left(\omega^2 - k_{||}{}^2 c_{Ai}{}^2\right)\mu_e\frac{H_m'^{(1)}(\mu_e a)}{H_m^{(1)}(\mu_e a)} \quad (3.22)$$

where $Z_m'(x) = \mathrm{d}Z_m(x)/\mathrm{d}x$. The integer m is the azimuthal mode number. It specifies the azimuthal structure: the oscillations with $m = 0$ are *sausage* (also

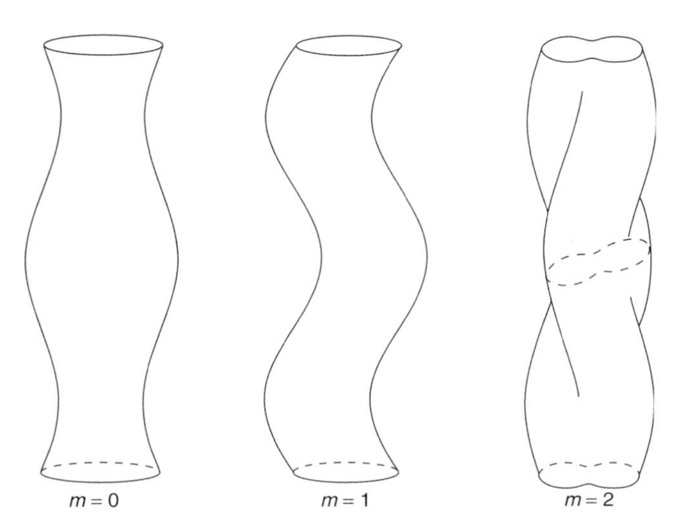

Figure 3.1 A sketch of eigenmodes of a slender flux tube: the sausage mode ($m = 0$), kink mode ($m = 1$), and ballooning mode ($m = 2$). (After Spruit [16].)

known as *radial, peristaltic,* or *fast magneto-acoustic* (FMA)) modes, those with $m = 1$ are *kink* modes, and those with $m \geq 2$ are referred as *flute or ballooning modes* (Figure 3.1).

Sausage modes ($m = 0$) are perturbations of the tube (loop) radius accompanied by perturbations of the plasma density and of the absolute value of the magnetic field. *Kink* modes ($m = 1$) are weakly compressible and represent displacements of the tube axis. Note that at $m = 0$ the solutions for Eq. (3.22) also describe *slow magnetosonic* and *torsional* oscillations. For a small plasma β, $\beta = 8\pi p/B^2 \ll 1$, *slow magnetosonic* oscillations almost do not disturb the magnetic field, and its behavior is the same as that for sound oscillations. In turn, *torsional* oscillations do not perturb the pressure of the plasma, being in fact Alfvén-type modes.

3.1.1
Trapped Modes

In order to solve Eq. (3.22), Edwin and Roberts [17] considered only the case $\mu_e^2 < 0$, which corresponds to the nonoscillatory evanescent behavior in the transverse direction outside the tube. The solution for Eq. (3.21) outside the cylinder is the McDonald function:

$$K_m(x) = (\pi/2) i^{m+1} H_m{}^{(1)}(ix) \tag{3.23}$$

therefore, this solution displays no acoustic damping. Hence, for trapped modes, instead of Eq. (3.22) one obtains

$$\rho_e \left(\omega^2 - k_{||}{}^2 c_{Ae}{}^2\right) \mu_i \frac{J'_m(\mu_i a)}{J_m(\mu_i a)} = \rho_i \left(\omega^2 - k_{||}{}^2 c_{Ai}{}^2\right) \mu_e \frac{K'_m(\mu_e a)}{K_m(\mu_e a)} \tag{3.24}$$

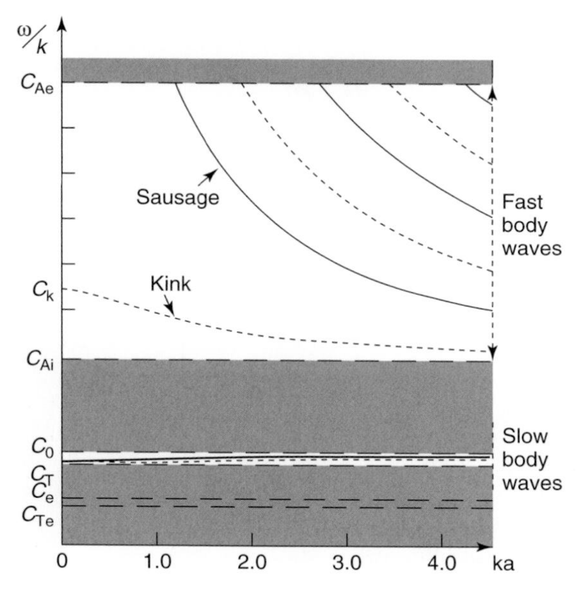

Figure 3.2 The dependence of the phase speed on the dimensionless wave number $k_{||}a$ for trapped fast and slow magnetoacoustic waves in a magnetic flux tube under the solar corona conditions: $c_{Ae} = 5c_{si}$, $c_{se} = 0.5c_{si}$, and $c_{Ai} = 2c_{si}$. (After Edwin and Roberts [17].)

Figure 3.2 shows the results for a numerical solution for dispersion Eq. (3.24), obtained in Ref. [17] under the solar corona conditions (i.e., c_{Ae}, $c_{Ai} > c_{se}$, c_{ci}). It is shown that three main modes exist: sausage, kink, and slow magnetosonic. Note that torsional modes, which are also described by Eq. (3.24), are not shown in Figure 3.2.

Besides, in the case of $m = 0$, an essentially incompressible dispersionless torsional (Alfvén) mode of MHD oscillations exists:

$$\omega_t = k_{||}c_{Ai} \tag{3.25}$$

The transverse wave number is determined by the formula $k_\perp = \lambda_j/a$, where λ_j are the roots of the equation $J_0(\lambda) = 0$. The longitudinal wave number $k_{||} = s\pi/l$ has discrete values ($s = 1, 2, 3, \ldots$), due to the frozen edges of the plasma cylinder, representing the line-tied boundary conditions at the loop footpoints. For $m = 1$, the kink mode arises in the spectrum of cylinder oscillations with the frequency

$$\omega_k = k_{||}\left(\frac{\rho_i {c_{Ai}}^2 + \rho_e {c_{Ae}}^2}{\rho_i + \rho_e}\right)^{1/2} \tag{3.26}$$

in the long-wavelength limit. The solutions are shown in Figure 3.2, which shows that the kink mode solutions extend up to the long-wavelength limit $k_{||}a \to 0$. In contrast, the trapped sausage mode has a cutoff at the phase speed $v_{ph} = \omega_+/k_c = c_{Ae}$ and does not exist for small wave numbers $k_{||}a < k_c a$. For the cutoff value of the longitudinal, $\mu_e = 0$, Eq. (3.24) yields $J_0(\mu_i a)|_{k=k_c} = 0$. The cutoff for the sausage mode occurs at the cutoff wave number k_c [17, 18]

$$k_{||} = k_c = \frac{\eta_j}{a}\left[\frac{(c_{si}^2 + c_{Ai}^2)(c_{Ae}^2 - c_{Ti}^2)}{(c_{Ae}^2 - c_{Ai}^2)(c_{Ae}^2 - c_{si}^2)}\right]^{1/2}, \quad j = 1, 2, 3, \ldots \tag{3.27}$$

where $\eta_j = (2.40, 5.52, 8.65, \ldots)$ are the zeros of the Bessel function J_0. Different zeros correspond to different radial modes. It follows from Eq. (3.27) that k_c does not depend on the wave frequency and tube length, but it does depend on the tube radius, characteristic speeds inside and outside the tube, and parameter η_j. Thus, trapped modes exist for $k_{||} \geq k_c$. For $k_{||} < k_c$, we have leaky modes.

3.1.1.1 Global Sausage Mode

An important case of the trapped modes is the global sausage mode. Its existence was confirmed by observations with high (~10 arcseconds) spatial resolution of oscillations in a single coronal loop in the solar microwave emission, made with the Nobeyama radioheliograph. In the 12 January 2000 flare event, the radio emission at 17 and 34 GHz displayed synchronous QPOs with the period ~15 s in different parts of the flare loop [19]. The period of the global sausage mode is $P_{GSM} = 2\,l/c_p$, where c_p is the phase velocity of the sausage mode corresponding to the wave number $k_{||} = \pi/l$, $c_{Ai} < c_p < c_{Ae}$. For a strong density contrast inside and outside the tube, the period of the global sausage mode should satisfy the following condition:

$$P_{GSM} < \frac{2\pi a}{\eta_0 c_{Ai}} \approx \frac{2.62a}{c_{Ai}} \tag{3.28}$$

From $P_{GSM} = 2\,l/c_p$ and Eq. (3.28), we obtain the necessary condition for the existence of the global trapped sausage mode [20]:

$$l/2a < \frac{\pi c_{Ae}}{2\eta_0 c_{Ai}} \approx 0.65\sqrt{\frac{\rho_i}{\rho_e}} \tag{3.29}$$

therefore, the loop should be sufficiently thick and dense. The same result can be obtained from the condition of the total internal reflection of the sausage mode [21] (see Section 3.1.2.1): $\rho_e/\rho_i < {k_{||}}^2/k^2_{\perp}$. For $k_{||} = \pi/l, k_{\perp} = \eta_0/a$, the latest inequality coincides with Eq. (3.29). Here, $k_{\perp}$ is the fundamental transverse wave number of the sausage mode. The density contrast in solar flaring loops can reach $\rho_i/\rho_e \approx 10^2$ [22]. In this case, a length-to-width ratio of the longest loop that can support a global trapped sausage mode is $l/2a < 6.5$.

3.1.1.2 Global Kink Mode

Dispersion equation (Eq. (3.24)) suggests that there are two kink ($m = 1$) modes in the limit $k_{||}a \ll 1$: slow and fast [23]. The slow kink mode displays the phase velocity roughly equal to the "tube" (cusp) speed:

$$\frac{\omega}{k_{||}} \approx c_T \equiv \frac{c_{si}c_{Ai}}{({c_{si}}^2 + {c_{Ai}}^2)^{1/2}} \tag{3.30}$$

The phase speed of the fast kink mode is

$$\frac{\omega}{k_{||}} \approx c_k \equiv \left(\frac{2}{1+\rho_e/\rho_i}\right)^{1/2} c_{Ai} \tag{3.31}$$

where c_k is the so-called kink speed. The speed of the slow wave c_T is below both the Alfvén c_{Ai} and sound c_{si} speeds. For the solar corona with the temperature $T = 10^6$ K, $c_s < 3 \times 10^7$ cm s^{-1} , in the 4 July 1999 event TRACE observations revealed the phase velocity $c_k \approx 10^8$ cm s^{-1}. Equation (3.31) is used for the determination of the absolute magnetic field in the loop [23]:

$$B = (4\pi\rho_i)^{1/2} c_{Ai} = \frac{\sqrt{2}\pi^{3/2} l}{P} \sqrt{\rho_i \left(1 + \frac{\rho_e}{\rho_i}\right)} \tag{3.32}$$

For example, for an oscillating loop with $l = 1.5 \times 10^8$ cm, $P = 360$ s, $\rho_e/\rho_i \approx 0.1$, for a sufficiently wide interval of plasma densities, $1–6 \times 10^9$ cm^{-3}, from Eq. (3.32) we estimate the magnetic field B to be within the interval 4–30 G.

3.1.2
Leaky Modes

It is natural to suppose that owing to the energy loss related to the emission of MHD waves into the surrounding medium, the eigenmode frequency ω and parameter μ are complex values. In this case, Sommerfeld's condition is satisfied, which is described by the Hankel function:

$$H_n^{(1)}(\mu_e r) \approx \left(\frac{2}{\pi \mu_e r}\right)^{1/2} \exp\left[i\left(\mu_e r - \frac{n\pi}{2} - \frac{\pi}{4}\right)\right] \tag{3.33}$$

that is, a cylindrical running wave should be formed outside the flux tube. Therefore, an oscillating plasma cylinder (a magnetic tube) can radiate MHD waves into the surrounding medium, thereby damping the oscillations. Such radiating (leaky) modes were first considered by Zaitsev and Stepanov [13], while Spruit [16] introduced the term *acoustic damping*.

3.1.2.1 Sausage Mode

In the simplest case of a slender ($a/l \ll 1$) and dense ($\rho_e/\rho_i \ll 1$) plasma cylinder in the axially symmetrical case ($m = 0$), instead of Eq. (3.22), we obtain the equation [13]:

$$\left(k_{||}a\right)^2 \ln \frac{1}{k_{||}a} = \frac{4\pi\rho_i}{B_e^2} \frac{a\left(\omega^2 - k^2 c_{Ai}^2\right)}{\mu_i} \frac{J_0(\mu_i a)}{J_0'(\mu_i a)} \tag{3.34}$$

which is approximately satisfied if $\mu_i a = \eta_j$ and η_j is determined from the equation $J_0(\eta_j) = 0$.

Equation 3.20) yields the frequencies of sausage and slow magnetosonic oscillations [13]:

$$\omega_+ = \left({k_\perp}^2 + {k_{||}}^2\right)^{1/2} \left({c_{si}}^2 + {c_{Ai}}^2\right)^{1/2} \tag{3.35}$$

$$\omega_- = \frac{k_{||}c_{si}c_{Ai}}{\left(c_{si}{}^2 + c_{Ai}{}^2\right)^{1/2}} \tag{3.36}$$

Here, the transverse wave number $k_\perp = \eta_j/a$; η_j are the zeros of the Bessel function $J_0(\eta_j) = 0$. Sausage modes, which are most effective in modulation of radiation from coronal loops, can undergo noticeable damping due to the fact that they radiate waves into the surrounding medium [13, 21]:

$$\gamma_a = \frac{\pi}{2}\omega_+ \left(\frac{\rho_e}{\rho_i} - \frac{k_{||}{}^2}{k_\perp{}^2}\right), \quad \frac{\rho_e}{\rho_i} > \frac{k_{||}{}^2}{k_\perp{}^2} \tag{3.37}$$

The physical mechanism of the "acoustic damping" is clear: loop oscillations are accompanied by excitation of waves in the ambient medium, which results in a decrease in the energy of the loop oscillations. As mentioned earlier, there is no acoustic damping for $\rho_e/\rho_i < k_{||}{}^2/k_\perp{}^2$, which is equivalent to the total internal wave reflection. Thus, a thick and dense loop is an ideal resonator for sausage modes.

To solve the transcendental equation (Eq. (3.22)) analytically, we will have to invoke various approximations and constraints, so let us use the numerical approach. We assume the frequency $\omega = \omega_0 - i\gamma$ to be a complex value, and the longitudinal wave number $k_{||}$ to be a real value, that is, in the adopted notation, ω_0 specifies the oscillation period, while γ is the wave decrement due to the emission of MHD waves by the oscillating tube. Thus, the arguments of the Bessel and Hankel functions in Eq. (3.22) can be complex, and the corresponding modes can be leaky.

When performing the numerical calculations for sausage modes ($m = 0$), we separate the imaginary and real parts in Eq. (3.22). Subsequently, following Cally [24, 25] for kink modes ($m = 1$), we solve the system of real equations for the unknown $\omega_0/k_{||}$ and $\gamma/k_{||}$. To restrict the infinite family of dispersion curves specified by the harmonic number j, we consider the most easily excited radial modes with $j = 1, 2, 3$.

Figure 3.3 shows the obtained dependencies of $\omega_0/k_{||}$ and $\gamma/k_{||}$ on the parameter $k_{||}a$. To compare these results with the numerical solutions obtained by Edwin and Roberts [17] for trapped modes, we took the same relations between characteristic speeds in the solar corona: $c_{se} = 0.5c_{si}$, $c_{Ai} = 2c_{si}$, $c_{Ae} = 5c_{si}$. The dispersion curves constructed for various radial harmonics j closely coincide with those for trapped modes studied in Ref. [17] (hatched area) in the interval $c_{Ai} < \omega_0/k_{||} < c_{Ae}$ and with their natural extensions for $\omega_0/k_{||} > c_{Ae}$. Figure 3.3 also shows the branches of slow magnetoacoustic ($\omega_0 = k_{||}c_{si}$) and Alfvén ($\omega_0 = k_{||}c_{Ai}$) modes.

The dispersion curves for sausage modes ($m = 0$) in the region $C_{Ai} < \omega_0/k_{||} < C_{Ae}$ completely coincide with those obtained by Edwin and Roberts [17] for trapped modes under the condition $\mu_e^2 < 0$. As a result, there is no acoustic damping in the hatched domain ($\gamma = 0$), and perturbations in the surrounding region are described by a real MacDonald function $K_m(\mu_e a)$. Meanwhile, the arguments of the cylindrical functions can be complex ($\mathrm{Im}\omega \neq 0$) and the corresponding modes can be leaky. Since $K_m(z) = (\pi/2)i^{m+1}H_m{}^{(1)}(iz)$, the solution for Eq. (3.22) includes trapped modes as a special case with $\mu_e^2 < 0$.

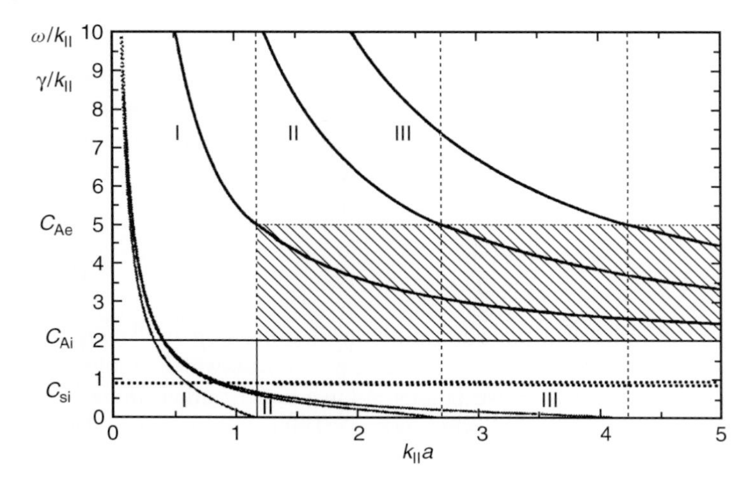

Figure 3.3 Dispersion curves of sausage oscillations for the first three harmonics ($j = 1, 2, 3$). The black (upper) and gray (lower) curves indicate the dependencies for $\omega_0/k_{||}$ and $\gamma/k_{||}$, respectively. The straight line $\omega_0/k_{||} = c_{Ai}$ represents the branch of Alfvén waves; the horizontal dashed lines correspond to slow magnetoacoustic waves. The hatched rectangle delineates the region considered in [17]. The vertical lines correspond to longitudinal wave numbers $k_{||} = k_c$ determined from Eq. (3.27) for $j = 1, 2, 3$.

The behavior of the quality factor $Q = \omega_0/\gamma$ for the sausage oscillations at various parameters $k_{||}a = ka$ is presented in Figure 3.4a. In this figure, the parameters are close to those of a solar flaring loop [26]: $c_{se} = 2 \times 10^7 \mathrm{cm\ s^{-1}}$, $c_{si} = 2c_{se}$, $c_{Ae} = 10c_{si}$, and $\rho_i/\rho_e = 30$. Figure 3.4 indicates that $Q \approx$ constant at $ka \ll 1$, in accordance with the previous results [13, 21].

Figure 3.4b plots the Q-factor for the calculated sausage mode versus the ratio of Alfvén speeds inside and outside the magnetic flux tube $c_{Ai}/c_{Ae} \approx \sqrt{n_e/n_i}$, $(B_i \approx B_e)$ for $\omega_0 =$ constant and $k_{||}a \ll 1$. It is shown here that the curve constructed with the use of the formula derived analytically from Eq. (3.22) by Zaitsev and Stepanov [13]

$$Q \approx \frac{2n_i}{\pi n_e} \tag{3.38}$$

is very close to that calculated numerically.

In Figure 3.5, the oscillation period $P = 2\pi/\omega_0$ of sausage oscillations is plotted against the parameter $k_{||}a$ for $j = 1, 2, 3$. For comparison, the period determined from Eq. (3.34) for slender ($k_{||}a \ll 1$) flux tube [13]

$$P = \frac{2\pi a}{\eta_j\sqrt{{c_{Ai}}^2 + {c_{si}}^2}} \tag{3.39}$$

is also presented in Figure 3.5. As it is easy to verify, Eq. (3.39) for leaky modes at $k_{||}a \ll 1$ is also consistent with the numerically calculated values.

It follows from the analysis presented here (Figure 3.3) that the phase speed of leaky modes at $k_{||}a \ll 1$ (a slender flux tube) is $v_{ph} \propto 1/k_{||}$. Therefore, the oscillation

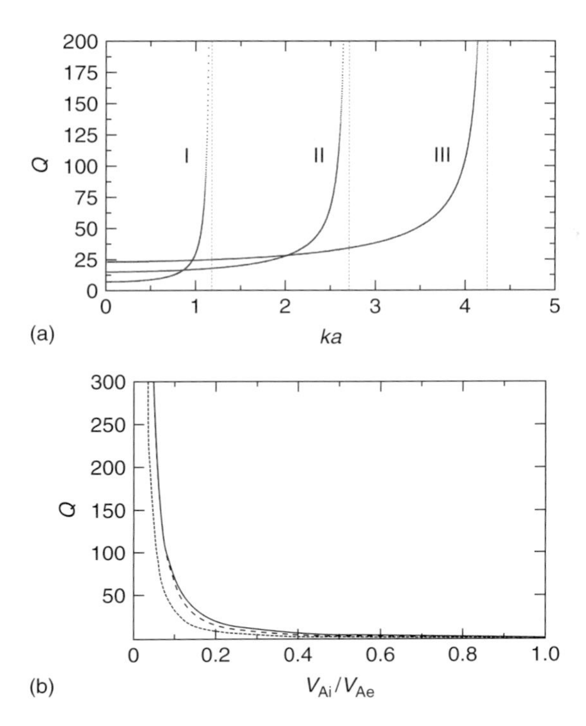

Figure 3.4 (a) The Q-factor of sausage oscillations versus the parameter $k_{||}a$. The vertical dashed lines correspond to longitudinal wave numbers $k_{||} = k_c$ determined from Eq. (3.23) for $j = 1, 2, 3$. (b) Q-factor of sausage oscillations versus the ratio of Alfvén speeds inside and outside the flux tube at $k_{||}a \ll 1$ and $j = 1$. The dashed line corresponds to Eq. (3.38).

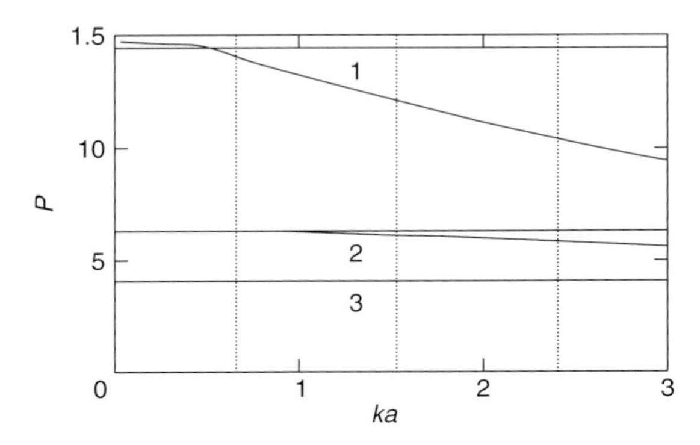

Figure 3.5 The period of the first three harmonics of sausage oscillations (in seconds) versus the parameter $k_{||}a$ for typical solar flaring loop parameters $c_{si} = 4 \times 10^7$ cm s^{-1}, $c_{Ai} = 10^8$ cm s^{-1}, and $a = 3 \times 10^8$ cm. The straight horizontal lines follow the approximate Eq. (3.39). The vertical lines correspond to the longitudinal wave numbers $k_{||} = k_c$ for $j = 1, 2, 3$.

period $P \propto 1/k_{||}v_{\rm ph}$ is most likely specified by the radius of the loop a rather than by its length $l \gg a$. In turn, the eigenfrequency ω_0 and the Q-factor of leaky sausage modes of a thin magnetic flux tube depend weakly on the wave number $k_{||}$. Note that Eq. (3.38) was confirmed recently by Jakimiec and Tomczak [27] on the basis of the analysis of hard X-ray solar pulsations: the pulse period is proportional to the flare loop radius a.

As an illustration for the proposed model for the pulsations in the emission produced by leaky sausage modes, let us consider the well-known radio flare event on 16 May 1973 at the frequency $\nu = 230$ MHz, described in detail by McLean and Sheridan [28] (Figure 3.6).

From Figure 3.6, one can estimate the Q-factor of the pulsating event $Q \approx \pi N$, where $N \approx 10$ is the number of pulses with the amplitudes that are smaller than the initial amplitude with the factor of e. Kopylova *et al.* [29] showed that since the minimum value of the plasma density in the pulsed source ($\nu \geq \nu_{\rm p} = \sqrt{4\pi e^2 n/m} = 230$ MHz) is about $n_{\rm i} \approx 6.6 \times 10^8$ cm^{-3} for the plasma temperature $T = 3 \times 10^6$ K, the dissipative processes inside the loop, including ion viscosity and electron thermal conductivity, are less important compared to acoustic damping with the Q-factor $Q \approx 2n_{\rm i}/\pi n_{\rm e}$ (Eq. (3.38)). Comparing both expressions for the Q-factor, we obtain

$$\frac{n_{\rm i}}{n_{\rm e}} \approx \frac{\pi^2 N}{2} {\rm cm}^{-3} \tag{3.40}$$

and for $N = 10$, the density ratio is $n_{\rm i}/n_{\rm e} \approx 50$. Therefore, the charged particle density outside the oscillating loop is $n_{\rm e} \approx 1.3 \times 10^7$ cm^{-3}. Following [30], we assume that the electron density outside the loop varies with height in accordance with the Baumbach–Allen formula:

$$n_{\rm e}(h) = 10^8 \left(\frac{2.99}{x^{16}} + \frac{1.55}{x^6} + \frac{0.036}{x^{3/2}} \right) {\rm cm}^{-3} \tag{3.41}$$

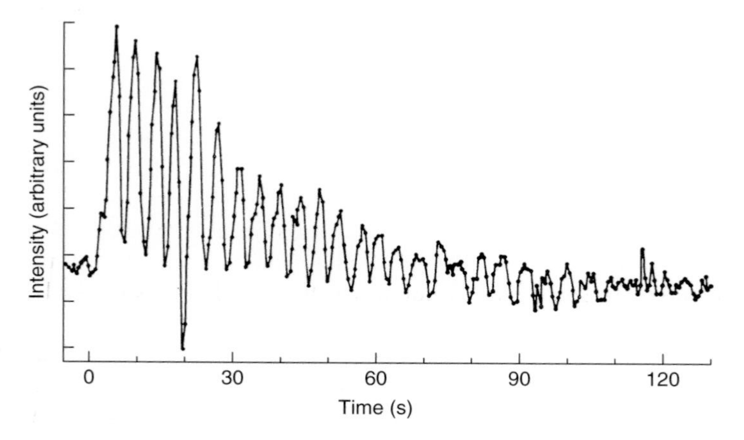

Figure 3.6 Pulsations of solar radio emission with the period of 4.3 s detected on 16 May 1973 (03:14 UT) at 230 MHz. (After McLean and Sheridan [28].)

where $x = 1 + h/R_\odot$. Substituting $n_e \approx 1.3 \times 10^7$ cm^{-3} into Eq. (3.41), we find that the oscillating meter-wavelength source was located at a height $h \approx 3.6 \times 10^{10}$ cm ($\approx 0.52R_\odot$).

3.1.2.2 Kink Modes

As shown earlier, Eq. (3.21) also contains kink modes ($m = 1$) with the phase speed determined by Eq. (3.26). At the first glance, kink modes are less important for plasma pressure deviations, and their contribution to the modulation of flaring emission is small compared to that of the sausage modes. Nevertheless, numerous direct observations from space (TRACE, Hinode, Solar Terrestrial Relations Observatory, and SDO) provide evidence for the existence of kink oscillations with substantial amplitudes in solar magnetic loops [3, 4, 31–33]. Therefore, kink oscillations in the loops can be also responsible for modulation of flare radiation. An example of loop kink oscillation observed by TRACE is shown in Figure 3.7.

Analyzing kink oscillations of solar flare loops, Nakariakov *et al.* [4] obtained a low Q-factor of such pulsations, $Q = \omega/\gamma \leq 10$, and proposed that the damping could be due to anomalously high values of plasma viscosity or resistivity. The paper by Nakariakov *et al.* [4] stimulated a huge flow of papers on coronal loop oscillations and the origin of the low Q-factor. Uralov [34] suggested that the character of the loop oscillations is specified by the medium, that is, by dispersion properties of the coronal plasma. Oscillations of the corona driven by a flare or a filament eruption also involve coronal loops into the process of oscillations. The low quality of kink oscillations was explained by dispersion blurring of the FMA wave impulse produced by a flare [34, 35]. Note that, according to [34], all loops in the flare region should oscillate; however, observations show that, in fact, only a few of them do, and they do it with different frequencies and phases. Arregui *et al.* [36] took into account the inhomogeneity of coronal loops and proposed the resonant mechanism of kink mode damping based on their interaction with Alfvén-type modes. A physically transparent mechanism for the observed rapid decay of loop oscillations was recently suggested by Chen and Schuck [37]: an oscillating loop spends its energy on surmounting the resistance of the ambient medium.

To explain the rapid decay of kink oscillations, Mikhalayev and Soloviev [38] considered oscillations of a double plasma cylinder in the form of a cord surrounded by a coaxial shell. In their model, the condition for rapid acoustic decay is satisfied if the plasma density in the shell is lower than that in the surrounding corona. This artificial approach can be useful in modeling the multifiber structure of a coronal loop. However, there are some reasons for which the simplest single-cylinder model may also illustrate the rapid damping of kink oscillations.

To clear up the situation, Kopylova *et al.* [39] analyzed Eq. (3.21) numerically for the same velocities c_{si}, c_{se}, C_{Ai}, and C_{Ae} as those in [17]. Since the number of types of kink modes substantially exceeds that of sausage modes, in Figure 3.8, we present the illustration for only three kink modes; as we can see, only mode "a" is leaky. The dependence of the damping rate of "a" mode on ka is the same as that obtained in [38] for the kink mode of a flux tube with a shell. There is no

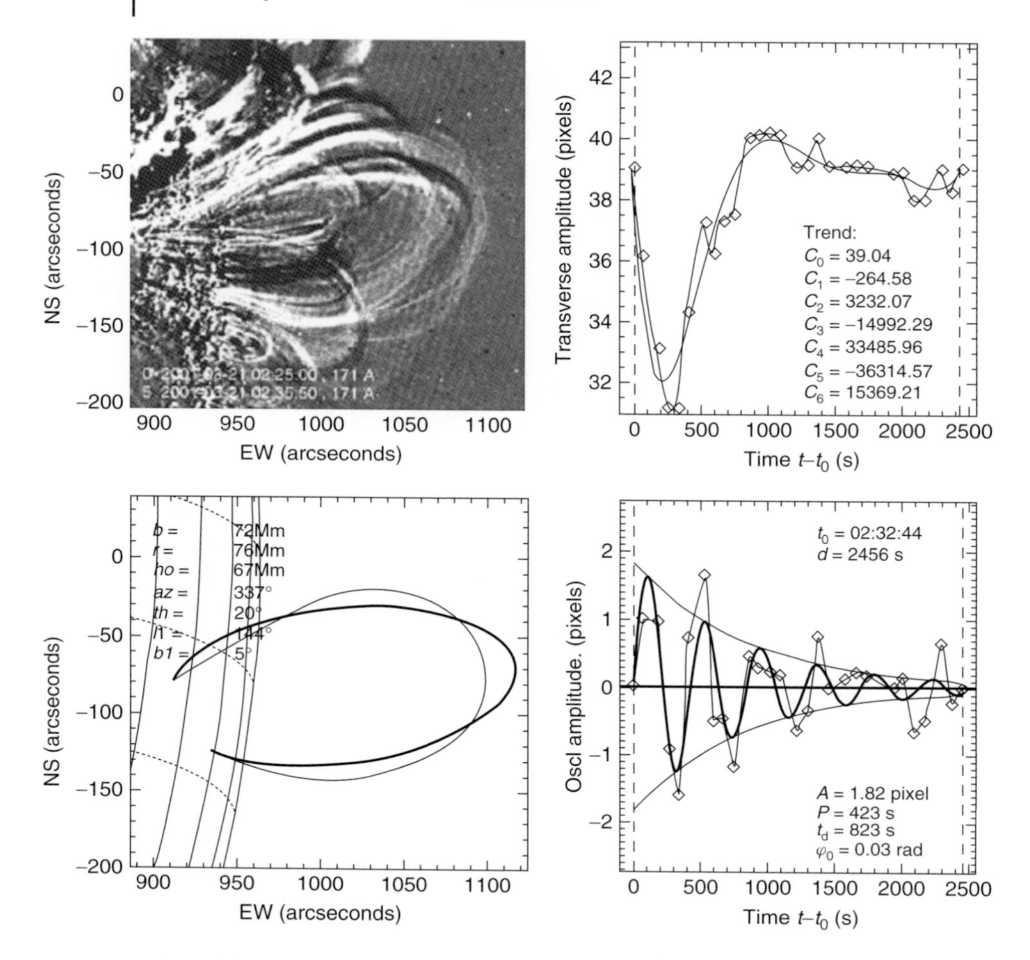

Figure 3.7 Displacement of an EUV loop during MHD kink mode oscillations in the solar event on 21 March 2001, 02:32 UT. Average pulse period is 423 s, and e-folding decay time is 823 s. The oscillation amplitude ~1000 km. (After Aschwanden *et al.* [31].)

acoustic damping for the Alfvén "b" mode, since this mode produces no deviation in the plasma pressure. The "c" mode, studied previously in [17], is also nonleaky. Figure 3.8 also presents two out of a number of slow magnetoacoustic modes.

Let us consider the behavior of "a" and "c" type modes at $k_{||}a \to 0$. For $\omega_0^2/{k_{||}}^2 > {c_{\mathrm{Ai}}}^2$ (curves above the Alfvén "b" mode in Figure 3.8), $|\mu_{\mathrm{i}}a| \approx \omega_0 a/c_{\mathrm{Ai}} \ll 1$, and the dispersion equation has the form of Eq. (3.26). Substituting $c_{Ai} = 2c_{si}$ and $c_{Ae} = 5c_{si}$ into Eq. (3.26), we obtain $\omega_0/k_{||} \approx 2.62c_{si}$, which is consistent with the behavior of "a" and "c" modes at $k_{||}a \to 0$ (Figure 3.8).

For the global kink mode, $k_{||}a \approx \pi a/L$. According to TRACE observations [30, 31], $k_{||}a \approx 0.2$, $P \approx 300$ s; thus, only the "a," "b," and "c" modes can be considered

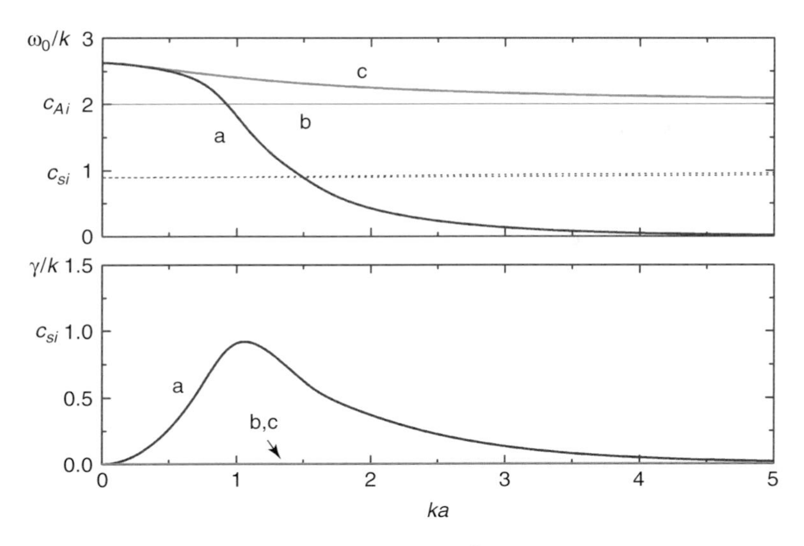

Figure 3.8 Dispersion curves for low-frequency fast magnetoacoustic kink modes. The dashed lines indicate two slow magnetoacoustic modes.

to explain the events, and only the "a" mode is leaky. Since the pulsation decay is quite rapid, one can suggest that the "a" mode is excited in coronal loops. Figure 3.8 shows that at $k_{||}a \approx 0.2$, the Q-factor $Q \approx 60$, that is, we expect about $N \approx Q/\pi \approx 20$ oscillations. However, TRACE data suggest $N = 4 - 5$. Perhaps, this discrepancy can be explained by nonlinear effects: indeed, in the linear theory, it is assumed that the displacement of the loop is much smaller than its minor radius, while the observed displacement often exceeds 3–4 minor radii. On the other hand, the finite amplitude effects do not seem to cause difficulties to the resonant absorption model [40].

Another mechanism for kink oscillation decay, based on the radiative damping, was proposed by Soloviev *et al.* [41]. They have shown that free kink oscillations with small but finite amplitude in an external plasma effectively generates FMA waves. As a result, the perturbation energy is completely dissipated during several periods of loop oscillations. Aschwanden *et al.* [42] proposed an observational test for different mechanisms for kink oscillations damping based on the analysis of scaling of the damping time with the oscillation period. However, the presently available resolution does not allow one to discriminate between different mechanisms.

3.1.3 Ballooning Modes

The theory of MHD modes of a straight magnetic cylinder does not take into account a very important aspect of coronal loop oscillations – the effect of a loop curvature. Several attempts were made to construct an analytical theory of MHD modes in a curved magnetic loop [43–46]. However, the theory is still not complete and a number of important questions are still open. Instead, one can use approximate results obtained for laboratory plasma devices, such as tokamaks.

The curvature of the magnetic field lines and the fairly large value of the plasma parameter $\beta = 8\pi n k_B T/B^2 > 0.1$–$0.3$ in flaring coronal loops provide favorable conditions for the excitation of the ballooning mode of flute instability. The conditions for ballooning instability were considered in Chapter 2 in the context of the formation of a magnetic loop above a sunspot and the flare energy release. However, ballooning instability may display both periodical and nonperiodical mode. Ballooning oscillations result from the joint action of a destabilizing force $F_1 \sim p/R_{\text{loop}}$ related to the pressure gradient and the curvature of the magnetic field and a restoring force $F_2 \sim B^2/R_{\text{loop}}$, resulting from the tension of magnetic field lines. Neglecting the effect of gravitational stratification, we will consider a semicircular loop with the major radius R. Assume that the equilibrium is perturbed and there is a plasma "tongue" with the scale $L_1 = L/s$, where $L = \pi R$ and s is the number of the "tongues" along the loop. The dispersion equation of the ballooning mode has the form [47]:

$$\omega^2 - k_{\|}^2 C_A^2 = -\frac{p}{R_{\text{loop}} \rho d}; \; d = \begin{cases} b, b \gg \lambda_\perp \\ \lambda_\perp, b \ll \lambda_\perp \end{cases}; \; d = \begin{cases} b, b \gg \lambda_\perp \\ \lambda_\perp, b \ll \lambda_\perp \end{cases} \tag{3.42}$$

Here, $b = \rho\,(\partial\rho/\partial r)^{-1}$ is the typical scale of plasma density inhomogeneity across the magnetic field in the loop, and $\lambda_\perp$ is the lateral size of the plasma "tongue." Since the loop footpoints are frozen in the photosphere, $k_{\|} = s\pi/L$, where s is equal to the number of oscillating regions along the loop length l, and the oscillation period is

$$P_1 = \frac{2L}{C_A}\left(s^2 - \frac{L\beta}{2\pi d}\right)^{-1/2} \approx \frac{2L}{C_A s} \tag{3.43}$$

Otherwise, Eq. (3.42) can be reduced to the expression derived in the straight cylinder approximation. For a typical solar coronal loop, $l = 10^{10}$ cm and $\beta \sim 0.1$, $L\beta/2\pi d \leq 1$, which justifies the use of Eq. (3.42).

3.2
MHD Resonator at $\sim 1 R_\odot$ in the Solar Corona

Along with coronal magnetic loops, other types of waveguides and resonators for gravitational, acoustic, and MHD oscillations and waves are possible in stellar atmospheres owing to their inhomogeneity. For example, resonant properties of the solar chromosphere and photosphere have been widely studied in the context of heating of the corona and the explanation for the 5 min and 3 min oscillations. There is a one-dimensional resonator for acoustic waves with the period of about 300 s, formed by the subphotospheric convection zone and the temperature minimum at the height of 700 km above the photosphere, exists on the Sun. Likewise, recent numerical modeling of the response of a sunspot atmosphere to the broadband

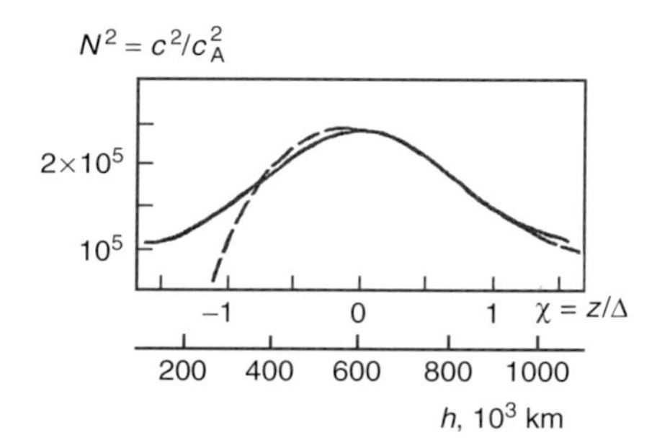

Figure 3.9 The height dependence of the square of the refractive index of FMA waves obtained from the data for type II radio bursts (dashed line). The approximation in Eq.(3.45) with $\delta = 2$, $N_1^2 = 8 \times 10^4$, $\Delta = 3 \times 10^{10}$ cm is shown by the solid line.

excitation demonstrated the generation of high-quality oscillations with the periods about 3 min [48].

From the analysis of type II solar radio emission generated by MHD shock waves, Zaitsev and Stepanov [49] found that FMA oscillations in a coronal resonator can be excited by beams of high-energy protons accelerated in the flare site under the condition of Cherenkov resonance at the height $\sim 1 R_\odot$ above the photosphere, with the minimum of the Alfvén speed c_A (Figure 3.9). This minimum corresponds to the maximum of the refractive index of FMA waves $N = c/c_A$; thereby, this region can present a resonator or a waveguide for FMA waves. Using the ideal MHD approach for $\beta \ll 1$, we can obtain the equation for FMA waves [49]:

$$\frac{d^2\psi}{dz^2} + \left[\frac{\omega^2}{c^2} N^2(z) - k_\perp{}^2\right]\psi = 0 \tag{3.44}$$

Here, the magnetic field B is uniform and directed along the z-axis of a Cartesian coordinate system, $\psi = \text{div}\zeta_\perp \propto \exp(-i\omega t + ik_\perp r_\perp)\varphi(z)$, ζ is the displacement of a volume element of plasma, and $k_\perp{}^2 = k_x{}^2 + k_y{}^2$. To simplify the calculations, let us represent the profile of the square of the refractive index in the following form:

$$N^2(z) = \frac{c^2}{c_A{}^2(z)} = N_1^2\left(1 + \frac{\delta}{\text{ch}^2\chi}\right) \tag{3.45}$$

where δ is a number, $\chi = z/\Delta$, Δ is the characteristic nonuniformity scale of the refractive index N, $N_1 = N\ (\pm\infty)$, and $z = 0$ at the point of the maximum N. In Figure 3.9, the dependence (Eq. (3.44)) is represented by the solid line.

We can see that the $N_1^2(z)$ curve obtained from the data for a type II burst is fairly closely approximated by Eq. (3.45) with $\delta = 2$, $N_1^2 = 8 \times 10^4$, $\Delta = 3 \times 10^{10}$ cm. Substituting Eq. (3.45) into Eq. (3.44), we obtain

$$\frac{d^2\psi}{d\chi^2} + \Delta^2\left(\frac{\omega^2}{c^2} N_1{}^2 - k_\perp{}^2 + \frac{2\omega^2}{c^2}\frac{N_1^2}{\text{ch}^2\chi}\right)\psi = 0 \tag{3.46}$$

Equation (3.46) is analogous to the Schrödinger equation with the potential

$$U = -\frac{2\omega^2 N_1^2}{c^2 \cos h^2 \chi} = \frac{U_1}{\cos h^2 \chi} \tag{3.47}$$

and with the energy $E = \frac{\omega^2 N_1{}^2}{c^2} - k_\perp{}^2$. The discrete spectrum (trapped modes) corresponds to $E < 0$. Moreover, the inequality $E > -U_1$ must be satisfied. Thus, we can find the region of existence for modes trapped in the waveguide, $p < q < 3p$, where $p = (\omega N_1 \Delta / c)^2$, $q = (k_\perp \Delta)^2$. The solution for the Schrödinger equation with the potential (Eq. (3.47)) is well known [50]. It is expressed in the form of hypergeometric functions. Using this solution, we find the discrete spectrum of oscillations in the potential well:

$$(p - q)_s = -\frac{1}{4}\left[-(2s-1) + \sqrt{1+8p}\right]^2 \tag{3.48}$$

where $s = 0, 1, 2, \ldots$; there are a finite number of levels $s < (-1 + \sqrt{1+8p})/2$ depending on the value p. The solution for the dispersion equation (Eq. (3.48)) is shown in Figure 3.10. Note that $p_{\min} = s(s+1)/2$.

FMA oscillations in such a resonator can be excited by beams of high-energy protons accelerated in the flare site under the condition of Cherenkov resonance. The instability threshold for FMA oscillations with a period 5–10 min for (0.1–1.0) MeV protons is determined from the equality of the growth rate of beam instability and Landau damping for electrons of thermal plasma [49]: $n_{\text{beam}}/n_0 \approx 10^{-4}$.

Oscillations of this coronal FMA resonator can lead to modulation of the radio emission propagating through the resonator due to variations of the optical thickness τ driven by periodical variations of the plasma density. In the case of free–free absorption, the modulation of intensity of the radio emission is

$$\frac{\delta I}{I} \approx \delta\tau(t) \approx \frac{\delta N}{N}\tau \tag{3.49}$$

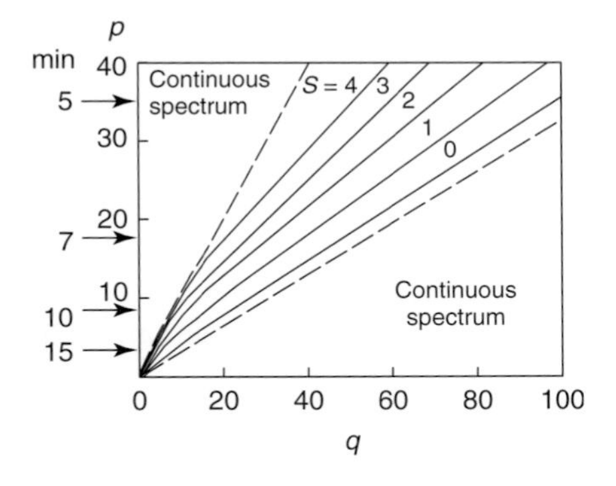

Figure 3.10 The spectrum of FMA waves in the coronal resonator. The arrows indicate the values of p corresponding to the given oscillation period. The value of p increases continually with an increase in q.

We can see from Eq. (3.49) that a resonator will effectively modulate radio emission passing through it if the optical depth $\tau \geq 1$. This is true under the solar coronal condition for the long-wavelength part of the decimeter and meter bands. Hence, such a resonator can be the reason for $\leq$10% modulation depth of the S-component and the noise storms with a period 3–15 min [49].

3.3 Excitation Mechanisms for Loop Oscillations

The excitation of loop oscillations can be due to internal sources, for example, self-excitation of RLC circuit, considered in Chapter 2, and/or the influence from outside (triggers, resonances).

3.3.1 External Triggers

Disturbances from flares, filament eruptions, and MHD waves can generate loop oscillations. There are several examples of such kind of excitation (Figure 3.11). Waves in coronal plasma involve magnetic loops into the oscillation process. Rapid damping of kink oscillations observed by TRACE can be due to dispersion blurring of the wave impulse (Figure 3.11a), produced, for example, by a flare [4, 35]. The ballooning oscillations of a loop can be driven by a flux of FMA waves from a source located under the loop [43]. The flux of FMA waves originated from ponderomotive interaction with a current-carrying loop produces effective acceleration $\mathbf{g}_{\text{eff}}$ (Figure 3.11b). The elastic force of the magnetic field lines is directed against $\mathbf{g}_{\text{eff}}$, and as a result, the ballooning mode of the oscillations appears. Taroyan *et al.* [51] discovered excitation of slow magnetoacoustic waves, $\omega = k_{\|}c_{\text{s}}$, in a solar coronal loop, as a consequence of heating in one of the loop footpoints during a microflare.

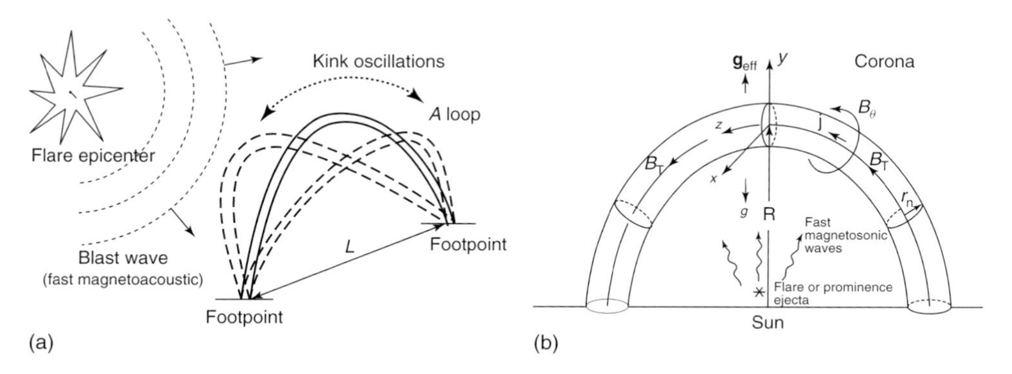

Figure 3.11 (a) The excitation of kink oscillations by a FMA wave from a flare. (b) Excitation of loop ballooning pulsations by FMS waves [43].

An interesting possibility for kink modes excitation in coronal loops by coronal mass ejection (CME) quasi-steady upflow was suggested by Nakariakov *et al.* [52]. The authors collected 17 oscillating solar events associated with a flare and CME and showed that periodic shedding of Alfvénic vortices that appear due to the upflow can generate kink oscillations under the condition of resonance with the loop eigenmodes. Indeed, in the presence of the force connected with the vortex shedding caused by the CME upflow, the dynamics of the apex of a coronal loop can be modeled by the equation of an oscillator with external influence:

$$\ddot{x} + 2\gamma \dot{x} + \omega_0^2 x = a(t) \cos\left[2\pi \nu_{\mathrm{a}}(t) t\right] \tag{3.50}$$

Equation (3.50) makes it possible to illustrate the selectivity of the generation mechanism (not all loops from the loop set are experience oscillation), the large growth rate for the global mode, and horizontally polarized oscillations.

On the other hand, ponderomotive interaction or the "leakage" of wave energy from the oscillating current-carrying loop may trigger periodic energy release in a flare arcade [53]. The same kind of interaction may be provided by the effect of mutual induction [54]. The influence of an oscillating "non-flaring" loop on a flaring loop arcade can induce current-driven plasma instabilities and, in turn, the anomalous resistivity. This can periodically trigger magnetic reconnection and, hence, quasi-periodic pulsations of X-ray, optical, and radio emission from the flaring arcade (Figure 3.12).

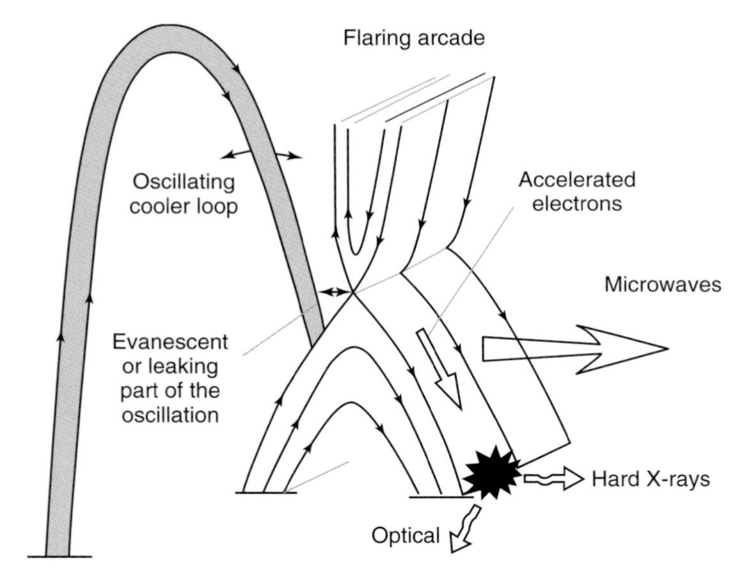

Figure 3.12 A sketch of the mechanism for generation of quasi-periodic pulsations in flaring emission. A cool (shaded) loop undergoes FMA oscillations. The external leaking part of the oscillations can reach magnetic null points in the arcade, inducing quasi-periodic modulation of the electric current. Periodically accelerated particles in the dense atmosphere will cause quasi-periodic microwave, optical, and X-ray emission.

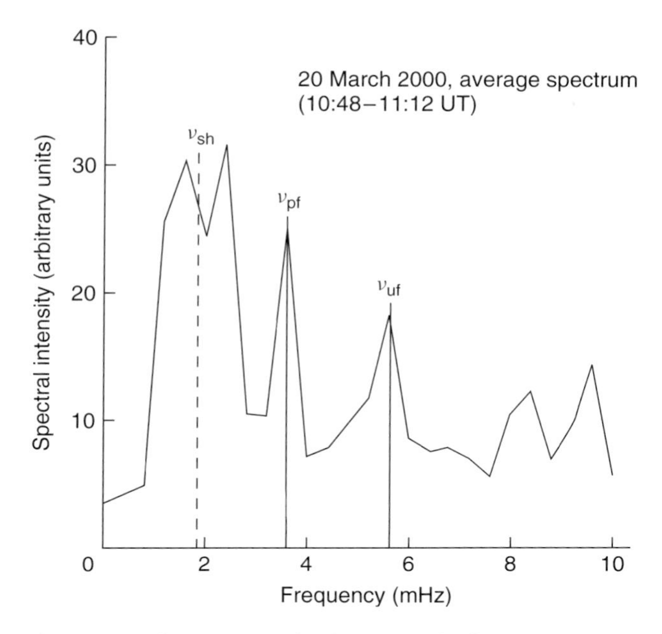

Figure 3.13 The spectrum of pulsations in the flare event on 20 March 2000 observed with Metsahovi radio telescope at 37 GHz. The dashed line shows the frequency of the 1.8 mHz (10 min period), while the solid lines indicate the frequencies of 5 min (3.6 mHz) and 3.3 min (5.6 mHz) oscillations.

3.3.2
Parametric Excitation of Loop Oscillations by p-Modes

TRACE and SOHO data [55, 56] and observations at 11 and 37 GHz using Finnish Metsähovi radio telescope [57, 58] provide numerous evidences for 5 min oscillations in the solar corona. Figure 3.13 presents the spectral analysis of 20 March 2000 solar flare event, which revealed 10 min, 5 min, and 3.3 min oscillations. It is well known that photospheric p-modes cannot directly penetrate the solar corona. Hence, the question arises: why the p-mode oscillations driven by photosphere convection observed at the coronal level?

Exploring kink oscillations of coronal loops with periods of about 300 s, Aschwanden *et al.* [3] conjectured that such oscillations can be initiated by 5 min oscillations of the photosphere (p-modes). De Pontieu *et al.* [56] suggested tunneling or direct propagation of photospheric oscillations into the outer atmosphere via slow MHD and shock waves. Such events can be interpreted using the electric circuit model for a coronal loop (Chapter 2) and the parametric resonance between acoustic eigenmodes of a loop and p-modes. Indeed, 5 min velocity oscillations of the photosphere modulate electric current in a coronal loop footpoints. Parametric resonance between the photosphere oscillations and acoustic oscillations $\omega_0 = k_{||}c_s$

of the loop appears. This effect is manifested in the excitation of loop oscillations with the periods of 5 min (pumping-up frequency ω), 10 min (subharmonic $\omega/2$), and 3.3 min (the first upper frequency of parametric resonance $3\,\omega/2$). The modulation of the electric current leads to the modulation of sound speed in the loop [58]:

$$c_s = \left(\frac{2\gamma k_B T}{m_i}\right)^{1/2}\left(1 + \frac{\Delta T}{2T}\right) = c_{s0}(1 + 0.5q\cos\omega t),$$

$$q = \frac{4}{3}\frac{\gamma - 1}{\gamma}\frac{I^2}{\pi c^2 r^2 p}\frac{I_\sim}{I} \tag{3.51}$$

Here, p is the gas pressure, r is the loop radius, $k_{||} = s\pi/l$, and l is the loop length, $s = 1, 2, 3, \ldots, I \gg I_\sim$.

Deviations of the loop plasma parameters can be written as

$$\frac{d^2 y}{dt^2} + \omega_0^2(1 + q\cos\omega t)y = 0 \tag{3.52}$$

Here, the parameter q determines the width of zones near the frequencies of a parametric resonance $\omega_n = n\omega/2$, $n = 1, 2, 3, \ldots$, where the parametric instability occurs [59]. The width of the first instability zone $(-q\omega_0/2 < \omega/2 - \omega_0 < q\omega_0/2)$, where ω_0 is close to $\omega/2$, is larger. Therefore, the amplitude of variations of the sound speed under the action of photospheric oscillations is higher compared to the second considerable narrowed zone $\omega \approx \omega_0$, if $q \ll 1$ (Figure 3.13). Since p-modes cannot directly penetrate into the corona, the parametric resonance may provide effective transfer for the energy of photosphere oscillations to the upper layers of the atmosphere. It offers the possibility for the understanding the heating mechanism of stellar coronae. Estimates made for solar loops showed that if the current exceeds 7×10^9 A, the energy flux of acoustic oscillations arising in a coronal magnetic loop as a result of the parametric resonance exceeds losses due to optical radiation.

Hindman and Jain [60] suggested that loop oscillations are generated due to buffeting excitation of loop magnetic fibrils by p-modes. They assumed, however, that magnetic fields are potential and no electric current exists. Our mechanism suggests the interaction of p-modes with electric currents in a coronal loop and the heating of loop footpoints due to the dissipation of the currents. Since p-modes do not penetrate into the corona, the parametric resonance can provide an efficient channel for transferring the energy of photosphere oscillations into stellar coronae.

3.3.3
Internal Excitation

In addition to the autoexcitation of a coronal loop as an equivalent RLC circuit (Chapter 2), some other ways of internal excitation of oscillations of coronal loops are possible.

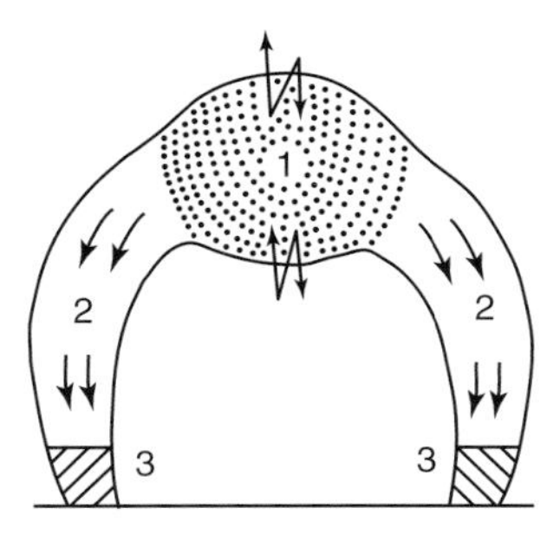

Figure 3.14 A model for the quasi-periodic modulation of hard X-rays in a flare loop [61]. 1: the volume of a rapid energy release (flare kernel), where sausage oscillations are excited; 2: modulated fluxes of energetic electrons; and 3: regions where hard X-ray emission is generated.

3.3.3.1 The Excitation of the Sausage Mode by Instantaneous Energy Release

Sausage oscillations of a coronal magnetic loop may occur due to rapid ($t_{\mathrm{imp}} < P$) energy release inside a loop [61]. The sharp plasma heating will increase the gas-kinetic pressure in the loop, thereby expanding the magnetic flux tube. These pulsations modulate the volume of the loss cone for high-energy electrons with the period described by Eq. (3.38). The modulated fluxes of >10 keV electrons injected into the loop footpoints produce pulsed hard X-ray emission with the same period P (Figure 3.14):

$$P \approx \frac{a}{\sqrt{c_{\mathrm{Ai}}{}^2 + c_{\mathrm{si}}{}^2}} \tag{3.53}$$

The emission flux of hard X-rays is proportional to the flux of energetic electrons ejected from a loop: $J \approx \ln_1/\sigma t_{\mathrm{D}}(\sigma)$. Here, n_1 is the number density of >10 keV electrons, $\sigma = B_{\mathrm{max}}/B_{\mathrm{min}}$ the loop mirror ratio, and $t_{\mathrm{D}}(\sigma)$ the time of electron diffusion into the loss cone, specified by the particular diffusion process involved. From these relations, one can see that, to the first approximation, the relative variations in the hard X-ray emission flux are proportional to the variations in the loop mirror ratio σ due to the pulsations of the magnetic field, $\Delta J/J \approx \Delta\sigma/\sigma \approx \Delta B/B$. To estimate ΔB, we should take into account that the oscillations are triggered by the work done on the loop magnetic field by additional gas pressure $\Delta p \approx nk_{\mathrm{B}}T$. Hence, $\Delta B/B \approx 4\pi k_{\mathrm{B}}T/B^2$, and the relative depth of the modulation of the hard X-ray emission is

$$M = \frac{\Delta J}{J} \approx \frac{4\pi k_{\mathrm{B}} T}{B^2} = \beta/2. \tag{3.54}$$

In other words, if the cool-plasma approximation ($\beta \ll 1$) is satisfied in a flare loop, hard X-ray pulsations should have small amplitude.

To determine the Q-factor of such sausage pulsations, we should note that the plasma density in the solar flare kernels exceeds that in the surrounding region by one to three orders of magnitude [62, 63]. Thus, there is a jump in the specific impedance of the sausage mode $Z = m_{\mathrm{i}}n(c_{\mathrm{A}}^2 + c_{\mathrm{s}}^2)^{1/2}$ at the "loop–environment" interface. In other words, the solar flare loops could provide a good resonator for sausage oscillations, so that the radiative (acoustic) losses may be neglected relative to dissipative losses. Other factors that may cause dissipative damping of loop sausage oscillations are ion viscosity, collisional damping, Landau damping, and electron thermal conductivity. It is essential for the diagnostics of plasma parameters of solar flare loops that, under the solar corona conditions, the most important factor for sausage oscillations is the electron thermal conductivity. Hence, the quality of oscillations $Q = \omega_+/\gamma$ may be expressed as [64]

$$Q = \frac{3}{\beta^2}\frac{m_e}{m_i}\nu_{ei}P, \quad \nu_{ei} = \frac{5.5n}{T^{3/2}}\ln\left(10^4\frac{T^{2/3}}{n^{1/3}}\right) \tag{3.55}$$

We can apply Eqs. (3.53)–(3.55) to the parameters of hard X-ray pulsations (P, Q, and M) to estimate the plasma number density, temperature, and magnetic field in the flare kernel:

$$\begin{aligned} n &\approx 4\times 10^{-11}\frac{Qa^3}{P^4}\frac{M^{7/2}}{(1+10M/3)^{3/2}}\ \mathrm{cm}^{-3} \\ T &\approx 10^{-8}\frac{a^2}{P^2}\frac{M}{(1+10M/3)}\ \mathrm{K} \\ B &\approx 3\times 10^{-17}\frac{Q^{1/2}a^{5/2}}{P^3}\frac{M^{7/4}}{(1+10M/3)^{5/4}}\mathrm{G} \end{aligned} \tag{3.56}$$

Equation (3.56) were applied for diagnostics of flare plasma using hard X-ray pulsations [65]. Three solar flares observed by *Venera* 12-M1 (90–700 keV) on 3 December 1978 20:30 UT and on 6 November 1979, 08:38 UT and by *Solar Maximum Mission* (30–286 keV) on 6 September 1981, 23:53 UT revealed low-quality ($Q \approx 10$) quasi-periodic pulsations with $M \leq 0.5$ and with the periods of 1.6, 2.3, and 6 s, respectively. This type of analysis for such three events with a between 10^8 and 10^9 cm indicates temperatures $T \approx 10^7$ K, magnetic field $B \approx 100$ G, and plasma densities n between 10^{10} and 10^{12} cm^{-3} [65]. These are in agreement with values of these parameters derived from other models.

3.3.3.2 The Excitation of the Global Kink Mode by Chromosphere Evaporation

Another internal way of generation of loop oscillations is related to heating of a loop footpoints by a solar flare [66]. As a result, hot ($\geq 10^7$ K) chromospheric plasma evaporates from the loop footpoints at the speed $v \geq 3\times 10^7$ cm s^{-1} [63]. Propagating along curved field lines, the evaporated matter induces the centrifugal force $F_c = nm_i v^2/R$, which stretches the magnetic loop upward, owing to the "freezing-in" condition. The elastic force of the magnetic field lines counteracts with the centrifugal force and "global" kink oscillations appear with the period determined by Eq. (3.26)

$$P_{\mathrm{GKM}} = 2l\left(\frac{\rho_i+\rho_e}{B_i^2+B_e^2}\right)^{1/2} \approx \frac{2l}{c_k} \tag{3.57}$$

Here, $R \approx l/\pi$ is the radius of the loop, c_k is the kink velocity (Eq. (3.21)), and we assume that the plasma $\beta \ll 1$.

3.3.3.3 The Excitation of the Sausage Mode by High-Energy Protons under the Bounce-Resonance Condition

The excitation of FMA oscillations in an extended ($\sim 1 R_\odot$) coronal resonator by beams of energetic protons under the Cherenkov resonance condition was considered in Ref. [49] (Section 3.3). However, in a coronal loop, which is in fact a magnetic bottle for energetic particles, another way of excitation of FMA oscillations is possible. A substantial part of accelerated protons is trapped in the loop and oscillates between the magnetic mirror points. The amplification of MHD oscillations is possible in the case of resonant interaction between the oscillations and magnetically trapped particles. Numerous resonant plasma instabilities arise due to the presence of nonthermal particles in the plasma [47]. If the wave frequency ω is comparable to the typical bounce frequency of the fast particles Ω, the bounce-resonance effects should be considered. These are very important for the problem of particle transfer in the Earth's magnetosphere [67–69].

Let us illustrate this mechanism of amplification of MHD oscillations by the following example [21]. Assume, for simplicity, that the stationary magnetic field in the loop depends on the longitudinal coordinate z:

$$B(z) = B_a \left(l + \frac{z^2}{l^2} \right) \tag{3.58}$$

B_a being the magnetic field at the loop apex. In this case, the undistributed motion of trapped particles is described in the drift approximation by the harmonic oscillator equation

$$\ddot{z} + \Omega^2 z = 0 \tag{3.59}$$

where $\Omega = (2\mu B_a/m_i)^{1/2}/l$ is the bounce frequency, $\mu = m_i v_\perp^2/2B(z)$ the magnetic moment of the particle, and $v_\perp$ the velocity component transversal to the magnetic field. If the sausage mode exists in the plasma, the force $F_{||} = -\mu \nabla B_\sim$ affects a particle, where $B_\sim$ is the component of the magnetic field of the wave parallel to $\boldsymbol{B}$. For the wave field $B_\sim(z,t) = B \sin(\omega t - k_{||} z)$, the general expression for the wave growth rate is [69]

$$\gamma_+/\omega_+ = 8\pi^4 n \sin^2\vartheta \sum_{s=1}^{\infty} s \int \mathrm{d}I \mathrm{d}\mu \mu^2 \frac{\partial f}{\partial I} J^2{}_s(k_{||} l)\delta(\omega - s\Omega) \tag{3.60}$$

Here, n denotes the density of rapid particles at $z = 0$, $I = m_i \Omega l^2/2$ is the longitudinal adiabatic invariant, l the amplitude of particle bounce, θ the angle between the wave vector and the stationary magnetic field, $J_s(w)$ is the Bessel function, and $\delta(w)$ the Dirac's δ-function. The distribution function $f(\mu, I)$ in Eq. (3.44) is related to the velocity distribution function as follows:

$$f(\mu, I) = \frac{2B_a F(v,s)}{n m_i^2 l} \tag{3.61}$$

Equation (3.60) shows that the contribution to the wave growth rate is due to the particles satisfying the bounce-resonance condition $\Omega(\mu) = \omega/s$, $s = 1, 2, 3, \ldots$ As it follows from Eq. (3.60) for the instability, the condition $\partial f/\partial I > 0$ must be

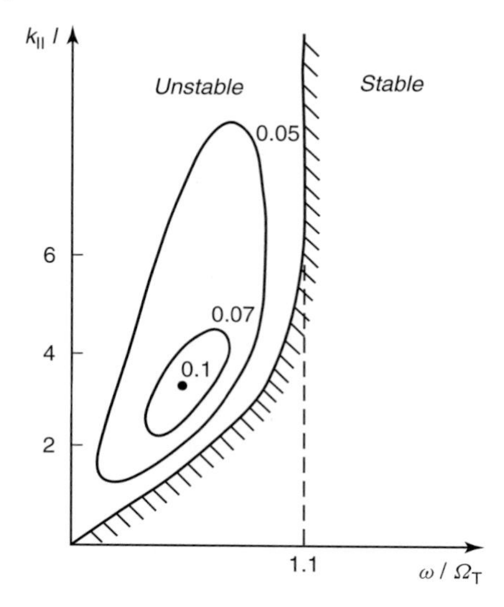

Figure 3.15 The unstable and stable regions and lines of constant growth rate in the plane (ω/Ω_T, $k_{||}l$) for $p = 2$.

fulfilled in some region (μ, I). Suppose that the distribution function F is isotropic, that is, it depends only on the total energy $\varepsilon = m_i v^2/2 = \mu B_a + \Omega I$:

$$F(v, s) \sim \varepsilon^p \exp\left[-\varepsilon(p + 3/2)/T_h\right] p > 0 \tag{3.62}$$

where T_h is the average energy of the rapid particles. Note that a loss cone exists in a loop, but for the mirror ratio $\sigma = B_{max}/B_{min} \geq 10$ it can be neglected. For a moderate p and $kl \approx 2\text{–}3$, the maximum growth rate is [21]

$$\gamma_+^{max}/\omega_+ \approx 0.1\beta_h \tag{3.63}$$

Here, $\beta_h = 8\pi n T_h/{B_a}^2$ is the "plasma beta" for energetic particles. The unstable region is shown in Figure 3.15, where $\Omega_T^2 = 2T_h/m_i l^2(p + 3/2)$.

For example, let us estimate the self-excitation criteria for the sausage mode under the bounce-resonance condition when acoustic damping is the most important factor in solar flare loops. It is inferred from Eqs. (3.36) and (3.58) that, when $\rho_e/\rho_i \approx 0.02$, $k_{||}^2/k_\perp^2 \approx a^2/l^2 \approx 10^{-3}$ the self-excitation condition of the sausage mode is fulfilled if the relative pressure of energetic protons $\beta_h > 0.2$.

References

1. Leighton, R.B. (1960) Aerodynamic phenomena in stellar atmospheres. Proceedings of the IAU Symposium No.12 on Cosmical Gas Dynamic (ed. R.N. Thomas), pp. 321–325.
2. Kurtz, D.W. (2001) *Sol. Phys.*, **220**, 123.
3. Aschwanden, M.J., Fletcher, L., Schrijver, C.J., and Alexander, D. (1999) *Astrophys. J.*, **520**, 880.
4. Nakariakov, V.M., Ofman, L., DeLuca, E.E., Roberts, B., and Davila, J.M. (1999) *Science*, **285**, 862.

5. Gershberg, R.E. (2005) *The Solar-Type Activity Amongst the Main Sequence Stars*, Springer.
6. Rodono, M. (1974) *Astron. Astrophys.*, **32**, 337.
7. Bastian, T.S., Bookbinder, J., Dulk, G.A., and Davis, M. (1990) *Astrophys. J.*, **353**, 265.
8. Mullan, D.J., Mathioudakis, M., Bloomfield, D.S., and Christian, D.J. (2006) *Astrophys. J. Suppl.*, **164**, 173.
9. Watts, A.L. and Strohmayer, T.E. (2007) *Adv. Space Res.*, **40**, 1446.
10. Ma, B., Li, X.-D., and Chen, P.F. (2008) *Astrophys. J.*, **686**, 492.
11. Stepanov, A.V., Zaitsev, V.V., and Valtaoja, E. (2011) *Astron. Lett.*, **37**, 276.
12. Ershkovich, A.I. and Nusinov, A.A. (1971) *Kosm. Issled.*, **9**, 470.
13. Zaitsev, V.V. and Stepanov, A.V. (1975) *Issled. Geomagn. Aeronomy. Fiz. Solntsa*, **37**, 3.
14. Hain, K. and Lüst, R. (1958) *Z. Naturforsch.*, **13a**, 936.
15. Appert, K., Gruber, R., and Vaclavik, J. (1974) *Phys. Fluids*, **17**, 1471.
16. Spruit, H.C. (1982) *Sol. Phys.*, **75**, 3.
17. Edwin, P.M. and Roberts, B. (1983) *Sol. Phys.*, **88**, 179.
18. Roberts, B., Edwin, P.M., and Benz, A.O. (1984) *Astrophys. J.*, **279**, 857.
19. Nakariakov, V.M., Melnikov, V.F., and Reznikova, V.E. (2003) *Astron. Astrophys.*, **412**, L7–L10.
20. Aschwanden, M.J., Nakariakov, V.M., and Melnikov, V.F. (2004) *Astrophys. J.*, **600**, 458.
21. Meerson, B.I., Sasorov, P.V., and Stepanov, A.V. (1978) *Sol. Phys.*, **58**, 165.
22. Doshek, G.A. (1994) Proceedings of the Kofu Symposium, NRO Report 360 (eds S. Enome and T. Hirayama), p. 173.
23. Nakariakov, V.M. and Ofman, L. (2001) *Astron. Astrophys.*, **372**, L53.
24. Cally, P.S. (1986) *Sol. Phys.*, **103**, 277.
25. Cally, P.S. (2006) *Sol. Phys.*, **233**, 79.
26. Aschwanden, M.J. (2003) NATO Advances Research Workshop, NATO Science Series II, p. 22.
27. Jakimiec, J. and Tomczak, M. (2010) *Sol. Phys.*, **261**, 233.
28. McLean, D.J. and Sheridan, K.V. (1973) *Sol. Phys.*, **32**, 485.
29. Kopylova, Y.G., Melnikov, A.V., Stepanov, A.V.*et al.* (2007) *Astron. Lett.*, **33**, 706.
30. Aschwanden, M.J. and Acton, L.W. (2001) *Astrophys. J.*, **550**, 475.
31. Aschwanden, M.J., De Pontieu, B., Schrijver, C.J., and Title, A.M. (2002) *Sol. Phys.*, **206**, 99.
32. Antolin, P. and Verwichte, E. (2011) *Astrophys. J.*, **736**, 121.
33. Aschwanden, M.J. (2009) *Space Sci. Rev.*, **149**, 31.
34. Uralov, A.M. (2003) *Astron. Lett.*, **29**, 486.
35. Terradas, J., Oliver, R., and Ballester, J.L. (2005) *Astrophys. J.*, **618**, L149.
36. Arregui, I., Anries, J., Van Doorsselaere, T., Goossens, M., *et al.* (2007) *Astron.Astrophys.*, **463**, 333.
37. Chen, J. and Schuck, P.W. (2007) *Sol. Phys.*, **246**, 145.
38. Mikhalayev, B.B. and Soloviev, A.A. (2005) *Sol. Phys.*, **227**, 249.
39. Kopylova, Y.G., Melnikov, A.V., Stepanov, A.V., and Tsap, Y.T. (2007) *Bull. Crimean Astrophys. Obs.*, **103**, 79.
40. Terradas, J., Goosssens, M., and Verth, G. (2010) *Astron. Astrophys.*, **524**, 23.
41. Soloviev, A.A., Mikhalayev, B.B., and Kiritchek, E.A. (2002) *Plasma Phys. Rep.*, **28**, 699.
42. Aschwanden, M.J., Nightingale, R.W., Andries, J. *et al.* (2003) *Astrophys. J.*, **598**, 1375.
43. Sakai, J.-I. (1982) *Astrophys. J.*, **263**, 970.
44. Cargill, P.J., Chen, J., and Garren, D.A. (1994) *Astrophys. J.*, **423**, 854.
45. van Doorsselaere, T., Debosscher, A., Andries, J., and Poeds, S. (2004) *Astron. Astrophys.*, **424**, 1065.
46. Tsap, Y.T., Kopylova, Y.G., Stepanov, A.V. *et al.* (2008) *Sol. Phys.*, **253**, 161.
47. Mikhailovsky, A.B. (1974) *Theory of Plasma Instabilities*, vols. **1 and 2**, Consultant Bureau, New York.
48. Botha, G.J.J., Arber, T.D., Nakariakov, V.M., and Zhugzhda, Y.D. (2011) *Astrophys. J.*, **728**, 84.
49. Zaitsev, V.V. and Stepanov, A.V. (1982) *Sov. Astron.*, **26**, 340.

50. Landau, L.D. and Lifshits, E.M. (1991) *Quantum Mechanics, Nonrelativistic Theory*, Butterworth-Heinemann, Oxford, Boston.
51. Taroyan, Y., Erdelyi, R., Wang, T.J., and Bradshaw, S.J. (2007) *Astrophys. J.*, **659**, L173.
52. Nakariakov, V.M., Aschwanden, M., and Van Dorsselaere, T. (2009) *Astron. Astrophys.*, **502**, 661.
53. Nakariakov, V.M., Fullon, C., Verwichte, E., and Young, N.P. (2006) *Astron. Astrophy.*, **452**, 343.
54. Khodachenko, M.L., Zaitsev, V.V., Kislyakov, A.G., and Stepanov, A.V. (2009) *Space Sci. Rev.*, **149**, 83.
55. De Moortel, I., Ireland, J., Hood, A.W., and Walsh, R.W. (2002) *Astron. Astrophys.*, **387**, L13.
56. De Pontieu, B., Erdelyi, R., and De Moortel, I. (2005) *Astrophys. J.*, **624**, L61.
57. Kislyakov, A.G., Zaitsev, V.V., Stepanov, A.V., and Urpo, S. (2006) *Sol. Phys.*, **233**, 89.
58. Zaitsev, V.V. and Kislyakov, A.G. (2006) *Astron. Rep.*, **50**, 823.
59. Landau, L.D. and Lifshitz, E.M. (1976) *Mechanics*, Pergamon Press, Oxford.
60. Hindman, B.W. and Jain, R. (2008) *Astrophys. J.*, **677**, 769.
61. Zaitsev, V.V. and Stepanov, A.V. (1982) *Sov. Astron. Lett.*, **8**, 132.
62. de Jager, C. (1979) *Sol. Phys.*, **64**, 135.
63. Doschek, G.A., Antiochos, S.K., Antonucci, E. *et al.* (1986) Chromospheric Explosion in Energetic Phenomena on the Sun, NASA Conference Publication 4-1.
64. Braginskii, S.I. (1965) *Rev. Plasma Phys.*, **1**, 205.
65. Desai, U.D., Kouveliotou, C., Barat, C. *et al.* (1987) *Astrophys. J.*, **319**, 567.
66. Zaitsev, V.V. and Stepanov, A.V. (1989) *Sov. Astron. Lett.*, **15**, 66.
67. Dungey, J.M. (1965) *Space Sci. Rev.*, **4**, 199.
68. Roberts, C.S. and Schulz, M. (1968,) *J. Geophys. Res.*, **73**, 7361.
69. Karpman, V.I., Meerson, B.I., Mikhailovsky, A.B., and Pokhotelov, O.A. (1977) *Planetary Space Sci.*, **25**, 573.

Further Heading

Stepanov, A.V., Kliem, B., Zaitsev, V.V. *et al.* (2001) *Astron. Astrophys.*, **374**, 1072.

Chiuderi, C. and Giovanardi, C. (1979) *Sol. Phys.*, **33**, 475.

4
Propagating MHD Waves in Coronal Plasma Waveguides

Recent progress in the imaging study of the solar corona allowed for the direct observation of magnetohydrodynamic (MHD) waves propagating along various plasma structures, resolving both the oscillation periods and wavelengths. The interest in coronal MHD waves is connected with several key challenges of modern solar physics: the problems of coronal heating and acceleration of the solar wind, possible triggering of coronal energy releases, that is, flares and coronal mass ejections (CMEs), and also the plasma diagnostics potential. As the observed wavelengths are comparable with the characteristic scales of the coronal plasma stratification and structuring (e.g., the density scale height is about 50 Mm and the typical minor radius of a coronal loop or a diameter of a prominence filament or a coronal jet is about 1 Mm), theoretical analysis of coronal waves must be based on the formalism developed in Chapter 3. Despite the abundant observational information provided by modern space and ground-based observational facilities, the study of coronal waves remains one of the most important research topics in solar and stellar physics.

As discussed in detail in Chapter 3, physical properties of MHD waves guided by plasma structures are quite different from their properties in a uniform medium, and this fact needs to be taken into account in the interpretation of observed phenomena and in theoretical estimations. In a low-β plasma, typical for the solar and stellar coronae, a plasma cylinder, representing coronal plasma structures, is known to have four main MHD wave modes: longitudinal, kink, sausage, and torsional. These modes correspond to the lowest values of the azimuthal wave number: $m = 0$ and 1. Waves with higher m have not yet been identified in the corona. The longitudinal wave is very similar to the familiar acoustic wave. It is essentially compressible, and it causes variations of the magnetic field, as the plasma flows in this wave are parallel to the magnetic field. The sausage mode is also essentially compressible, and it is accompanied by significant variation of the magnetic field, as the plasma flows are mainly transverse and cause variations of the cylinder cross section. The kink wave resembles an Alfvén wave of a nonuniform medium, while it is weakly compressible and hence is locally an oblique fast magnetoacoustic wave guided along the magnetic field by reflection or refraction on the plasma nonuniformity (e.g., the boundary of the cylinder). The true Alfvén wave in this geometry is the torsional wave that is essentially incompressible and transverse. In torsional waves, perturbations of the magnetic field are situated at

Coronal Seismology: Waves and Oscillations in Stellar Coronae, First Edition.
A. V. Stepanov, V. V. Zaitsev, and V. M. Nakariakov.

a magnetic surface corresponding to a constant value of the Alfvén speed. In a nonuniform medium, such as the solar corona, torsional waves situated at different magnetic surfaces are independent of each other. The four main MHD waves can couple with each other because of the curvature of the cylinder (e.g., in coronal loops), nonlinearity, and the equilibrium twist of the magnetic field. However, in observations, these waves are usually observed to be distinct, and hence, it is convenient to use this terminology in the interpretation of the observations.

4.1 MHD Waves in Vertical Coronal Magnetic Flux Tubes

A common feature of open magnetic configurations in the solar corona is polar plumes. They are observed as bright quasi-radial rays in coronal holes at heights of up to 10 or more solar radii by various coronal instruments in the extreme ultraviolet (EUV) and visible light (VL) channels (Figure 4.1). According to the theory presented in Chapter 3, polar plumes can act as waveguides for magnetoacoustic waves.

The first observational indication of compressible perturbations in polar plumes was found in the data obtained with the Harvard Skylab experiment [2]. Statistically significant short-period variations of Mg x emission intensity were detected, with amplitudes of about 10% and possibly moving at speeds lower than 100–200 km s^{-1}. Systematic observational study of waves guided by plumes became possible with the use of imaging and spectral instruments in the Solar and Heliospheric Observatory (SOHO). Analysis of data obtained in the white-light channel of SOHO/UVCS (Ultraviolet Coronagraph Spectrometer) revealed the presence of polarized brightness (density) fluctuations with periods of about 7–10 min at

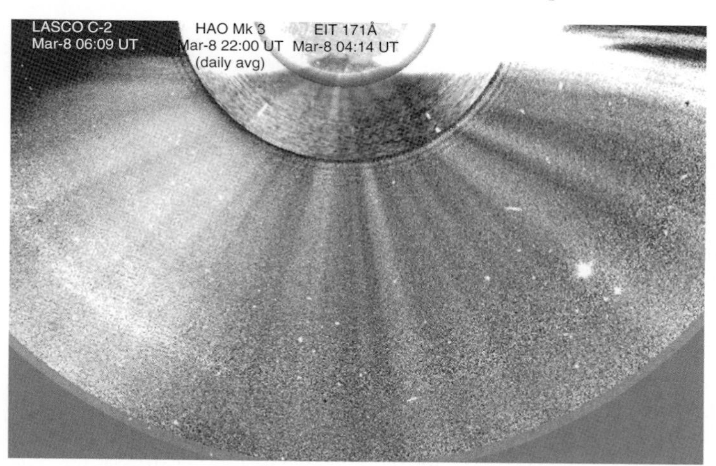

Figure 4.1 Embedded images of a coronal hole as seen by SOHO/EIT (171 Å), the Mauna Loa Solar Observatory (MLSO; Mark-III K-coronameter [Mk3] (white light pB corona)) Mk-3 white-light coronameter, and SOHO/LASCO (Large Angle and Spectrometric Coronagraph) (C2, white light). The bright radial structures are polar plumes. (From [1].)

a height of about 1.9 $R_\odot$ [3–5]. The estimation of the propagation speed of the fluctuations gave values in the range of 160–260 km s^{-1}. Statistically significant correlations were found between oscillation patterns at neighboring pixels along a vertical slit, suggesting that the variations were positioned along the plumes.

Time–distance maps, constructed for vertically oriented observational slits over several plumes at heights of 1.01–1.2$R_\odot$, observed in SOHO/EIT (Extreme ultraviolet Imaging Telescope) 171 Å datasets, showed the presence of diagonal stripes [6]. Such a diagonal stripe exhibits an EUV brightness disturbance, which changes its position in time and, consequently, propagates along the slit. The variation of the intensity in the disturbances suggests that they are perturbations of density and, consequently, the disturbances are compressive. (Another possibility for the variable intensity is the variation of the emitting column depth, e.g., by a kinklike perturbation [7]. However, the geometry of the plume in combination with the long wavelength of the perturbations ruled out that interpretation.) The observed amplitude of the variations in the intensity was estimated to be 10–20%. Assuming that the emission intensity is proportional to the density squared, it was obtained that the amplitude of the density perturbations was about 5–10%. Only the outwardly propagating waves were observed. The wave periods were 10–15 min, gathered in envelopes of 3–10 periods. The duty cycle of the waves was roughly symmetric. The projected speeds were about 75–150 km s^{-1}. Further analysis revealed that the wave amplitude grows with height.

The observational manifestation of incompressible or weakly compressible waves, such as torsional and kink waves, is usually indirect, and is associated with nonthermal broadening of coronal emission lines. The relative width of the spectral line λ is estimated to be

$$\frac{\Delta\lambda}{\lambda} \infty \left(\frac{kT}{m_\mathrm{i}} + \phi + \zeta^2\right)^{1/2} \tag{4.1}$$

where the first term on the right-hand side represents thermal broadening caused by the Doppler shift caused by thermal motion of the emitting ions of mass m_i, the second term represents various instrumental effects, and the third term accounts for the Doppler shift caused by various unresolved bulk movements of the emitting plasma along the line of sight. These motions can be connected with plasma turbulence and/or waves. There is an abundant evidence of nonthermal broadening of the emission lines observed in both plumes and interplume regions (see e.g., Refs. [8–11]). The typical nonthermal velocities are about 20 km s^{-1} and are usually observed to increase with height.

Theoretical modeling of the observed phenomena can be based on the model developed in Chapter 3.

4.1.1 Effects of Stratification

As the wavelengths of the wave phenomena observed in polar plumes are comparable with the scale height of the density stratification by gravity, its effects should be taken into account. Moreover, the adequate description of long-period waves requires inclusion of the effects of spherical geometry. In this section, we briefly overview these effects.

Consider an open magnetic flux tube, diverging in the radial direction. The magnetic field can be assumed to be purely radial,

$$B_0(r) = \frac{B_{00} R_\odot^2}{r^2} \tag{4.2}$$

where B_{00} is the magnetic field strength at the base of the corona ($r = R_\odot$) and r is the radius vector. Such a configuration corresponds to a magnetic monopole and is obviously not correct for modeling a global structure of the solar magnetic field. However, this configuration models the local magnetic structure of a plume over the lowest 20–30 Mm very well. In addition, we assume that the plasma is isothermal, with constant temperature T and sound speed C_s. The vertical dependence of the gravitational acceleration is $g = GM_\odot/r^2$, where $M_\odot$ is the mass of the Sun and G is the gravitational constant.

The hydrostatic equilibrium profile of the gravitationally stratified density can be readily obtained from the balance between the gas pressure gradient and gravity forces along the radial magnetic field,

$$\rho_0(r) = \rho_{00} \exp\left[-\frac{R_\odot}{H}\left(1 - \frac{R_\odot}{r}\right)\right], \tag{4.3}$$

where H is the scale height, $H/\mathrm{Mm} \approx 50\ T/\mathrm{MK}$, and ρ_{00} is the density at $r = R_\odot$. It is clear that Eq. (4.3) fails at large values of r, as it does not tend to zero. However, this expression works well for heights up to several solar radii, where the solar wind speeds are sufficiently low (i.e., subsonic). In the considered equilibrium, the sound speed C_s is constant, while the Alfvén speed varies with height as

$$C_A(r) = \frac{B_0(R_\odot) R_\odot^2}{r^2[4\pi\rho_0(R_\odot)]^{1/2}} \exp\left[\frac{R_\odot(r - R_\odot)}{2Hr}\right]. \tag{4.4}$$

The Alfvén speed initially grows with height, reaching a maximum at several solar radii, and then decreases. The specific position of the maximum is determined by the density scale height.

All four main MHD waves of a magnetic flux tube, the longitudinal, torsional, sausage, and kink, can propagate in the radial direction in the presence of the horizontal gradients of the equilibrium physical quantities (see Chapter 3). Here we consider two simplest, but important, cases of longitudinal and torsional waves.

We begin with the consideration of spherically symmetric longitudinal waves, which perturb the plasma density $\tilde{\rho}$ and the radial component of the plasma velocity, V_r. The governing wave equation is

$$\frac{\partial^2 \tilde{\rho}}{\partial t^2} - \frac{C_s^2}{r^2}\frac{\partial}{\partial r}\left(r^2 \frac{\partial \tilde{\rho}}{\partial r}\right) - g\frac{\partial \tilde{\rho}}{\partial r} = \mathrm{RHS} \tag{4.5}$$

where nonlinear and dissipative (e.g., connected with the finite viscosity or thermal conduction of the plasma) terms are gathered on the right-hand side and are represented as RHS. When the considered wavelengths are much less than both the scale height and the radius of the Sun, Eq. (4.5) can be solved in the Wentzel–Kramers–Brillouin (WKB) (or the single wave) approximation. Considering the nonlinear and dissipative to be of the same order as the ratio of the wavelength to the scale height, and passing to the frame of reference, co-moving with the upwardly propagating perturbation, we obtain from Eq. (4.5):

$$\frac{\partial \tilde{\rho}}{\partial r} + \left(\frac{1}{r} + \frac{g(r)}{2C_s^2}\right)\tilde{\rho} + \frac{1}{\rho_0(r)}\tilde{\rho}\frac{\partial \tilde{\rho}}{\partial \xi} - \frac{2\eta_0}{3C_s\rho_0(r)}\frac{\partial^2 \tilde{\rho}}{\partial \xi^2} = 0, \tag{4.6}$$

where $\xi = r - C_s t$ is the running coordinate (for details, see [12]) and η_0 is the dissipation coefficient, for example, the bulk viscosity. Accounting for thermal conduction changes, the qualitative value of the dissipative coefficient, but not the structure of the equation. Equation (4.6) is a spherical Burgers equation. The evolution of the waves is then determined by the competition of three effects: the stratification and increase in the wave front area with height, nonlinearity, and dissipation. According to Eq. (4.6), in the linear and dissipationless regime, when the third and the fourth terms are neglected, the relative perturbation of the density grows with height as

$$\frac{\tilde{\rho}}{\rho_0} \propto \frac{1}{r}\exp\left[\frac{R_\odot}{2H}\left(1 - \frac{R_\odot}{r}\right)\right]. \tag{4.7}$$

The growth of the relative amplitude of the acoustic waves with height, together with the observed propagation speed (that is, subsonic, as it should be because of the projection effect in the case of a wave guided along a thin flux tube in comparison with the wave length), suggests that the compressible propagating disturbances reported in [3–6] are likely to be acoustic waves. If one takes the two-dimensional effects that are associated with the finite width of the waveguiding flux tube into account, the waves should be rather slow magnetoacoustic. But, in a low-β plasma of the solar corona, slow magnetoacoustic waves are very similar to pure acoustic waves, and the above theory is quite adequate. The main difference is that the propagating speed is the tube speed introduced in Eq. (3.16). The observed vertical growth of the amplitude of compressible propagating disturbances is consistent with the theoretical model (Figure 4.2).

At certain heights, the growth of the amplitude requires accounting for nonlinear effects, which leads to the deformation of the wave shape and creation of a shock. As in a uniform medium, the higher density part of the wave perturbation propagates faster, and hence, an initially harmonic wave results in a sawtooth-shaped wave.

In the plasma equilibrium, given by Eqs. (4.2) and (4.3), linearly polarized (e.g., in the θ-direction, locally perpendicular to radial coordinate) Alfvén waves with spherical wave fronts are governed by the equation

$$\frac{\partial^2 V_\theta}{\partial t^2} - \frac{B_0(r)}{4\pi\rho_0(r)r}\frac{\partial^2}{\partial r^2}[rB_0(r)V_\theta] = \text{RHS} \tag{4.8}$$

where again the RHS represents nonlinear and dissipative terms. In this case, in contrast with Eq. (4.5), the speed of the waves, according to Eq. (4.4), varies with

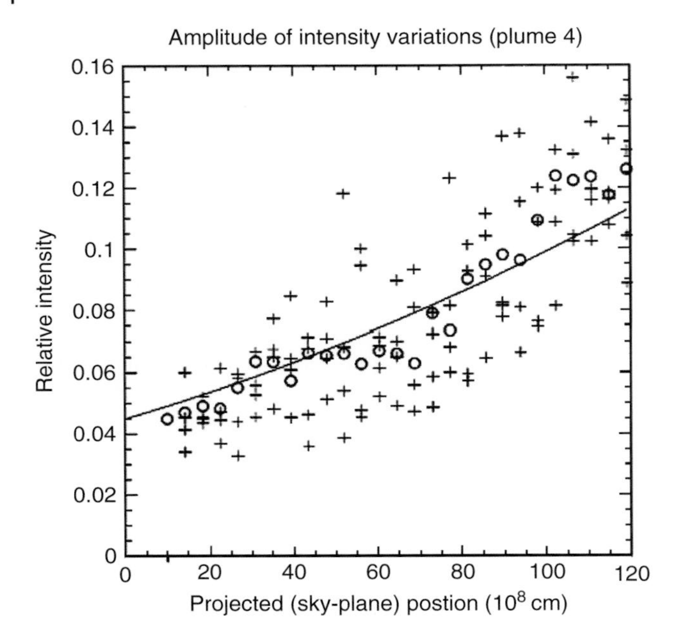

Figure 4.2 Vertical dependence of the upwardly propagating compressible wave observed in a polar plume with SOHO/EIT. The solid curve shows the theoretical solution, and various symbols correspond to different measurements. (From [13].)

height. Thus, in the WKB analysis of Eq. (4.7), it is convenient to not work in the spatial domain, as in Eq. (4.5), but in the time domain [13]. The evolution of a weakly nonlinear upwardly propagating Alfvén wave is then described by the equation

$$\frac{\partial V_\theta}{\partial r} - \frac{R_\odot^2}{4Hr^2} V_\theta - \frac{1}{4C_\mathrm{A}(C_\mathrm{A}^2 - C_\mathrm{s}^2)} \frac{\partial V_\theta^3}{\partial \tau} \frac{v_0}{2C_\mathrm{A}^3} \frac{\partial^2 V_\theta}{\partial \tau^2} = 0, \tag{4.9}$$

where $\tau = t - \int C_\mathrm{A}^{-1}(r)\mathrm{d}r$ and v_0 is a coefficient of dissipation, representing resistivity or shear viscosity. Equation (4.9) is a spherical analog of the scalar Cohen–Kulsrud–Burgers equation. The main difference between this equation and the spherical Burgers equation (Eq. (4.6)) is the order of the nonlinearity, which affects the characteristic wave signature. In particular, in contrast with an initially harmonic acoustic wave that develops into a saw-tooth wave, the Alfvén wave develops into a square wave ("meander"): the shocks are formed at both positive and negative half-periods.

In the linear and dissipationless regime, the third and the fourth terms of Eq. (4.9) can be neglected, and the height evolution of the wave amplitude is given by the linear solution

$$V_\theta \infty \exp\left[\frac{R_\odot (r - R_\odot)}{4Hr}\right]. \tag{4.10}$$

The amplitude of an upwardly propagating linear Alfvén wave grows with height. We must note that Eq. (4.9) was derived under the assumption that the scale height is much smaller than the solar radius [13], and, consequently, consideration of the limiting case of the nonstratified atmosphere, $H \to \infty$, is meaningless here. The growth of the Alfvén wave amplitude with height can be easily explained by simple geometrical reasoning. In the absence of dissipation and backward waves, the radial component of the *Poynting* flux of a spherical wave, $\boldsymbol{S} \propto \boldsymbol{B} \times \boldsymbol{V} \times \boldsymbol{B}$, is proportional to $1/r^2$. Hence, its radial component is given by the expression $S_r \propto B_0 V_\theta B_\theta$. Taking into account that in a running Alfvén wave the perturbations of the magnetic field and the plasma velocity are connected with each other as $B_\theta / B_0 = -V_\theta / C_A$, we get $S_r \propto B_0^2 V_\theta^2 / C_A$. Using Eq. (4.2), we obtain that $B_0 V_\theta^2 / C_A$ is constant, which gives us $V_\theta \propto \left[\frac{C_A(r)}{B_0(r)}\right]^{1/2} = \rho_0^{-1/4}(r)$. Using Eq. (4.3), we observe that this expression coincides with Eq. (4.10). The dependence given by Eq. (4.10) is used for the confirmation that the spectral nonthermal broadening could indeed be caused by upwardly propagating Alfvén waves (see e.g., [8–11]).

The third and fourth terms in Eq. (4.9) describe the nonlinear distortion of the wave profile through the generation of higher harmonics and the dissipation. At a height where the sound speed reaches the Alfvén speed, the discussed formalism fails, as the coefficient of the nonlinear term in Eq. (4.9) becomes infinite. The nature of the nonlinearity is the perturbation of the density, induced by the magnetic pressure perturbation in a linearly polarized Alfvén wave of finite amplitude. If the Alfvén wave is harmonic, $V_\theta = V_{\theta a} \cos(\omega t - kr)$, the induced density perturbation is

$$\rho = \frac{V_{\theta a}^2}{16\pi (C_A^2 - C_s^2)} \cos(2\omega t - 2kr) \tag{4.11}$$

The excitation of the compressible perturbation can be considered to be the nonlinear coupling of Alfvén and slow magnetoacoustic waves, with resonance occurring when their speeds coincide.

Torsional waves, which are guided by a spherically divergent magnetic flux tube, have similar properties: the growth of the amplitude with height and nonlinear steepening, with one important exception. It is connected with the fact that in flux tubes slow magnetoacoustic waves become longitudinal modes, and their propagation speed is different from the sound speed. Moreover, in the long-wavelength limit, these waves propagate at the tube speed, that is, subsonic and sub-Alfvénic. Hence, the resonance between the torsional waves, propagating at the Alfvén speed (they are situated at specific surfaces of the constant Alfvén speed and hence "do not know" about transverse structuring) and the longitudinal modes, affected by the structuring, which propagate at the tube speed, does not occur [14]. Another interesting property of weakly nonlinear, long-wavelength torsional waves is that they do not perturb the boundary of the magnetic flux tube, as the centrifugal and magnetic tension forces in these waves cancel out each other.

4.2
Propagating Waves in Coronal Loops

4.2.1
Propagating Compressible Waves in Coronal Loops

One of the most common examples of coronal waves is propagating compressible disturbances observed near footpoints of coronal loops, often in structures known as *magnetic fans* – divergent magnetic flux tubes situated over sunspots. The waves are essentially longitudinal, and are the perturbations of the emission intensity, propagating along bright filaments in the fans. In the late 1990s, the commissioning of high-resolution coronal EUV imagers SOHO/EIT and Transition Region and Coronal Explorer (TRACE) revealed the omnipresence of such waves [15–18]. More recently, signatures of these waves were detected with spectral [19, 20] and stereoscopic instruments [21] and in microwave [22] and soft X-ray [23] bands.

The propagating EUV disturbances in magnetic fans are very similar to the compressible waves detected in polar plumes (see Section 4.4): in a time–distance map, constructed by taking a sequence of EUV images along a certain selected slit along the assumed magnetic field in the image, there are periodic diagonal strips (Figure 4.3). Typical speeds of the propagating EUV disturbances observed in coronal loops vary from a few tens to 200 km s^{-1}. The large range of observed speeds is possibly connected with the projection effect. The typical periods are from 3 to 12 min. In the distribution of the events over the period, there are two clear peaks, corresponding to 3 min and 5 min bands. Typically, loops situated above sunspots show oscillations with periods of about 3 min, whereas nonsunspot loops (above plague regions) show oscillations closer to 5 min. The typical amplitudes of the EUV intensity perturbations are present in small percentages, reaching, in some cases, 10–14%. The waves have pronounced amplitude modulation in the form

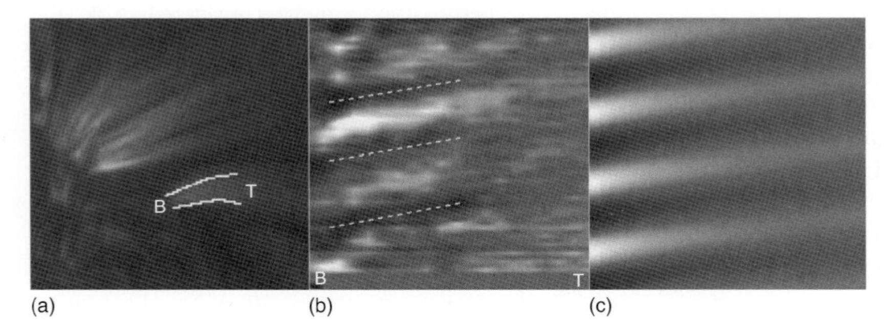

Figure 4.3 An example (TRACE 171 Å, 7 April 2000, 12:42 UT) of a coronal loop footpoint (a) supporting an oscillatory signal and the observed (b) and theoretically predicted (c) running difference between the intensity time series, in which the propagating intensity disturbances are seen as bright and dark, diagonal bands. (From [18].)

of intermittent quasi-monochromatic trains lasting several periods. At the same location, sequences of wave trains can be seen for several hours and even days. (Such behavior is similar to the time evolution of 3-min oscillations in sunspot umbrae, seen at the chromospheric heights [24].) Stereoscopic measurements performed with the EUV imagers on Solar TErrestrial RElations Observatory (STEREO) A and B spacecraft estimated wave path inclination angle as $37^\circ \pm 6^\circ$, giving the absolute value of the phase speed as 132 ± 11 km s^{-1} [20]. This value corresponds to the adiabatic sound speed of plasma with the temperature 0.84 ± 0.15 MK, which is consistent with the bandpass of the imager. Spectroscopic observations [19] showed an approximate in-phase relation between Doppler shift and intensity oscillations of the coronal emission lines. Both the intensity and Doppler shift amplitudes were found to decrease with the line formation temperature. An interesting feature of the waves is their detection length [25]: the waves are typically seen in the lower parts of the loops, up to the (projected) heights of 20 Mm. Recent spectroscopic observations performed with Hinode/EIS (Extreme-ultraviolet Imaging Spectrometer on board of *Hinode* (SOLAR-B)) in the Fe xii 195 Å emission line showed the presence of the propagating disturbances at distances up to 70–90 Mm [26]. This property, together with the shorter periods, suggests that there is some difference between the EUV intensity waves observed in polar plumes (Section 4.4) and that observed in loops. Analysis of 2D information showed that different strands, forming the magnetic fan, oscillate at different times, but generally show similar wave properties (periods, amplitude, and detection lengths).

The first model describing the propagating longitudinal waves was developed in [27]. In the model, the propagating disturbances are considered to be acoustic waves guided along the magnetic field in a looplike magnetic flux tube of constant cross section. Only field-aligned motions were taken into account; hence, the balance between the gravitational and gas pressure forces determined the equilibrium. The main difference between this model and the model discussed in the previous section is the curvature of the waveguiding field line. Let the shape of the flux tube be semicircular with the major radius being R_L, representing a coronal loop. Consider a curvilinear coordinate system, with the coordinate s directed along the magnetic field, with $s = 0$ at one of the footpoints. As the typical major radii of coronal loops are less than the radius of the Sun, we can neglect the effects of spherical geometry and the variation of the gravitational acceleration with height. Hence, the projection of the gravitational acceleration on the magnetic field line is

$$g(s) = \frac{GM_\mathrm{e}}{R_\odot^2}\left(1 + \frac{R_\mathrm{L}}{R_\odot}\sin\frac{s}{R_\mathrm{L}}\right)^{-2}\cos\frac{s}{R_\mathrm{L}} \tag{4.12}$$

Considering the isothermal plasma, from the balance of the gas pressure and gravitational forces along the field line, we obtain the profile of the plasma density,

$$\rho_0(s) = \rho_0(0)\exp\left(-\frac{\gamma g(0)}{C_\mathrm{s}^2}\frac{R_\mathrm{L}\sin(s/R_\mathrm{L})}{1 + (R_\mathrm{L}/R_\odot)\sin(s/R_\mathrm{L})}\right) \tag{4.13}$$

where $\rho_0(0)$ and $g(0)$ are the equilibrium plasma density and the gravitational acceleration at the footpoints, respectively. The profile is symmetric with respect to the loop apex. It is also convenient to introduce a function $H_P(s) = C_s^2/\lambda g(s)$ that represents the projected scale height.

Following the formalism introduced in the previous chapter, we consider the wavelength of the perturbations to be much shorter than the major radius of the loop and the scale height. Hence, applying the WKB approximation to the waves propagating in the positive direction of s, we obtain the evolutionary equation describing the parallel flows in the wave,

$$\frac{\partial V}{\partial s} - \frac{1}{2H(s)}V + \frac{\gamma+1}{2C_s}V\frac{\partial V}{\partial \xi} - \frac{1}{2\rho_0 C_s}\left[\frac{4\eta_0}{3} + \frac{\kappa_P(\gamma-1)^2}{\gamma R}\right]\frac{\partial^2 V}{\partial \xi^2} = 0 \quad (4.14)$$

where the last term represents dissipation due to finite viscosity (η_0) and parallel thermal conduction (κ_P), and $\xi = s - C_s t$ is the running coordinate. Perturbations of the plasma density, gas pressure, and temperature in the longitudinal wave can be obtained with the use of phase relations

$$\frac{\tilde{\rho}}{\rho_0} = \frac{V}{C_s}, \frac{\tilde{p}}{p_0} = \gamma\frac{V}{C_s}, \frac{\tilde{T}}{T_0} = (\gamma-1)\frac{V}{C_s}, \quad (4.15)$$

respectively, where the quantities with the index zero denote equilibrium values. Phase relations (Eq. (4.15)) are consistent with the observed relationship between the intensity and Doppler shift variations in propagating EUV disturbances observed in coronal loops [26].

Equation (4.14) gives that the amplitude of linear longitudinal waves in an ideal medium depends on the coordinate along the magnetic field line as

$$V(s) \propto \exp\left[\frac{\gamma g(0) R_L}{2C_s^2}\frac{\sin(s/R_L)}{1 + R_L/R_\odot \sin(s/R_L)}\right] \quad (4.16)$$

Together with Eq. (4.13), this expression corresponds to the conservation of the energy $\rho_0 V^2/2$ in the wave. Dependence (Eq. (4.16)) shows that the amplitude of the waves should grow up to the apex of the loop and then gradually decrease, returning to the initial value at the opposite footpoint. Such a behavior is clearly different from that observed in the presence of the upwardly propagating waves near footpoints only. A possible explanation for this discrepancy is connected with the combined action of the effects of the flux tube divergence with height and finite thermal conduction [18]. In addition, shorter-period and higher-amplitude waves are affected by nonlinear steepening, described by the third term of Eq. (4.14), and dissipate below the loop apex. Quantitative analysis of these effects is a subject of ongoing active research.

The simultaneous observation of the propagating coronal disturbances with imaging and spectral instruments shows another interesting feature: simultaneous imaging and spectral observations, performed with Hinode/XRT (The *Hinode* X-Ray Telescope) and EIS, showed that the periodic propagating disturbances were accompanied by upward Doppler velocities of $\approx$50 km s^{-1} in a coronal emission line in the magnetic fan. This suggested that the propagating disturbances were interpreted as upflows, rather than acoustic waves. Moreover, EIS observations of asymmetries in line profiles, with faint blue-wing excess in the order of 1–5% core intensity, seem to support the interpretation in terms of upflows [28]. However,

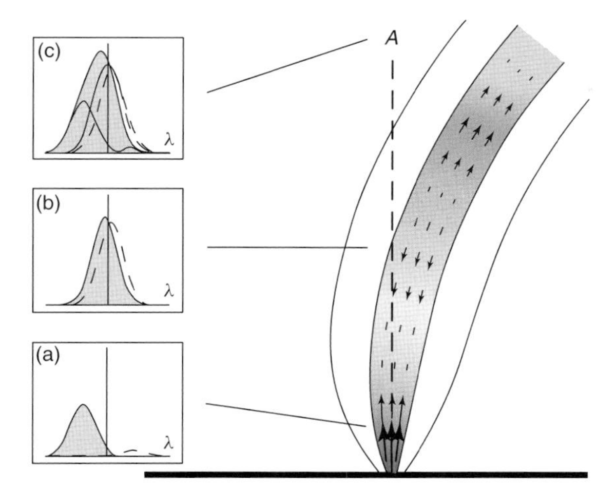

Figure 4.4 Illustration of the formation of the systematic Doppler shift and asymmetry of a coronal emission line from a propagating acoustic wave in a coronal loop. The velocity field and intensity perturbation of the wave are shown. The inset panels illustrate the emission line at various locations. The solid and dashed curves are for oscillation phase $\varphi = 0$ and $\varphi = \pi$, respectively. (a) Emission line from the loop with a large-amplitude slow wave. (b) Emission line from loop with small-amplitude slow wave or from a static background plasma in the line of sight. (c) Total emission line integrated over the line of sight. The thin lines show the two plasma components contributing to the full profile. (From [29].)

recent detailed modeling of observational spectral signatures of coronal acoustic waves demonstrated that the observations are consistent with the slow wave interpretation [29]. Indeed, an upwardly propagating acoustic wave with phase relations (Eq. (4.15)), when averaged over its period of oscillation, produces nonthermal line broadening that is correlated with the Doppler velocity, and line asymmetry, as found in observations (Figure 4.4). On the other hand, the upflow interpretation has several intrinsic difficulties: it is not clear why the plasma flows with speeds close to that of sound do not yet result in shocks. What is the source of the periodicity and modulation? And where does the matter brought up by the upflows go?

Another interesting recent development in the study of coronal longitudinal waves was the apparent relationship between the waves observed in the magnetic fan above a sunspot and quasi-periodic pulsations in flaring energy releases at an active region (AR) near the sunspot [30]. It was suggested that propagating disturbances could periodically trigger magnetic reconnection at a remote coronal magnetic structure, causing periodic bursts of emission. Potentially, this study opens up interesting perspectives for the understanding of the phenomenon of sympathetic flares, when flares induce each other, forming chains of energy releases in the corona. Also, the above may be relevant to the stellar coronal seismology, when stellar coronal oscillations are observed in the light curves of stellar flares. In that case, the oscillatory patterns representing low-amplitude oscillations are dramatically amplified by the energy releases they trigger.

4.2.2
Transverse Waves in Coronal Loops

Since their discovery in 1999 with the EUV imager TRACE [31, 32], transverse oscillations of solar coronal loops have been subject to intensive study. The oscillations are usually seen as periodic displacements of coronal loops on the plane of sky, with the periods ranging from a few to several minutes, different for different loops. The displacement amplitude is typically a few megameters, which corresponds to several minor radii of the oscillating loops. In most of the observed cases, the spatial structure of the oscillations corresponds to a global standing mode of the loop, with the nodes at the footpoints and the maximum displacement at the loop apex. In some cases, a second standing harmonics is also observed. In almost all the cases, the oscillations have horizontal polarization. After excitation by a flare or a coronal mass ejection, the oscillations decay very rapidly, in several periods. The observed geometry of standing transverse oscillations is consistent with the $m = 1$ (or kink) mode, discussed in Chapter 3. The kink mode is locally a fast magnetoacoustic wave, propagating obliquely to the magnetic field, and being guided along the field by a field-aligned plasma structure by reflection ofrefraction.

Recent observations of the dynamics of coronal loops with the ground-based spectrometer Coronal Multi-Channel Polarimeter (CoMP) showed the presence of propagating waves of the Doppler shift of the coronal emission line Fe xiii, along the magnetic field lines [33]. The period of the Doppler shift variations was several minutes. The direction of the magnetic field was inferred from the linear polarization measurements. The amplitude of the observed Doppler shift variations was low, $<1\,\mathrm{km\,s^{-1}}$, which corresponded to the displacement amplitude of about 50 km. As the plane of the observed loops was approximately perpendicular to the line of sight, the Doppler shift was caused by the periodic, propagating displacements of the plasma in the direction perpendicular to the magnetic field. On the other hand, this periodicity was not detected either in the emission intensity or in the line broadening. The projected speeds of the perturbations were 0.6–$4\,\mathrm{Mm\,s^{-1}}$, in the range of the expected Alfvén speed. The waves seem to be ubiquitous, filling up the entire observed active region. There is a disparity between the upward-propagating and downward-propagating wave power, the outward power being about a factor of 2 larger. The observed correlation length of the perturbations in the direction perpendicular to the magnetic field is about 9 Mm, and along the field 45 Mm. In contrast with the standing kink waves, the propagating waves do not seem to follow a certain plasma nonuniformity, such as an individual loop.

Initially, the transverse waves observed with CoMP were interpreted as Alfvén waves, as the observations showed incompressible waves propagating along the field at a speed close to the expected Alfvén speed. However, in the observations, the perturbed volume of plasma (e.g., a magnetic flux tube) was displaced in the transverse direction as a whole, and the transverse size of the perturbation was at least an order of magnitude shorter than the wavelength along the field. Hence, without a waveguiding field-aligned plasma nonuniformity, such a perturbation would not propagate along the field, as observed, but across it [34]. This is connected with the competition of two restoring forces: one the magnetic tension force and the other the gradient of the total pressure. Both the forces affect a volume of the plasma with the frozen-in magnetic field, displaced in a transverse direction. The magnetic tension force drives the perturbation along the field, as the Alfvén wave, while the gradient of the total pressure generates an oblique fast magnetoacoustic wave. The ratio of the energy partition between these two modes depends on the aspect ratio between the perpendicular and parallel spatial scales of the perturbation. If its parallel size is much shorter than the perpendicular scale, the perturbation can be considered to be quasi-plane and perpendicular to the magnetic field. It develops mainly in the Alfvén waves. In contrast, the perturbation localized in the perpendicular direction, such as that observed with CoMP, develops as fast waves. Such a scenario becomes clearer if one considers a limiting case, when the perturbation is infinitely extended along the magnetic field. In this case, a transverse displacement would simply change the density of the magnetic field lines and the plasma density and hence create a gradient of the total pressure. The magnetic tension force does not appear, as the field lines remain straight. Hence, the only excited wave would be a fast wave, propagating across the magnetic field. The CoMP waves are localized across the field, while propagating along it. Such a behavior is possible only in the presence of a field-aligned plasma nonuniformity, for example, of the temperature or density, when the periodic Doppler shift is produced by kink modes of the nonuniformity. Torsional waves should be excluded from consideration, as the induced rotation of the perturbed flux tube causes the Doppler shift simultaneously in both blue and red directions on the different sides of the flux tube. This leads to a periodic nonthermal broadening of the emission line because of spatial integration [35], and not because of the observed periodic Doppler shift. Consequently, CoMP observations show the ubiquity of kink waves, which are the propagating analog of the kink oscillations observed with TRACE. Moreover, their observation provides us with an important seismological evidence of the presence of unresolved, field-aligned filamentation of the active regions.

Recent 3D MHD numerical simulations supported the interpretation of the propagating transverse waves in terms of kink waves [36]. It was shown that the observed plasma motions could not exist in uniform plasma, although they could propagate along the field in the presence of the field-aligned plasma nonuniformity (it was a plasma cylinder with the density contrast about 2 in the simulations). In addition, the observed excess of the upward-propagating wave power in comparison with the downward-propagating wave power was interpreted in terms of resonant absorption: linear coupling of the collective kink mode to unresolved torsional or Alfvén motions (see, e.g., [37] and references therein). The coupling occurs in a narrow resonant layer inside the oscillating magnetic flux tube, where the phase speed of the traveling kink wave coincides with the local Alfvén speed. As the coupling takes energy from the kink wave, its amplitude decreases with the damping length

$$\frac{L_{\mathrm{D}}}{\lambda_{\|}} = \frac{2}{\pi} \frac{a}{l_{\perp}} \frac{\rho_{\mathrm{i}} + \rho_{\mathrm{e}}}{\rho_{\mathrm{i}} - \rho_{\mathrm{e}}}, \tag{4.17}$$

where $\lambda_{||}$ is the wavelength, a is the radius of the waveguide, and $l_{\perp}$ is the characteristic spatial scale of the variation of the Alfvén speed (or density) in the radial direction [38]. If the transverse profile of the Alfvén speed is a step function, corresponding to the infinitely sharp jump in the Alfvén speed at the cylinder boundary, the decay by resonant absorption is negligible. For a smooth profile of the Alfvén speed, the kink wave decays in a few wavelengths from the source, which is consistent with observations.

4.3
Waves in Coronal Jets

Jets are a frequently observed, dynamical feature of the solar corona. In particular, in the soft X-ray band, jets are seen as transient, collimated features with apparent high velocity outflows in the direction of collimation (Figure 4.5). According to a statistical study based on the analysis of SOHO/SXT (soft X-ray telescope, a telescope that views soft X-rays an instrument on the Yohkoh (Solar A) space probe) and Hinode/XRT data [39, 40], most of the observed jets are associated with small flares and microflares. The typical observed lengths of hot jets are in the range from a few tens to several hundred megameters, and their diameters are a few megameters, which gives the typical aspect ratio of about 10–100. The jets are typically observed for several tens of minutes. About half of the jets have a constant width, and about a third have a width decreasing with height. The projected outflow speeds are measured in the range of several hundred kilometers per second. In some cases, the speed reaches about a thousand kilometers per second. Most of the jets are situated over regions of mixed magnetic polarity, suggesting that they are likely to be connected with magnetic reconnection. Recently, transverse oscillations or coronal jets were detected [40] in the form of periodic displacements in the jet axis with a period of about 200 s and displacement amplitude of about 4000 km.

The theoretical description of this phenomenon requires the generalization of dispersion relation (Eq. (3.21)) on accounting for the field-aligned bulk flow. The main new feature introduced by the flow was the modification of the dispersion relations [41, 42] and wave–flow interaction effects associated with negative energy [43]. The latter can cause instabilities at steady flow speeds well below the Kelvin Helmholtz Instability (KHI) threshold, which occurs when the flow-speed shear exceeds the Alfvén speed by a factor of about 2 [44].

Following the formalism developed in Chapter 3 and [45], we consider a hot jet as a uniform plasma cylinder of radius a with the total pressure balance with the external medium at the boundary. Inside the cylinder, there is a field-aligned steady flow with constant speed U_{i}. In both external and internal media, the plasma beta is assumed to be less than unity. Also, in contrast with the previous subchapters, this time we can ignore the effect of gravitational stratification of the plasma. Indeed, for plasmas of temperature, for example, 10 MK, the density scale height is about 500 Mm, while the jets are observed up to heights of 100 Mm only.

Consider only linear transverse perturbations to the steady flow equilibrium of azimuthal wave number $m = 1$, describing the perturbation of the jet axis. The

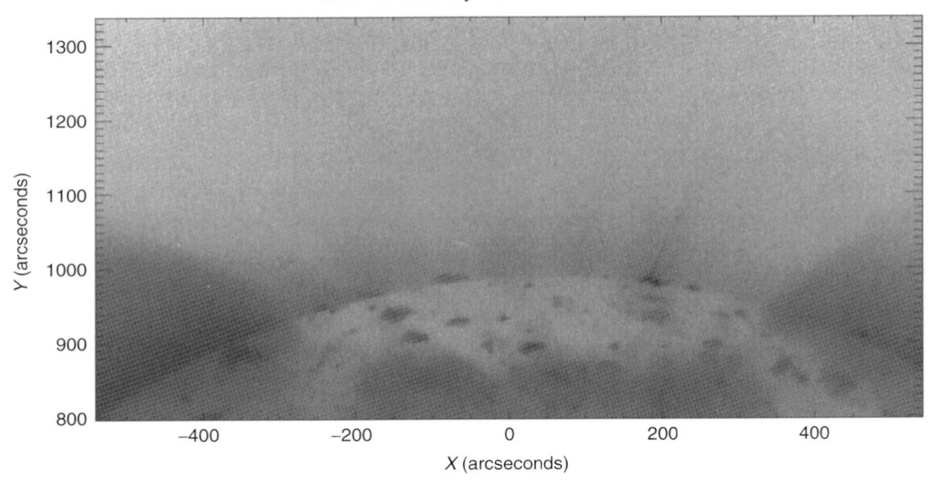

Figure 4.5 Hinode X-Ray Telescope (XRT) false-color image of the north polar coronal hole. A typical jet is seen in the center of this image as a thin dark outwardly stretched feature. (From [40].)

presence of the internal flow causes the Doppler shift of the quantities, describing the internal medium in Eq. (3.21): in the expression μ_i and in the second factor on the right-hand side, the frequency ω becomes $\Omega_\mathrm{i} = \omega - U_\mathrm{i}k$. Since the wavelengths of transverse perturbations observed by Cirtain *et al.* [40] are significantly longer than the radius a of the jet, it is sufficient to consider the limiting case of $|ka| = 1$. From Eq. (3.21) with the modifications accounting for the Doppler shift, the phase and group speeds of upwardly propagating (with $k > 0$) kink waves can be approximated by the explicit expressions,

$$\begin{aligned} \omega/k &\approx V_\mathrm{G} + \alpha(ka)^2 \ln(ka) \\ d\omega/dk &\approx V_G + 3\alpha(ka)^2 \ln(ka) \end{aligned} \tag{4.18}$$

where

$$V_\mathrm{G} \approx U_\mathrm{i} + C_\mathrm{Ai}\sqrt{1 + (B_\mathrm{e}/B_\mathrm{i})^2}, \alpha \approx \frac{C_\mathrm{Ai}}{2} \frac{(B_\mathrm{e}/B_\mathrm{i})^2}{\sqrt{1 + (B_\mathrm{e}/B_\mathrm{i})^2}}$$

assuming that the plasma density in the jet is much higher than that outside it. Both asymptotic expressions for the phase and the group speeds are independent of the sound speeds inside and outside the jet.

The perturbations of the velocity in this mode is given by the expressions

$$\begin{aligned} v_\mathrm{r} &= \frac{C_\mathrm{A0}^2}{\left(k^2 C_\mathrm{A0}^2 - \Omega_\mathrm{i}^2\right)} \frac{\mathrm{d}}{\mathrm{d}r} \Delta(r) \cos\Theta \\ v_\theta &\propto \exp\left[\frac{R_\odot(r - R_\odot)}{4Hr}\right]. \end{aligned} \tag{4.19}$$

where $\Theta = \omega t - kz$ is the phase of the oscillation, and $\Delta(r) = A_0 J_1(\mu_i r)$ inside the jet and $\Delta(r) = A_0 J_1(\mu_i a) \frac{K_1\left(\sqrt{-\mu_e^2}\, k\right)}{K_1\left(\sqrt{-\mu_e^2}\, a\right)}$ outside it, with the expressions for the $\mu_{i,e}$ being given by Eq. (3.16) and A_0 being the amplitude of the perturbation. The function $\Delta(r)$ describes the transverse structure of the perturbation. The z-component of the velocity perturbations is zero. Also, the kink wave perturbs the magnetic field, including its absolute value, and the density of the plasma, $\tilde{\rho} = -\rho_0 \Delta(r) \sin\Theta / \Omega_i$, where the equilibrium density ρ_0 should be assumed to be either inside or outside the jet. Thus, the transverse waves guided by coronal jets are *essentially* compressible, since they perturb the plasma density and the divergence of the velocity. Also, these waves are collective, since they are coherent perturbations of all magnetic surfaces inside and outside the jet (given by the function $\Delta(r)$).

The origin of the transverse oscillations of soft X-ray coronal jets has not yet been understood. One possible candidate mechanism could be the Kelvin–Helmholtz instability [46], connected with the generation of the perturbation due to the shear flow at the boundary of the jet. For a plasma cylinder of the observed parameters, with the typical values of the Alfvén speed ($C_{Ai} = 3C_{A0}$, which correspond to the density contrast $\rho_e/\rho_i = 0.13$), the instability threshold value of the steady flow speed is $U_i = 4.47 C_{Ai}$. Since the observed values of the jet speeds do not exceed the Alfvén speed inside the jet, the instability threshold is not reached and this possibility should be excluded. Another option is related to negative energy wave instabilities, when some modes grow if dissipative processes or energy leakage decreases the total energy of the considered system. In that case, the system is essentially nonconservative, and the wave takes the energy from the driver of the steady flow. However, in [43], it was shown that, in the considered situation, sub-Alfvénic flow speeds could lead to the instability of longitudinal modes only, which does not explain the generation of the transverse perturbations. Consequently, the observed transverse waves are likely to be excited somewhere at the origin of the jet.

As observable parameters of kink waves guided by coronal jets are linked with the physical parameters of the internal and external media and the speed of the jet through relations [13, 47], it is possible to use the observations of the waves for seismological diagnostics of the plasma and determination of the physical mechanisms operating in jets. This is an interesting and promising topic for future research.

4.4
Evolution of Short-Wavelength, Fast Magnetoacoustic Waves

Wave dispersion is a property of a medium when different spectral components of a wave propagate at different phase and group speeds, connected with the presence of a characteristic spatial scale in the medium. In a uniform medium, ideal MHD waves are dispersionless, as the ideal MHD equations do not contain a characteristic scale. On the other hand, in MHD waveguides, the spatial scale is the width of the waveguide. Because of this, fast magnetoacoustic waves of both kink and sausage symmetry are dispersive (see Chapter 3). Dispersion is most pronounced for the wavelengths comparable with the width of the waveguiding structure (e.g., the minor radius of a loop). In waveguides, evolution of broadband MHD perturbations is accompanied by the appearance of wave patterns with typical, quasi-periodic dispersive signatures. Obviously, as the formation of such a pattern takes some time, the wavelength of the perturbation should be much smaller than the distance to the observation point from the wave source. In other words, this effect can be observed for the perturbations with the wavelengths much shorter than the length of the waveguide, for example, the length of the coronal loop.

In a pioneering work [48], qualitative analysis of propagating sausage modes of a plasma cylinder demonstrated that, at some distance from the initial perturbation, a broadband pulse develops into a characteristic quasi-periodic wave train. The waves were considered to be in a straight cylinder with the step-function transverse profile of the equilibrium physical quantities. Three distinct phases of the wave train were distinguished: the low-amplitude periodic phase, the subsequent high-amplitude quasi-periodic phase, and the decay phase. Such an evolution scenario is determined by the presence of a minimum of the group speed dependence upon the wave number. The periodic phase should be observed from the time h/C_{Ae} to time h/C_{Ai}, where h is the distance between the initial excitation and the observation point. Then the observed signal becomes quasi-periodic, lasting till time $h/\min(V_{\mathrm{g}})$, where V_{g} is the group speed of the mode. An estimation of the characteristic period of the periodic phase, provided in [48], is

$$P_{\mathrm{prop}} \approx \frac{2.6a}{C_{\mathrm{Ai}}}\sqrt{1-\frac{\rho_{\mathrm{e}}}{\rho_{\mathrm{i}}}} \tag{4.20}$$

In loops with large density contrast, $\rho_{\mathrm{i}} \gg \rho_{\mathrm{e}}$, the observed period is $P_{\mathrm{prop}} \approx 2.6a/C_{\mathrm{Ai}}$. This value is approximately the transverse fast wave transit time in the loop, which is of the order of 1 s. The corresponding wavelength is about the width of the waveguide. It is important to keep in mind that Eq. (4.20) provides us only with the estimation of the mean period, as the intrinsic feature of the discussed effect is the variation of the period with time. The initial stage of the fast sausage pulse evolution in a straight magnetic slab was numerically modeled in [49], and was found to be consistent with the analytical prediction. Wave trains with signatures qualitatively similar to those theoretically predicted were found in 303 and 343 MHz coronal data recorded by the Icarus spectrometer [48].

The transverse profile of the Alfvén and fast speeds in the loop, directly connected in a low-β coronal plasma with the density profile, affects the fast wave dispersion and, consequently, the wave train signature. This effect was studied in [50] by considering waves in a zero-β plasma slab with the density profile given by the generalized symmetric Epstein function, $\rho_0 = \rho_{\max}\mathrm{sech}^2\left[(x/w)^p\right] + \rho_\infty$, where x is the transverse Cartesian coordinate, and $\rho_{\max}, \rho_\infty, w$ are the values of the density at the center of the slab and at infinity, and the slab width, respectively. The power index p determines the steepness of the profile. The cases when the power index p equals either unity or infinity correspond to the symmetric Epstein profile or to the step-function profile, respectively. In both limiting cases, exact dispersion relations for kink and sausage modes can be obtained analytically. The case of the symmetric Epstein profile is particularly interesting; the dispersion relations are not transcendental, such as in the case of the step-function profile, but algebraic, allowing for the exact explicit solution. It was established that the group speed of both kink and sausage modes has a minimum for all profiles with the power index being greater than unity, which, in other words, is steeper than the symmetric Epstein profile. Thus, the properties of the profile affect the dispersion relation and can be estimated by the analysis of wave trains formed by the dispersion.

Numerical simulations of the developed stage of the dispersive evolution of a fast wave train, propagating along the magnetic field in a straight slab of a low-β plasma with a smooth transverse profile of the plasma density, confirmed the qualitative prediction of [48]. It was found that development of an impulsively generated pulse leads to the formation of a quasi-periodic wave train with the mean wavelength comparable with the slab width [51]. In agreement with the analytical theory, wave trains have pronounced period and amplitude modulation. In this situation, a convenient analytical tool is the wavelet transform technique [52]. In particular, evolution of a broadband sausage pulse leads to the formation of a wave train with a characteristic "tadpole" wavelet signatures (or, rather a "crazy tadpole" as it comes tail-first, see Figure 4.6a).

A striking evidence of this effect came from the white-light observations of the solar corona with the Solar Eclipse Coronal Imaging System (SECIS), which discovered propagating short-period compressible waves in coronal loops [53, 54]. The waves were observed to travel along the assumed magnetic field lines in an off-limb active region at the projected speed of $\sim$2100 km s^{-1}. The waves were observed to have a quasi-periodic wave train pattern with the mean period of about 6 s. The wavelet spectrum of a typical wave train observed with SECIS is shown in Figure 4.6b. There is an obvious similarity between the theoretically predicted, "crazy tadpole" signature and the observational detections. Indeed, both the quasi-harmonic tail and the broadband head are clearly seen in the wavelet spectrum. This supports the interpretation of the propagating disturbances observed by SECIS as fast magnetoacoustic wave trains, excited by a broadband impulsive driver.

An important parameter that determines the specific wavelet signature of the fast magnetoacoustic wave train is the initial spectrum of the perturbation. This spectrum is determined by the duration of the initial perturbation, or its spatial localization. Figure 4.7 shows wave trains, generated by initial Gaussian-shaped sausage perturbations of different widths, in a plasma slab with the symmetric Epstein profile of the density. The wave trains are recorded as the density perturbations at a distance of $\sim$15 slab widths from the initial position along the slab. Different panels of Figure 4.7 demonstrate the wave trains developed from the initial Gaussian perturbations of different longitudinal widths and their wavelet

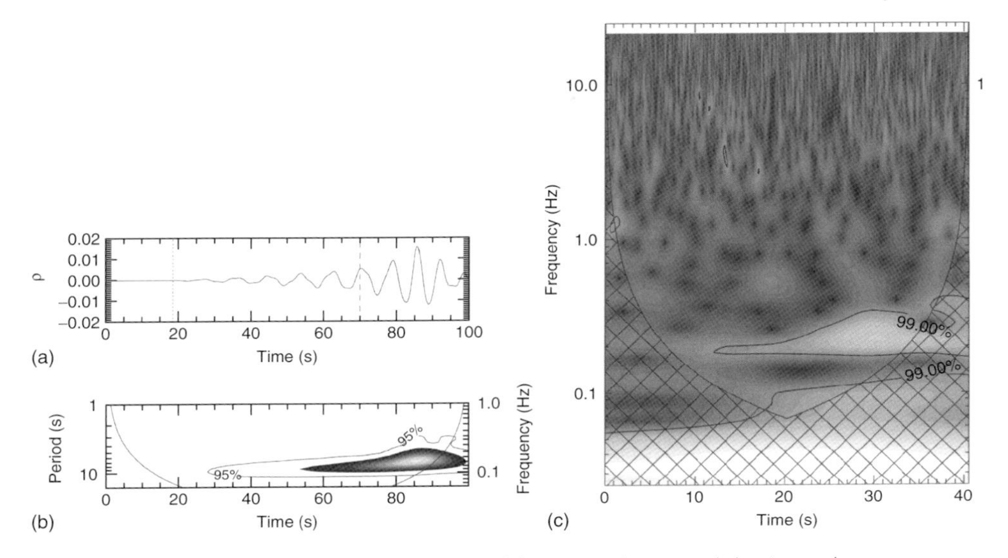

Figure 4.6 (a,b) Numerical simulation of an impulsively generated fast magnetoacoustic wave train guided by a slab of enhanced plasma density. (a) The wave train of the plasma density perturbation, measured at the distance of 35 slab widths from the wave source. The vertical lines show the pulse arrival time if the density was uniform: the dotted line using the external density, and the dashed line the density at the center of the structure. (b) Morlet wavelet spectrum of the signal, demonstrating the characteristic "crazy tadpole" signature. (From [51]) (c) A Morlet wavelet spectrum of a wave train observed with SECIS instrument. (From [53].)

spectra. In all the cases, the spectral energy of the wave trains is situated in the region corresponding to the periods shorter than the cutoff period of the guided sausage mode. Hence, a part of the long-period part of the initial energy leaves the slab via the sausage wave leakage, confirming the discussion in Chapter 3. It is also clear that the wave trains generated by longer initial pulses have less pronounced "crazy tadpole" spectra and are more monochromatic. Longer initial pulses have spectra with the main part of the energy localized at the wave numbers smaller than the cutoff wave number. Consequently, for longer pulses, the guided part of the initial spectral energy is localized just above the cutoff wave number, giving more monochromatic spectra of the developed wave trains.

Observational detection of short-period fast wave trains and the study of the amplitude and spatial modulation provide us with an interesting perspective for probing fine transverse structuring of the coronal plasma. However, full-scale implementation of this seismological method will be possible only when the necessary spatial and time resolution is achieved. The time resolution needs to be better than 1 s, allowing for the confident identification of the fast wave transverse transit time (about 5 s or shorter), while the spatial resolution has to be at least five times better than the projected wave length. Moreover, this should be achieved in both time and spatial domains simultaneously; otherwise unresolved positive and negative perturbations would cancel out each other.

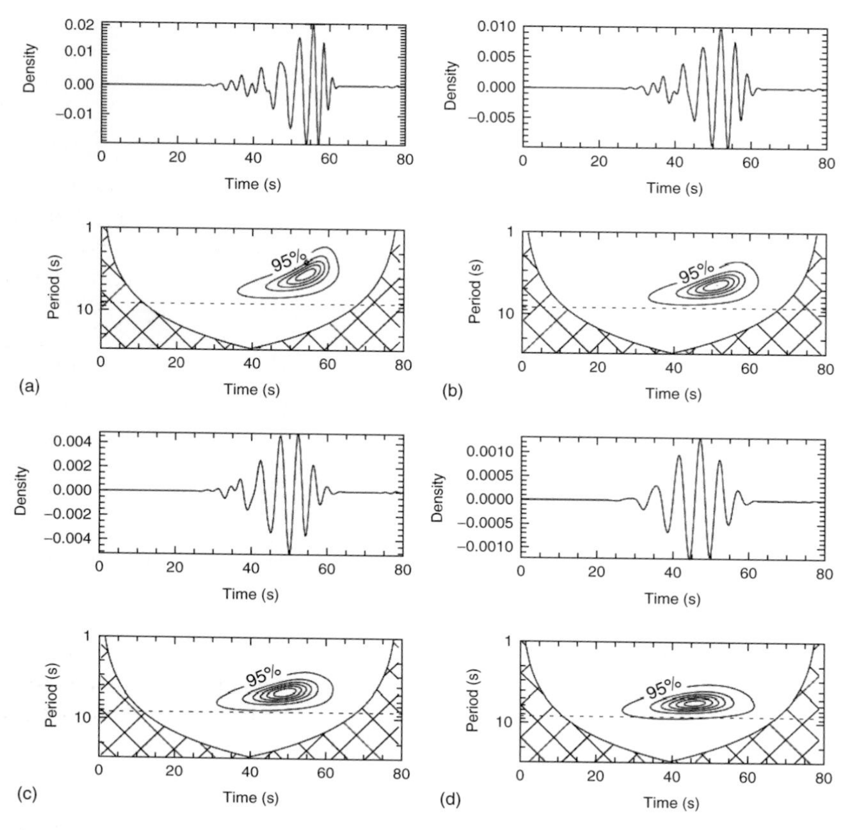

Figure 4.7 Density perturbations in impulsively generated fast wave trains guided by a plasma slab and their Morlet wavelet spectra, for different widths of the initial sausage pulse: (a) $w/2$, (b) w, (c) $3w/2$, and (d) $5w/2$, where w is the slab half-width. The horizontal dotted line shows the value of the longest possible period of a trapped sausage mode, which, for the plasma profile taken in the simulations, is 8.2 s. (From [52].)

Similar tadpole patterns were also detected in solar decimetric type IV radio events recorded by the Ondřejov radio spectrograph [55]. The characteristic features are usually observed simultaneously at all radio frequencies in the 1.1–4.5 GHz frequency range. The characteristic period of the wavelet tadpole patterns was about 70 s, which is significantly longer than the periods detected with SECIS. A possible interpretation is that the guiding plasma structure in this case is not a loop, but a current sheet. In the vicinity of the current sheet, the Alfvén speed is reduced because of the decrease in the magnetic field. Hence, the fast wave transverse transit time is longer. Numerical simulations of fast magnetoacoustic modes guided by current sheets confirmed this interpretation [56]. Moreover, estimations of the current sheet's width, derived at by using Eq. (4.20), gave the value of 1 Mm that is consistent with the estimations made by other techniques.

4.5 Alfvén Wave Phase Mixing

As discussed earlier, in a nonuniform medium, perturbations of the magnetic field Alfvén waves are situated at the surfaces of constant Alfvén speed and are polarized parallel to these surfaces. In an Alfvén wave, the transverse perturbation of the magnetic field does not change the absolute value of the field, keeping a constant distance between the perturbed field lines. Because of the frozen-in effect, such a behavior does not perturb the plasma density, making the Alfvén wave essentially incompressible.

In an ideal plasma, Alfvén waves situated at neighboring Alfvén surfaces do not "feel" with each other. This can be illustrated by a set of mechanical pendulums attached to the same axis, oscillating in the plane perpendicular to the axis. In an ideal case, when the pendulums are not coupled with each other, oscillations of the pendulums are completely independent. If the pendulums have different resonant frequencies, the oscillations have different instant phases, and one does not see any collective motion of the system. If one excites the pendulums, giving them the same initial phase, the difference in the frequencies will lead to phase mixing of the oscillations. However, if the difference in frequencies is not large, there will be a certain period of time when the oscillations are almost in phase and the behavior of the system will seem to be coherent. The pendulums can couple with each other, for example, via the air between them or through the deformation of the axis, while the efficiency of this coupling may be very low. This analogy can be used for the understanding of the Alfvén wave dynamics in a smoothly nonuniform plasma. In particular, Alfvén wave phase mixing occurs when the waves propagate in a medium with a smooth transverse profile of the Alfvén speed.

Consider a 1D smooth nonuniformity of the Alfvén speed in the direction across the magnetic field. Let the field be straight and directed along the z-axis, while the Alfvén speed gradient is in the x-direction. Linear plane Alfvén waves, which are polarized in the y-direction, are described by the wave equation

$$\frac{\partial^2 V_y}{\partial t^2} - C_A^2(x)\frac{\partial^2 V_y}{\partial z^2} = 0 \tag{4.21}$$

with the solution $V_y = \Psi(x)f\left(z \pm C_A(x)t\right)$, where $\Psi(x)$ and $f(z)$ are the functions describing the initial profile of the wave, and the sign of the argument determines the direction of the wave propagation along the z-axis. We note that plasma flows in two other directions, in the plane x–z, corresponding to magnetoacoustic collective modes. Equation (4.21) shows that Alfvén waves propagating on different Alfvén surfaces, which are in the considered geometry (the planes $x = \text{const}$), have different speeds coinciding with the local values of the Alfvén speed. Hence, the waves are subject to phase mixing, accompanied by the generation of increasingly steep gradients in the x-direction in the wave fronts. These growing transverse gradients cause several interesting physical effects.

4.5.1 Damping of Alfvén Waves because of Phase Mixing

Sufficiently steep transverse gradients eventually result in dissipative effects, such as finite viscosity and resistivity, coming into play. Accounting for these effects requires an additional term, $\overline{\eta}(\partial^2/\partial x^2 + \partial^2/\partial z^2)\partial V_y/\partial t$, where $\overline{\eta}$ represents the dissipation (e.g., the shear viscosity), on the left-hand side of Eq. (4.21). In the developed stage of phase mixing, the transverse spatial scales of the Alfvénic perturbation are much smaller than the longitudinal wavelength; hence, the dissipative terms can be rewritten as $\overline{\eta}\partial^3 V_y/\partial x^2\partial t$. Considering the wave traveling in the positive direction of z, we introduce a running coordinate $\xi = z - C_A(x)t$. We get that $\partial/\partial t = -C_A(x)\partial/\partial\xi$, $\partial/\partial z = \partial/\partial\xi$, and $\partial/\partial x = -t\left(dC_A/dx\right)\partial/\partial\varsigma$. Hence, the slow evolution of the phase-mixed Alfvén wave in the presence of finite dissipation is described by the diffusion equation

$$\frac{\partial V_y}{\partial t} + \overline{\eta}t^2\left(\frac{\mathrm{d}C_A}{\mathrm{d}x}\right)^2\frac{\partial^2 V_y}{\partial\xi^2} = 0 \tag{4.22}$$

Taking that $\partial/\partial\xi \approx k_P$, where k_P is the longitudinal wavelength, we can readily integrate Eq. (4.22), getting

$$V_y \approx V_y(0)\exp\left[-\frac{\overline{\eta}}{3}\left(\frac{dC_A}{dx}\right)^2 k_P^2 t^3\right] \tag{4.23}$$

where the first factor on the right-hand side represents the amplitude of the Alfvén wave at $t = 0$. Thus, phase-mixed Alfvén waves decay superexponentially [57]. A similar dependence can be obtained in the spatial domain, considering the decrease in the wave amplitude with the distance from its source. In addition, the efficiency of phase mixing damping can be increased by nonlinear steepening because of the decrease in the longitudinal wavelength [58].

Dependence of the Alfvén speed on the longitudinal coordinate (e.g., in the case of waves guided by open coronal magnetic structures, it is the stratification and the vertical change of the magnetic flux tube diameter) modifies the longitudinal wavelength and hence affects the efficiency of Alfvén wave phase mixing [59]. Depending on the specific geometry, the efficiency of wave damping can either be higher or lower in comparison with 1D nonuniformity. For example, in the equilibrium with uniform density and exponentially diverging flux tubes, the wave amplitude has an exotic $\exp[-\exp(z)]$-dependence [60].

In the rarefied parts of the solar corona, the decrease in the transverse wavelength in phase-mixed Alfvén waves can reach the collisionless scale. One of the important new effects appearing in this case is the wave–particle interaction, connected with the generation of a parallel electric field [61]. Consideration of this effect in terms of kinetic description of the plasma established that the parallel electric field increases with a decrease in the transverse wavelength by phase mixing,

$$E_P \approx k_P^2\rho_s^2\left(\frac{\mathrm{d}C_A}{\mathrm{d}x}\right)t\exp\left[-\frac{\sqrt{\pi}}{6}\frac{C_A^2}{V_{Te}}\left(\frac{\mathrm{d}C_A}{\mathrm{d}x}\right)^2 k_P^2\rho_s^2 t^3\right] \tag{4.24}$$

where V_{Te} is the electron thermal speed and $\rho_s = C_s/\omega_{ci}$ with ω_{ci} being the ion cyclotron frequency [62]. Equation (4.25) demonstrates that, in the collisionless regime, as well as in the collisional regime, phase-mixed Alfvén waves experience superexponential damping. The parallel electric field, generated by phase mixing, causes electron acceleration and Landau damping of the waves.

4.5.2
Enhanced Nonlinear Generation of Oblique Fast Waves by Phase-Mixed Alfvén Waves

In a nonlinear regime, transverse gradients in an Alfvén wave of finite amplitude cause coupling to oblique fast magnetoacoustic waves. This effect is connected with the ponderomotive force appearing because of the wave-induced temporal and spatial gradients of the magnetic pressure. In the 1D smooth nonuniformity of the Alfvén speed in the direction across the magnetic field and in the low-β regime, coupling of weakly nonlinear fast and Alfvén waves is described by the equation

$$\frac{\partial^2 V_x}{\partial t^2} - C_A^2(x)\left(\frac{\partial^2 V_x}{\partial x^2} + \frac{\partial^2 V_x}{\partial z^2}\right) = -\frac{1}{8\pi\rho_0}\frac{\partial^2}{\partial t \partial x}\left(B_y^2\right) \tag{4.25}$$

The right-hand-side term depends quadratically on the perturbation of the magnetic field in the Alfvén wave, and hence generates fast waves with double the frequency of the Alfvén wave. Also, it increases with an increase in the Alfvén wave amplitude, and grows with a decrease in the time and spatial scales in the Alfvén wave. Interestingly, the back reaction of the fast wave on the Alfvén wave can be neglected, as Alfvén waves are subject to higher-order, cubic nonlinearity (see, e.g., Eq. (4.8) and its discussion). Moreover, in contrast with the Alfvén waves that are confined to the magnetic field lines, fast waves readily travel across the field.[1] Hence, once excited by Alfvén waves, the fast waves leave the region of coupling, carrying out the energy from the system. Consequently, Alfvén waves can be considered to be a source of fast waves that propagate obliquely from the excitation region.

Phase mixing of Alfvén waves, caused by the transverse smooth inhomogeneity of the Alfvén speed, leads to the generation of progressively smaller transverse scales in the waves, increasing the right-hand-side term in Eq. (4.25). Hence, development of phase mixing should be accompanied by the enhanced generation of fast waves, with continuously growing efficiency. For a harmonic, initially plane Alfvén wave, the right-hand-side term in Eq. (4.25) becomes proportional to $A_a^2(\mathrm{d}C_A/\mathrm{d}x)z$ [63]. In other words, the driving term grows secularly with the distance from the Alfvén wave source. Thus, phase mixing of Alfvén waves dramatically amplifies the right-hand side of Eq. (4.25). This should result in enhanced generation of fast waves.

1) In a more rigorous analysis, refraction of fast waves caused by the nonuniformity of the Alfvén speed profile should be taken into account.

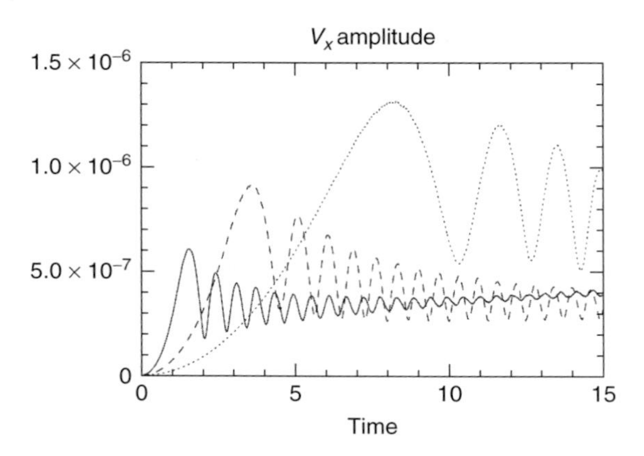

Figure 4.8 Excitation of fast magnetoacoustic waves by phase-mixed Alfvén waves of finite amplitude, and its saturation because of the destructive interference. Different curves correspond to different gradients of the Alfvén speed in the equilibrium. The solid curve corresponds to the steeper profile of the Alfvén speed. (From [64].)

Numerical simulations of the initial phase of the fast wave generation confirmed this theoretical prediction: indeed, fast waves appear in the region of the transverse gradient of the Alfvén speed during the passage of an initially plane Alfvén wave [64]. The amplitude of the fast waves grows linearly with time during the initial stages of the Alfvén wave evolution. However, the fast waves were found to saturate at certain low levels that were dependent on the amplitude and wavelength of the initial Alfvén wave, as well as the scale length of the gradient of the Alfvén speed in equilibrium (Figure 4.8). The saturation level is typically proportional to the square of the Alfvén wave amplitude.

An interpretation of the saturation of fast wave generation by phase-mixed Alfvén waves is perhaps connected with wave interference. Consider an initially plane Alfvén wave. Phase mixing distorts the wave front, generating transverse gradients. At an early stage, the transverse structure of the Alfvén wave is not very developed, and the gradients grow linearly in time. It generates obliquely propagating fast waves whose amplitudes also grow linearly in time. However, at later stages, there are many peaks in the transverse structure of the Alfvén wave, and while each of these is growing in amplitude neighboring peaks are not in phase, that is, their separation is not an integer multiple of the generated wave's wavelength. Indeed, since the number of maxima in the wave source term on the right-hand side of Eq. (4.25) grows and their separation decreases, it is impossible for these to be coherent emitters of fast waves. Thus, while phase mixing continues to generate shorter and shorter scale lengths, the generated fast wave amplitude saturation occurs soon after the phase difference across the Alfvén speed nonuniformity becomes sufficient. An interesting consequence of this interpretation is that in a longitudinally stratified medium. If the transverse Alfvén speed gradients decrease with height, phase mixing can remain "underdeveloped" and the destructive interference does not appear. In this case, the generation of oblique fast waves may be not limited by the saturation.

References

1. Deforest, C.E., Hoeksema, J.T., Gurman, J.B. *et al.* (1997) *Sol. Phys.*, **175**, 393.
2. Withbroe, G.L. (1983) *Sol. Phys.*, **89**, 77.
3. Ofman, L., Romoli, M., Poletto, G., Noci, G., and Kohl, J.L. (1997) *Astrophys. J.*, **491**, L111.
4. Ofman, L., Romoli, M., Poletto, G., Noci, G., and Kohl, J.L. (2000) *Astrophys. J.*, **529**, 592.
5. Morgan, H., Habbal, S.R., and Li, X. (2004) *Astrophys. J.*, **605**, 521.
6. Deforest, C.E. and Gurman, J.B. (1998) *Astrophys. J.*, **501**, L217.
7. Cooper, F.C., Nakariakov, V.M., and Williams, D.R. (2003) *Astron. Astrophys.*, **409**, 325.
8. Doschek, G.A. and Feldman, U. (1977) *Astrophys. J.*, **212**, 143.
9. Doyle, J.G., Banerjee, D., and Perez, M.E. (1998) *Sol. Phys.*, **181**, 91.
10. Dolla, L. and Solomon, J. (2008) *Astron. Astrophys.*, **483**, 271.
11. Banerjee, D., Teriaca, L., Gupta, G.R. *et al.* (2009) *Astron. Astrophys.*, **499**, 29.
12. Ofman, L., Nakariakov, V.M., and Sehgal, N. (2000) *Astrophys. J.*, **533**, 1071.
13. Nakariakov, V.M., Ofman, L., and Arber, T.D. (2000) *Astron. Astrophys.*, **353**, 741.
14. Vasheghani Farahani, S., Nakariakov, V.M., Van Doorsselaere, T., and Verwichte, E. (2011) *Astron. Astrophys.*, **520**, 80.
15. Berghmans, D. and Clette, F. (1999) *Sol. Phys.*, **186**, 207.
16. De Moortel, I., Ireland, J., and Walsh, R.W. (2000) *Astron. Astrophys.*, **355**, L23.
17. Robbrecht, E., Verwichte, E., Berghmans, D. *et al.* (2001) *Astron. Astrophys.*, **370**, 591.
18. De Moortel, I. (2009) *Space Sci. Rev.*, **149**, 65.
19. Wang, T.J., Ofman, L., and Davila, J.M. (2009,) *Astrophys. J.*, **696**, 1448.
20. Marsh, M.S. and Walsh, R.W. (2009) *Astrophys. J.*, **706**, L76.
21. Marsh, M.S., Walsh, R.W., and Plunkett, S. (2009) *Astrophys. J.*, **697**, 1674.
22. Gelfreikh, G.B., Tsap, Y.T., Kopylova, Y.G. *et al.* (2004) *Astron. Lett.*, **30**, 489.
23. Sakao, T., Kano, R., Narukage, N. *et al.* (2007) *Science*, **318**, 1585.
24. Bogdan, T.J. and Judge, P.G. (2006) *Phil. Trans. R. Soc. Lond. Ser. A*, **364**, 313.
25. De Moortel, I., Hood, A.W., Ireland, J., and Walsh, R.W. (2002) *Sol. Phys.*, **209**, 61.
26. Wang, T.J., Ofman, L., and Davila, J.M. (2009) *Astron. Astrophys.*, **503**, L25.
27. Nakariakov, V.M., Verwichte, E., Berghmans, D., and Robbrecht, E. (2000) *Astron. Astrophys.*, **362**, 1151.
28. De Pontieu, B., McIntosh, S.W., Hansteen, V.H., and Schrijver, C.J. (2009) *Astrophys. J.*, **701**, L1.
29. Verwichte, E., Marsh, M., Foullon, C. *et al.* (2010) *Astrophys. J.*, **724**, L194.
30. Sych, R., Nakariakov, V.M., Karlicky, M., Anfinogentov, S. *et al.* (2009) *Astron. Astrophys.*, **505**, 791.
31. Aschwanden, M.J., Fletcher, L., Schrijver, C.J., Alexander, D. *et al.* (1999) *Astrophys. J.*, **520**, 880.
32. Nakariakov, V.M., Ofman, L., DeLuca, E.E., Roberts, B., and Davila, J.M. (1999) *Science*, **285**, 862.
33. Tomczyk, S., McIntosh, S.W., Keil, S.L. *et al.* (2007) *Science*, **317**, 1192.
34. Van Doorsselaere, T., Brady, C.S., Verwichte, E., and Nakariakov, V.M. (2008) *Astron. Astrophys.*, **491**, L9.
35. Van Doorsselaere, T., Nakariakov, V.M., and Verwichte, E. (2008) *Astrophys. J.*, **676**, L73.
36. Pascoe, D.J., Wright, A.N., and De Moortel, I. (2010) *Astrophys. J.*, **711**, 990.
37. Ruderman, M.S. and Erdélyi, R. (2009,) *Space Sci. Rev.*, **149**, 199.

38. Terradas, J., Goossens, M., and Verth, G. (2010) *Astron. Astrophys.*, **524**, 23.
39. Shimojo, M., Hashimoto, S., Shibata, K. *et al.* (1996) *Publ. Astron. Soc. Jpn.*, **48**, 123.
40. Cirtain, J.W., Golub, L., Lundquist, L. *et al.* (2007) *Science*, **318**, 1580.
41. Goossens, M., Hollweg, J.V., and Sakurai, T. (1992) *Sol. Phys.*, **138**, 233.
42. Nakariakov, V.M. and Roberts, B. (1995) *Sol. Phys.*, **159**, 213.
43. Joarder, P.S., Nakariakov, V.M., and Roberts, B. (1997) *Sol. Phys.*, **176**, 285.
44. Ruderman, M.S., Verwichte, E., Erdelyi, R., and Goossens, M. (1996) *J. Plasma Phys.*, **56**, 285.
45. Vasheghani Farhani, S., Van Doorsselaere, T., Verwichte, E., and Nakariakov, V.M. (2010) *Astron. Astrophys.*, **498**, L29.
46. Ferrari, A., Trussoni, E., and Zaninetti, L. (1981) *Mon. Not. R. Astron. Soc.*, **196**, 1051.
47. Ofman, L., Nakariakov, V.M., and Deforest, C.E. (1999) *Astrophys. J.*, **514**, 441.
48. Roberts, B., Edwin, P.M., and Benz, A.O. (1984) *Astrophys. J.*, **279**, 857.
49. Murawski, K. and Roberts, B. (1993) *Sol. Phys.*, **143**, 89.
50. Nakariakov, V.M. and Roberts, B. (1995) *Sol. Phys.*, **159**, 399.
51. Nakariakov, V.M., Arber, T.D., Ault, C.E. *et al.* (2004) *Mon. Not. R. Astron. Soc.*, **349**, 705.
52. Nakariakov, V.M., Pascoe, D.J., and Arber, T.D. (2005) *Space Sci. Rev.*, **121**, 115.
53. Katsiyannis, A.C., Williams, D.R., McAteer, R.T.J. *et al.* (2003) *Astron. Astrophys.*, **406**, 709.
54. Williams, D.R., Mathioudakis, M., Gallagher, P.T. *et al.* (2002) *Mon. Not. R. Astron. Soc.*, **336**, 747.
55. Mészárosová, H., Karlický, M., Rybák, J., and Jiricka, K. (2009) *Astrophys. J.*, **697**, L108.
56. Karlický, M., Jelínek, P., and Mészárosová, H. (2011) *Astron. Astrophys.*, **529**, A96.
57. Heyvaerts, J. and Priest, E.R. (1983) *Astron. Astrophys.*, **117**, 220.
58. Ofman, L. and Davila, J.M. (1998), *J. Geophys. Res.*, **103**, 23677.
59. De Moortel, I., Hood, A.W., and Arber, T.D. (2000) *Astron. Astrophys.*, **354**, 334.
60. Ruderman, M.S., Nakariakov, V.M., and Roberts, B. (1998) *Astron. Astrophys.*, **338**, 1118.
61. Tsiklauri, D., Sakai, J.-I., and Saito, S. (2005) *Astron. Astrophys.*, **435**, 1105.
62. Bian, N.H. and Kontar, E.P. (2011) *Astron. Astrophys.*, **587**, A130.
63. Nakariakov, V.M., Roberts, B., and Murawski, K. (1997) *Sol. Phys.*, **175**, 93.
64. Botha, G.J.J., Arber, T.D., Nakariakov, V.M., and Keenan, F.P. (2000) *Astron. Astrophys.*, **363**, 1186.

5
Prominence Seismology

Among oscillating solar events, prominences and filaments excite special interest, since they play an important role in high-energy processes: flares and coronal mass ejections (CMEs). Prominences, for example, can cause injections of partially ionized plasma into a current-carrying coronal loop and, as a result, the flare energy is released (Chapter 2). There are various forms of prominences: hedgerows, arches, funnels, surges, sprays, loops, and so on [1]. The solar paradigm is also used to study stellar prominences [2].

5.1
Prominence Models

Here, we consider two most widely known models of a prominence constructed within the framework of magnetohydrodynamic (MHD): the Kippenhahn–Schlüter model [3] and the model of Kuperus and Raadu [4]. In the Kippenhahn–Schlüter model [3], the prominence has the shape of a thin vertical isothermal layer, which hangs on magnetic lines bent downward. The magnetic tension forms a force directed upward, which counterbalances the downward gravitation force (Figure 5.1), while the magnetic pressure, which increases with the distance from x-axis, is responsible for the transverse force, which counterbalances the plasma pressure gradient.

In this case, the equations of MHD have the following solution:

$$B_x = B_{x0} \tanh(y/d), \quad B_y = B_{y0}, \quad B_z = B_{zo}, \quad p = p_0 \operatorname{sech}^2(y/d) \tag{5.1}$$

where

$$d = \frac{2B_{y0}}{B_{x0}} \frac{k_B T}{m_i g} \tag{5.2}$$

is the thickness of the prominence. The plasma pressure in the center of the prominence is equal to the outer magnetic pressure related to the vertical component of the magnetic field, while the half-width of the prominence, that is, the distance at which the plasma pressure decreases appreciably compared to the maximum pressure in the center, is specified with d.

Coronal Seismology: Waves and Oscillations in Stellar Coronae, First Edition.
A. V. Stepanov, V. V. Zaitsev, and V. M. Nakariakov.

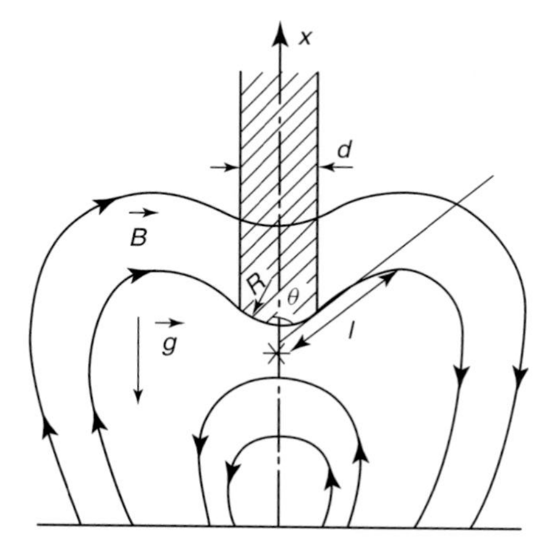

Figure 5.1 The Kippenhahn–Schluter model of a homogeneous prominence supported by magnetic field.

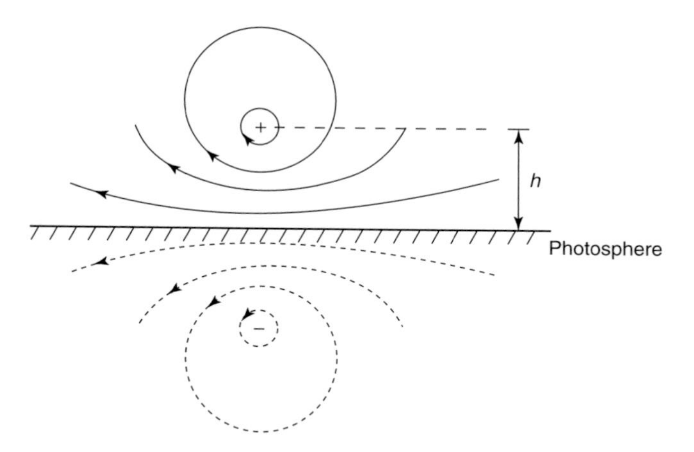

Figure 5.2 The magnetic field of a current-carrying filament at a distance h above the photosphere, and the induced current system in the photosphere in the Kuperus–Raadu model. The induced currents are equivalent to a virtual filament with the opposite but equal current at a depth h below the photosphere. (From [4].)

In the Kuperus–Raadu model [4], the prominence is considered as a horizontal filament swinging in the corona (Figure 5.2), along which a linear current flows. If we approximate the photosphere with a sharp boundary with infinite conductivity, then the magnetic field of the current-carrying filament may be considered as superposition of the fields of two equal linear currents, one of which flows at the height h above the photosphere, while the other in the opposite direction at the depth h below the photosphere. In this case, the filament undergoes the action of the Ampere force, directed upward and equal to the repulsion force between two linear currents. The condition of equilibrium for the Ampere force and gravity may be written as

$$\frac{I^2}{c^2 h} = \pi R^2 \rho g \tag{5.3}$$

where I is the current flowing through the cross section of a filament with the radius R, ρ the plasma density inside the filament, and g the gravitational acceleration. The equilibrium condition may be written in another form if the current is expressed through the magnetic field B_φ at the distance R from the axis of the current-carrying filament: $B_\varphi = 2I/cR$. In this case, Eq. (5.3) will take the following form:

$$\frac{B_\varphi^2}{4\pi h} = \rho g \tag{5.4}$$

Assuming, for example, $B_\varphi = 10\text{G}$, $h = 10^9\text{cm}$, $\rho = 10^{-13}\text{g cm}^{-3}$, and $R = 2 \times 10^8$ cm, we see that such a filament may be maintained in the corona by the Ampere force, if the current $I = 3 \times 10^{19}\text{statampere} = 10^{10}\text{ampere}$ flows along the filament. This is a reasonable value for currents that may exist in magnetic filaments.

5.2
Prominence Oscillations

Several comprehensive reviews are devoted to prominence oscillations [5, 6]. Let us start from the original simple models for prominence oscillations. In a number of cases, the observed oscillations are associated with oscillations of individual filaments, and hence their theory can be readily obtained from the generalization of the magnetic cylinder model, discussed in Chapter 3. In this chapter, we concentrate on large-scale, global oscillations of prominences.

More than four decades ago, Hyder [7] gave a description of prominence oscillations vertical with respect to the solar surface in terms of the simple harmonic oscillator. A vertical deviation of a prominence from the equilibrium induces a supplementary force of a magnetic field tension, which brings the prominence back to the Kippenhahn–Schlüter equilibrium. Thereby, effective magnetic viscosity was considered as the primary reason for the damping of the oscillations. Vertical oscillations are also easily obtained within the framework of the Kuperus–Raadu model [3]. Suppose, for example, that a prominence shifted vertically by $z \ll h$ from the equilibrium. Then, the restoring force appears

$$m\frac{\mathrm{d}^2 z}{\mathrm{d}t^2} = \frac{R^2 B_\varphi^2}{4(h+z)} - \frac{R^2 B_\varphi^2}{4h} \tag{5.5}$$

Taking into account Eq. (5.4) and the fact that the mass of the filament per unit length is $m = \rho\pi R^2$, we obtain the oscillator equation $\ddot{z} + (g/h)z = 0$ with the oscillation period

$$P = 2\pi\sqrt{\frac{h}{g}} \tag{5.6}$$

For $h = 10^9$cm, the period 20 min is obtained. Indeed, oscillations of prominences with such periods are sometimes observed.

Horizontal oscillations were studied subsequently by Kleczek and Kuperus [8], while prominence oscillations in a current sheet by Kuperus and Raadu [4]. In the case of horizontal oscillations, the viscosity effect is substantially smaller. However, during horizontal oscillations of a prominence, the coronal plasma is alternately compressed and rarified, and compression waves can be generated. The period of the waves is equal to the oscillation period of the prominence. The situation is similar to that of the radiation of acoustic waves from a piston source. This problem was solved by Rayleigh and later applied to the case of a circular piston by Lindsay [9]. Indeed, the horizontal displacement y in the middle of the prominence after a circular piston, which radiates hemispherical waves of the type $\sim\exp[i(\omega t - kr)]/r$, can be expressed as [8]

$$y'' + \xi \left[1 - \frac{J_1(2ka)}{ka}\right] y' + \omega^2 y = 0 \qquad (5.7)$$

Here, $k = 2\pi/\lambda$, where λ is the wavelength, a the radius of the piston (the prominence radius), and $J_1(z)$ the Bessel function. The second term in Eq. (5.7) describes the "acoustic resistance" of the prominence oscillations. The period of horizontal oscillations of a prominence in the case when the ends of the filament are rigidly fixed in the photosphere is [8]

$$P = 2\pi \frac{L}{c_A} \qquad (5.8)$$

where L is the length of the filament and $c_A = B/\sqrt{4\pi\rho}$ the Alfvén speed in it. For a filament with the length $L = 5 \times 10^9$cm and density $\rho = 10^{-13}$g cm^{-3}, we obtain the period of horizontal oscillations $P = 20$ min, provided the magnetic field that maintains the prominence is $B \approx 28$G. The quality factor of the oscillations appears to be low: the characteristic decay time is equal to several periods only [8].

The study of waves and oscillations in prominences may give a clue to the determination of parameters of prominences from a comparison between observations and the theory. Currently, however, the seismology of prominences is still far from completion due to the fact that the theory lags behind observations. Basically, the studies are restricted with the analysis of low-amplitude oscillations and waves and of the efficiency of their decay resulted from various dissipation processes and from radiation into the surrounding medium (see a detailed review [5]). Along with that, some attempts are made to take into account the effect of heating of a prominence on the period of its oscillations [10].

A good example of prominence oscillations can be found in the observations of prominence using SOHO/EIT (SOlar and Heliospheric Observatory/Extreme ultraviolet Imaging Telescope) (195 Å) data. It was two successive trains of transverse oscillations triggered by EIT waves from two flares (X1.3 class flare at 06:17 UT and C8.9 class flare at 16:39 UT) on 30 July 2005 in the same remote active region [11]. Figure 5.3 illustrates prominence oscillations in the event of 30 July 2005.

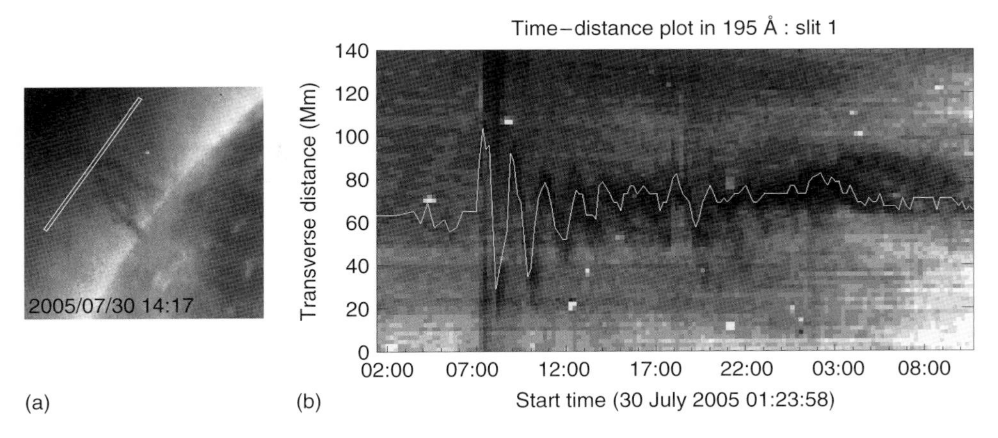

Figure 5.3 (a) Oscillating prominence near the solar limb driven by two flares. (b) Time–distance plot of prominence oscillations with an average period of about 100 min [11].

5.3 The Heating Effect

To illustrate how the heating of a prominence affects its oscillation period, let us consider a simple model. Suppose that a homogeneous prominence is maintained by the magnetic field of a coronal loop, as shown in Figure 5.1. We analyze vertical oscillations of the prominence. There are evidences that a prominence may undergo such oscillations before a flare [12, 13]. The equation of plasma motion subjected to a gravitational field and a nonuniform magnetic field may be written in the following form:

$$\rho \frac{\mathrm{d}\vec{v}}{\mathrm{d}t} = \frac{1}{4\pi}\left[\mathrm{curl}\vec{B} \times \vec{B}\right] - \nabla p + \rho\vec{g} + \frac{k_\mathrm{B}\rho(T_\mathrm{e} + T_\mathrm{i})}{m_\mathrm{i} R^2}\vec{R} \tag{5.9}$$

Here, $\vec{g}$ represents the free-fall acceleration and $\vec{R}$ is the mean radius of curvature of the field lines. The last term in the right-hand side in Eq. (5.9) expresses the influence on the particles of both the centrifugal force ($\sim\rho v_{\|}^2/R$) and the force associated with the nonuniformity of the magnetic field ($\propto \rho v_{\perp}^2|\nabla_{\perp}B|/2B \approx \rho v_{\perp}^2/2R$). When studying vertical oscillations of a prominence, one may neglect the pressure gradient in Eq. (5.9): the thickness $d \sim R$ of the prominence is generally substantially smaller than its vertical extent l_z, so that $\nabla p \approx 2\rho k_\mathrm{B} T/m_\mathrm{i} l_z \ll 2\rho k_\mathrm{B} T/m_\mathrm{i} d$. Here, it is assumed that the plasma is isothermal ($T_\mathrm{i} = T_\mathrm{e} = T$). In this case, the equilibrium state of the prominence is described by

$$\rho g_\mathrm{eff} d = 2\frac{B^2}{4\pi}\cos\theta \tag{5.10}$$

Here, θ is the angle between the vertical direction and the magnetic field: $g_\mathrm{eff} = g + 2\rho k_\mathrm{B} T/m_\mathrm{i} R$. In the vicinity of the equilibrium state (Eq. (5.10)), the equation of small oscillations will take the following form:

$$\ddot{z} + 2\frac{B^2}{4\pi}\frac{\sin\theta}{\rho l d}z = 0 \tag{5.11}$$

where l is the characteristic size of the "sag" in the magnetic lines. Equation (5.11) yields the expression for the squared oscillation frequency:

$$\omega^2 = \frac{B^2}{2\pi}\frac{\sin\theta}{\rho l d} \tag{5.12}$$

The oscillations may be excited through an onset of the ballooning mode of the flute instability [14], if the prominence has a critical thickness $d > d_*$, where

$$d_*^2 \approx \frac{B^2 R}{4\pi \rho g_{\text{eff}}} \tag{5.13}$$

Using the value of ρ given in Eq. (5.13) and substituting it into Eq. (5.12), we can determine the squared oscillation frequency near the instability threshold:

$$\omega^2 \approx 2\frac{g_{\text{eff}} d}{Rl}\sin\theta \tag{5.14}$$

Evaluating the angle θ from the equilibrium condition (Eq. (5.10)) and setting $d \approx R$, we obtain the expression for the prominence oscillation period:

$$P \approx \frac{2\pi}{3^{1/4}}\left(\frac{g}{l} + \frac{2k_{\text{B}}T}{m_{\text{i}} l d}\right)^{-1/2} \tag{5.15}$$

Equation (5.15) implies that if, for example, $d = 2 \times 10^8$ cm, the force associated with the curvature of the magnetic field will dominate over the gravitational force, provided the prominence temperature $T > 3 \times 10^4$ K. Let us also take $l = 5 \times 10^9$ cm and set $g = 2.74 \times 10^4$ cm s^{-2}; then as the active region heats up from 10^4 K to (5–10) $\times 10^6$ K, the prominence oscillation period $P \sim T^{-1/2}$ will shorten from 60 to 3–5 min, which approximately corresponds to the values observed by Bashkirtsev and Mashnich [15]. For $T > 3 \times 10^4$ K, Eq. (5.15) yields a simple expression for the plasma temperature in the prominence region:

$$T \approx \frac{2\pi^2}{\sqrt{3}}\frac{m_{\text{i}} l d}{k_{\text{B}} P^2} \tag{5.16}$$

Figure 5.4 [16] shows an example for the heating effect on the oscillation period. Supposing [16] $P \approx 10^4$ s, $ld \approx 10^{20}$ cm^2, from Eq. (5.16) we obtain $T \approx 10^5$ K.

The prominence matter might be heated due to quasi-periodic interaction between the magnetic field of the loop (this field maintains the prominence) and underlying fields of opposite polarity (marked by the asterisk in Figure 5.1). The heating causes the prominence to disappear in the H_α line; such apparent disappearance is occasionally observed before powerful flares [17]. The partly ionized matter of a prominence may also efficiently be heated due to dissipation of the electric current [18]. In Chapter 2, we presented Eq. (2.61) for generalized

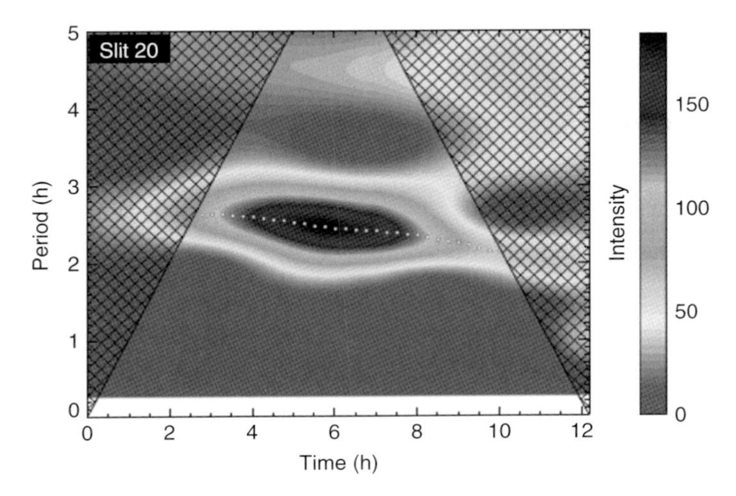

Figure 5.4 The wavelet spectrum of an oscillating prominence. The oscillation period slowly decreases until the filament erupts. (Figure taken from [16].)

Ohm's law, from which the expression for Joule dissipation of the electric current due to excitation of the oscillations may be derived

$$q = \frac{j^2}{\sigma} - \frac{F^2 \rho}{c n m_{\mathrm{i}} \nu'_{\mathrm{ia}}} \left[\frac{\mathrm{d}\vec{V}}{\mathrm{d}t} \times \vec{B} \right] \vec{j} \tag{5.17}$$

For the inner region of the prominence, for $\beta \ll 1$, the basic contribution to the energy release is related to the second term on the right-hand side of Eq. (5.17), that is, to the current dissipation due to ion–atom collisions in partly ionized plasma under nonsteady-state conditions.

5.4 Nonlinear Oscillations: Dynamical Modes

According to observations [19–21], quiescent prominences may display ascending vertical flows of matter with the velocity $\sim$1 km s^{-1}, accompanied by inflow of matter into the prominence from both sides with the horizontal velocity $\sim$5 km s^{-1}. Sometimes these motions of plasma in a quiescent prominence result in activation of the latter. This process starts with a slow rise of matter with the velocity of several kilometers per second and may last for a few hours; the prominence brightens, and eruption of matter occurs, which is sometimes accompanied with the appearance of flare ribbons in H_α. Thereby, plasma motion apparently plays a substantial role in evolution and energy release of the prominence. The other noticeable detail is that most studies of prominences are based on linear approximation, which is justified by the fact that the amplitude of oscillations in prominences is generally small (1–2 km s^{-1}). However, events with substantially higher (up to 90 km s^{-1} [6])

oscillation amplitudes are observed, which implies that nonlinear effects should also be taken into account in models of prominences.

An attempt to consider plasma motions and nonlinear effects within the framework of Kippenhahn and Schlüter's model was made in study [22]: for a certain structure of the velocity field and the magnetic field and for a specific pressure and plasma density distribution, a self-similar solution is found, which in the general case has the following form:

$$V_i(\vec{r}, t) = u_i(t) + \omega_{ij}(t) r_j \tag{5.18}$$

$$B_i(\vec{r}, t) = b_i(t) + a_{ij}(t) r_j \tag{5.19}$$

$$p(\vec{r}, t) = \frac{1}{2} P_{ij}(t) r_i r_j + P_k(t) r_k + P(t), \quad P_{ij} = P_{ji} \tag{5.20}$$

$$\rho(\vec{r}, t) = \rho(t) \tag{5.21}$$

These expressions represent exact self-similar solutions to MHD equations taking into account dissipative terms. They may be considered as the general local approximation of an inhomogeneous plasma flow in an inhomogeneous magnetic field in the case of spatially nonuniform temperature [23, 24].

In the study [25], the self-similar solution was generalized for partially ionized plasma in a prominence. In this case, the nonstationarity of the problem causes strong anisotropy in conductivity; as a result, the equation for the magnetic field obtained from the generalized Ohm's law $\vec{j} = \sigma \vec{E}$ cannot be applied. Owing to taking into account nonstationarity and partial ionization of plasma, supplementary terms caused by ion–atom collisions and specifying different dynamical modes in the prominence appear in the magnetic induction equation.

Let us consider a prominence as a vertical current sheet supported by a magnetic field due to stretching of its lines (Figure 5.5).

We assume the plasma confinement in the current sheet to be due only to the $(1/c)\left[\vec{j} \times \vec{B}\right]$ force and neglect the effect of thermal instability. Besides, we consider the prominence plasma to be heat isolated from the background coronal plasma; in other words, we assume that the system state varies adiabatically (with adiabatic index $\gamma = C_P/C_V$). This is possible because the characteristic times of the prominence evolution are sufficiently low. The density, ρ, and the gravitational acceleration are considered homogeneous. The MHD equations describing such a model are [25]

$$\frac{\partial \rho}{\partial t} + \rho \operatorname{div} \vec{V} = 0 \tag{5.22}$$

$$\rho \frac{d\vec{V}}{dt} = -\nabla p + \frac{1}{4\pi}\left[\operatorname{rot}\vec{B} \times \vec{B}\right] - \rho g \vec{x}_0 \tag{5.23}$$

$$\frac{\partial \vec{B}}{\partial t} = \eta \nabla^2 \vec{B} + \operatorname{rot}\left[\vec{V} \times \vec{B}\right] + \frac{F^2 \rho}{n_i m_i \nu'_{ia}} \operatorname{rot}\left[\frac{d\vec{V}}{dt} \times \vec{B}\right] - \frac{F}{n_i m_i \nu'_{ia}} \operatorname{rot}\left[\vec{f}_a \times \vec{B}\right] - \frac{1}{e n_i} \operatorname{rot}\left[\vec{j} \times \vec{B}\right] \tag{5.24}$$

$$\frac{dp}{dt} + \gamma p \operatorname{div} \vec{V} = 0 \tag{5.25}$$

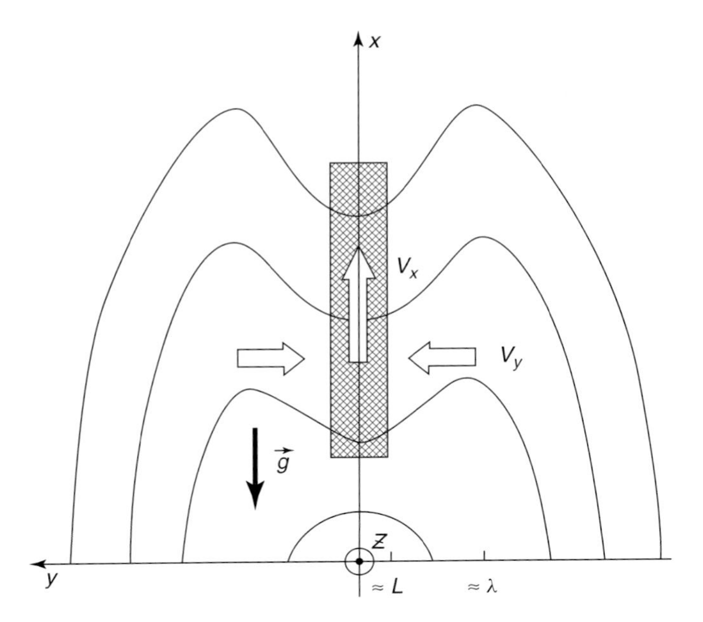

Figure 5.5 A solar prominence as a vertical current sheet supported by a magnetic field due to line stretching (cross section). The vectors $V_x(t), V_y(t)$ are the velocity components of the prominence matter [25].

In Eq. (5.24), terms of the order of $(L/\lambda)^2$, where λ is the characteristic spatial scale of the magnetic field (Figure 5.4), related to the temperature inhomogeneity, are neglected. According to Eq. (5.18), the velocity of the upward plasma flow in the prominence is assumed to be spatially homogeneous,

$$V_x = V_x(t) \tag{5.26}$$

and the components of the horizontal flow velocity are considered equal to

$$V_y = \frac{a'}{a}y,\ V_z = V_z(t) \tag{5.27}$$

Here, a' means the time derivative, and the dimensionless function $a(t)$ characterizes the degree of current-sheet compression. On the basis of Eq. (5.19), the magnetic field component expressions are assumed to have the following form:

$$B_x = B_{x0}(t)y/\lambda,\ B_y = B_{y0}(t),\ B_z = B_{z0}(t) \tag{5.28}$$

From Eq. (5.22), we obtain

$$\frac{\rho'}{\rho} + \frac{a'}{a} = 0 \tag{5.29}$$

The solution to this equation is

$$\rho(t) = \rho_0/a(t) \tag{5.30}$$

where ρ_0 is constant. Taking into account Eq. (5.20), we consider the pressure $p(y)$ to have the following form:

$$p(y,t) = p_{00}(t) - p_{y0}(t)y^2/\lambda^2 \tag{5.31}$$

From Eq. (5.25), we find

$$p_{y0}(t) = P_0/a^{\gamma+2}, p_{00}(t) = P_{00}/a^{\gamma} \tag{5.32}$$

where P_0 and P_{00} are constant. Since $p = p^{(i)} + p^{(e)} + p^{(a)}$, and we assume the thermal equilibrium $T_i = T_e = T_a = T$ and homogeneity of the densities n_i, n_a, time dependencies similar to Eq. (5.32) can be written for each plasma component. From induction equation (Eq. (5.24)), we obtain

$$\begin{aligned} B_{x0}(t) &= \frac{B_{01}}{a^2}\exp\left(-\frac{D}{2}a' + \frac{D^*}{2}\int a^{-\gamma}\,\mathrm{d}t\right) \\ B_{y0}(t) &= B_{02} \\ B_{z0}(t) &= \frac{B_{03}}{a}\exp\left(-\frac{D}{4}a' + \frac{D^*}{2}\int a^{-\gamma}\,\mathrm{d}t\right) \end{aligned} \tag{5.33}$$

Here, $B_{01}, B_{02},$ and B_{03} are constants,

$$D^* = 2P_0^{(a)}\frac{D}{F\lambda^2\rho_0} \tag{5.34}$$

where $P_0^{(a)}$ is a constant similar to P_0 in Eqs. (5.31) and (5.32) in the expressions for the neutral plasma component, and

$$D = \frac{4Fm_a}{(1-F)\rho_0 V_i \Sigma} \tag{5.35}$$

where Σ is the cross section of the ion-neutral collision, V_i is the average thermal speed of ions, and F is a numerical coefficient related to the presence of neutrals. Finally, projecting Eq. (5.23) on the axes, we find the equations for the velocity components:

$$V_x' = \frac{B_{x0}(t)B_{y0}(t)}{\lambda\rho(t)} - g \tag{5.36}$$

$$V_z' = 0 \tag{5.37}$$

and for the dimensionless function $a(t)$:

$$a'' = \frac{c_S^2}{\lambda^2}a^{-\gamma} - \frac{c_A^2}{\lambda^2}a^{-2}\exp\left(-Da' + D^*\int a^{-\gamma}\,\mathrm{d}t\right) \tag{5.38}$$

where $c_S^2 = 2P_0/\rho_0, c_A^2 = B_{01}^2/4\pi\rho_0$

In fact, Eq. (5.38) describes the dynamics of the prominence model, since all its parameters, such as the density and velocity of matter, the magnetic field parameters, and the pressure can be expressed in terms of $a(t)$. With certain values of B_{01}, ρ_0, and T, when the condition $D^*\tau^2/D \ll 1$ is fulfilled, where τ is the characteristic time of the system evolution, we can neglect the term proportional to

$\int a^{-\gamma} dt$ in the exponent in Eq. (5.38). Taking τ equal to λ/c_A (our further analysis proves this choice to be correct), we find

$$\frac{D^* \tau^2}{D} = \frac{c_S^2}{c_A^2} = \frac{8\pi P_0}{B_{01}^2} = \beta \tag{5.39}$$

If constants P_0 and B_0 are such that $\beta \ll 1$, Eq. (5.38) can be simplified:

$$a'' = \frac{c_S^2}{\lambda^2} a^{-\gamma} - \frac{c_A^2}{\lambda^2} a^{-2} \exp(-Da') \tag{5.40}$$

Further analysis is given for this simplified case. Introducing a new variable $b = a'$, we can write Eq. (5.40) in the following form:

$$a' = b \tag{5.41}$$

$$b' = \frac{c_S^2}{\lambda^2} a^{-\gamma} - \frac{c_A^2}{\lambda^2} a^{-2} \exp(-Db) \tag{5.42}$$

The condition $a' = 0, b' = 0$ yields the states of equilibrium in the phase plane (a, a'). In this case, the state of equilibrium is

$$\begin{aligned} b_0 &= 0, \\ a_0 &= (c_S/c_A)^{2/(\gamma-2)} = \beta^{1/(\gamma-2)} \end{aligned} \tag{5.43}$$

Let us assume that, in the equilibrium, the Kippenhahn–Schlüter model describes state the system, by Eqs. (5.1) and (5.2) with the equilibrium pressure P_{eq} and magnetic field B_{eq}. Therefore, relations (Eq. (5.43)) are equivalent to the balance between the equilibrium magnetic pressure (for the vertical component of the magnetic field outside the prominence ($\gamma \approx \lambda$)) and the kinetic pressure inside the prominence ($\gamma = 0$), that is, $\beta_{eq} = 8\pi P_{eq}/B_{xeq}^2 = 1$, where $P_{eq} = P_0/a_0^{\gamma+2}$, $B_{xeq} = B_{01}/a_0^2$.

Assuming $a = a_0 + \xi, b = b_0 + \eta$, where $\xi(t)$ and $\eta(t)$ are minor increments, we linearize the system (Eq. (5.41), Eq. (5.42)) in the vicinity of the equilibrium state (Eq. (5.43)):

$$\begin{aligned} \xi' &= \eta \\ \eta' &= B\xi + A\eta \end{aligned} \tag{5.44}$$

where

$$A = C_A^2 \frac{D}{\lambda^2} \beta^{2/(2-\gamma)}, \quad B = C_A^2 \frac{2-\gamma}{\lambda^2} \beta^{3/(2-\gamma)} \tag{5.45}$$

Note that the coefficient A in front of the variable η in Eq. (5.44) is always positive. This means that the equilibrium state Eq. (5.43) is unstable. Its type is determined by the value of γ and the relation between the coefficients A and B:

1) $\gamma > 2, (B < 0), A^2 < 4|B|$ – unstable focus equilibrium;
2) $\gamma > 2, (B < 0), A^2 > 4|B|$ – unstable node equilibrium;
3) $\gamma < 2, (B > 0)$ – saddle equilibrium.

The unstable focus equilibrium corresponds to the oscillations of the mass density in the prominence. In such a dynamical regime, the difference between the maximum and minimum density ($\rho_{\max} - \rho_{\min}$) per period becomes larger with each oscillation cycle. The characteristic buildup time is

$$\tau = \frac{2}{A} = \frac{2\lambda^2}{c_A^2 D}\beta^{2/(\gamma-2)} \tag{5.46}$$

If $(A/B\tau) \ll 1$, the number of oscillations will be large and phase trajectories of the system will be close to ellipses. In this case, the oscillation period is

$$P = 4\pi \frac{\lambda}{c_A}\beta^{3/2(\gamma-2)}\left[4(\gamma-2) - \frac{c_A^2 D}{\lambda^2}\beta^{1/(2-\gamma)}\right]^{-1/2} \tag{5.47}$$

If the relative density of neutrals F decreases, so does the period of oscillations, and the buildup time $\tau \to \infty$. The case $F = 0$ was studied in Ref. [22]. For totally ionized plasma, the focus-type equilibrium is transformed into the center-type equilibrium and for the prominence we obtain stable infinite (in time) oscillations, or rest, if the system is exactly at the equilibrium point $a = a_0, a' = 0$. Figure 5.6 shows a schematic phase image of the system described by Eq. (5.40) when $\gamma > 2, (B < 0), A^2 < 4|B|$.

There are two basic types of solutions. The first describes the formation of density oscillations in the prominence, resulting then in a continuous decrease in density. The second type corresponds to the so-called collapse processes, when the matter

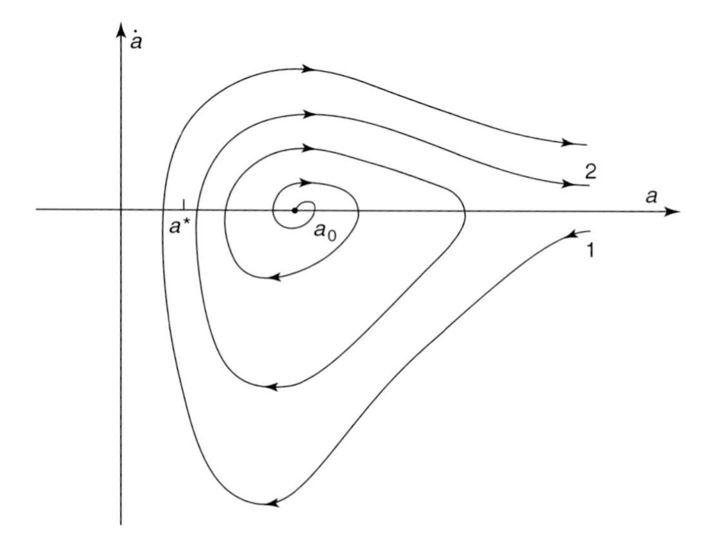

Figure 5.6 Phase trajectories of the system with $\gamma>2, (B<0), A^2 < 4|B|$. The prominence collapse coming then to a monotonic expansion (trajectory 1) and a quasi-periodic expansion (trajectory 2) are possible [25].

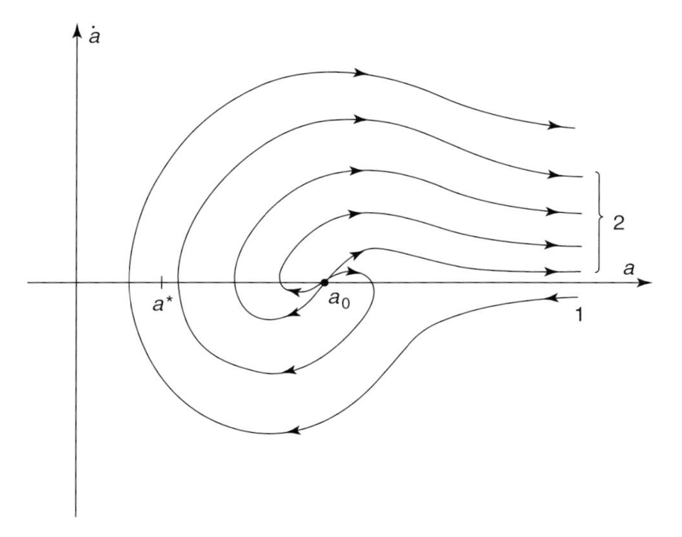

Figure 5.7 Phase trajectories of the system with $\gamma > 2$, $(B < 0)$, $A^2 \geq 4\,|B|$. The prominence collapse results then in the monotonic expansion (trajectory 1) and aperiodic expansion (trajectory 2) are possible [25].

is condensed to some finite density

$$\rho > \rho^* = \rho_0 \left(\frac{\beta}{\gamma - 1} \right)^{1/(2-\gamma)} \tag{5.48}$$

after which the density decreases monotonically. In the course of this process, the vertical component of the magnetic field decreases, and the prominence matter is pushed away and falls down.

When the equilibrium state of the system is of the node type, the basic features of the phase portrait are the same as those for the focus-type equilibrium (Figure 5.7).

However, in this case, instead of gradual buildup of density oscillations, the density decreases aperiodically. In the region of large and small values of $a(t)$, the system behavior is the same as that in the case of the focus-type equilibrium. For this reason, Eq. (5.48) is also valid for the node-type equilibrium.

When $\gamma < 2$, the equilibrium is of the saddle type (Figure 5.8).

In this case, along with the collapse followed by the prominence expansion in the system with certain initial conditions, slow condensation of the matter to the equilibrium density $\rho_{\text{eq}} = \rho_0 / a_0$ is also possible. In the phase portrait, this process corresponds to the motion along the saddle separatrix. Besides, monotonic collapse and monotonic expansion are possible, as well as the density decrease to a certain value $\tilde{\rho} = \rho_0 / \tilde{a}$ larger than the equilibrium value ρ_0 / a_0, which results in subsequent compression of the matter. The corresponding phase trajectories lie in the plane (a', a) to the left from the saddle point a_0. The duration of such processes

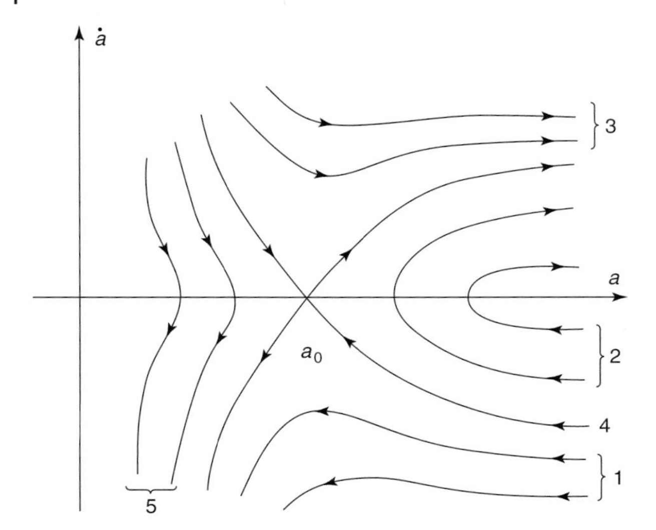

Figure 5.8 Phase trajectories of the system for $\gamma < 2$. The regimes of monotonic collapse (trajectory 1), collapse followed by expansion (trajectory 2), monotonic expansion (trajectory 3), and slow condensation to the equilibrium state density (trajectory 4) can exist, as well as the processes of the density decrease to some values larger than that in the equilibrium, which results in matter compression (trajectory 5) [25].

$\tilde{\tau}$ for $\gamma = 5/3$ can be determined by the following equation [25]:

$$\tilde{\tau} = \frac{35\pi}{216\sqrt{3}} \frac{\lambda}{C_A} \beta^{-9/2} \tag{5.49}$$

In the case of the oscillatory regime of the prominence, the oscillation buildup time τ and the oscillation period can vary in a broad interval, depending on the transverse component of the magnetic field $B_{y\,\mathrm{eq}}$, relative neutral density F, temperature T, and the equilibrium mass density ρ_{eq}. Tables 5.1 and 5.2 list some of these dynamic parameters as a function of $B_{y\,\mathrm{eq}}$ and the relative density of neutrals F.

In the case of the saddle-type equilibrium ($\gamma = 5/3$), we can estimate the collapse time $\tilde{\tau}$ using Eq. (5.49). Table 5.3 lists some values of $\tilde{\tau}$.

Tables 5.4 and 5.5 list the dependences of the vertical and horizontal components of the velocity of the prominence matter on the relative density of neutrals F and

Table 5.1 Oscillation buildup time τ for $\gamma = 3$, $T = 10^4$ K, and $n_\Sigma = 10^{12}$ cm^{-3}.

F / $B_{y\,\mathrm{eq}}$	0.1	0.3	0.5	0.9
0.1 G	76 h	22 h	10 h	1.6 h
1 G	7.6×10^3 h	2.2×10^3 h	1.1×10^3 h	163.5 h

Table 5.2 Period of oscillation P for $\gamma = 3$, $T = 10^4$ K, $n_\Sigma = 10^{12}$ cm^{-3}, and $F = 0.9$.

$B_{y\,eq}$	0.1 G	1 G	10 G	100 G
P	1.1 min	10 min	1.8 h	17 h

Table 5.3 Characteristic collapse time for $\gamma = 5/3$, $T = 10^4$ K, $n_\Sigma = 10^{12}$ cm^{-3}, and $F = 0.9$.

$B_{y\,eq}$	0.1 G	1 G	10 G	100 G
P	6 s	1 min	10 min	1.6 h

Collapse time $\tilde{\tau}$ depends weakly on the relative density of neutrals F.

Table 5.4 Vertical component of material velocity $V_{x\,\max}$ (km s^{-1}) for $T = 10^4$ K and $n_\Sigma = 10^{12}$ cm^{-3}.

F $B_{y\,eq}$	0.1	0.3	0.5	0.9
0.1 G	6	5.5	5.1	4
1 G	60.6	55.7	50.7	40
10 G	606.3	557.8	507.8	402.4

the magnetic field. In our calculations, we assumed λ (the characteristic spatial scale of the magnetic field structure) to be about 10 L (where L is the transverse scale of the prominence).

The one-dimensional model (in which parameters depend only on y) considered earlier, which assumes the processes to be adiabatic (we used a strictly simplified form of the energy equation (5.25)), and is described by self-similar solutions of the MHD equations, is, of course, greatly simplified. The obtained solutions merely represent local approximations, and they are not valid in the region $y \geq \lambda$. In spite of this, however, the model analysis shows that partial ionization of plasma causes the instability of the solar prominence considered in the Kippenhahn–Schlüter magnetohydrostatic model. Oscillations of density, magnetic field, and matter velocity, which are capable of destroying the prominence, are excited (prominence activation). The reason for this instability is that the neutral component of the plasma cannot be supported by the magnetic field and begins to fall down to the photosphere. The motion of neutrals leads, mainly owing to ion-neutral collisions, to the motion of the bulk plasma. The plasma motion in the magnetic field produces additional current along the filament, and the total current in the current sheet increases. Thus, the Ampere force $(1/c)\left[\vec{V} \times \vec{B}\right]$, which affects the current in the magnetic field and provides support for the prominence against gravity, increases and makes the prominence matter move upward. The upward motion of

Table 5.5 Horizontal component of material velocity $V_{y\,max}$ (km s^{-1}) for $T = 10^4$ K, $n_\Sigma = 10^{12}$ cm^{-3}.

F	0.1	0.3	0.5	0.9
$V_{y\,max}$ (km s^{-1})	1.15	1.08	1.01	0.87

the plasma reduces the current in the current sheet, the Ampere force decreases, and the prominence matter falls down again; the process reiterates. The energy of prominence oscillations comes from the potential energy of the prominence matter in the gravitational field. In addition, the collapse is possible, that is, the compression of the matter to a certain density, subsequently switching to expansion; also in addition, under certain conditions, plasma can condense to the state with the equilibrium density.

We would like to point out that ultralong-period oscillations, predicted by this model for large values of the magnetic field, have been recently found observationally [26, 27].

5.5 Flare Processes in Prominences

For inner regions of a prominence ($y \ll \lambda$) and in the case when the condition of low plasma $\beta \ll 1$ ($c_S^2 \ll c_A^2$) is fulfilled, the Joule heating of a prominence is primarily related to the last term in Eq. (5.17), that is, to dissipation that occurs in nonstationary conditions and that is caused by ion–atom collisions in partly ionized plasma. This term, which greatly increases the energy release compared to that for common Joule heating, is connected with the relative motion of neutral and ionized components of the plasma. Integrating Eq. (5.17) over the prominence volume and taking into account the expressions for $\mathrm{d}\vec{V}/\mathrm{d}t$, $\vec{B}$, and $\vec{j}$, specified by the nonlinear model, we find the energy release power in the form of Joule heating [18]:

$$W' = W_0 x \left(x^2 + Ax - G\right), x = \exp(-0.5 Da'/a) \tag{5.50}$$

where

$$W_0 = w \frac{F(2-F)}{1-F} \frac{8\pi k_B T g^2 m_i n_\Sigma}{3 B_{y0}^2 V_i \Sigma} \left(\frac{a}{a_0}\right)^3 \tag{5.51}$$

$$A = \frac{3 B_{y0}^2}{2 n_\Sigma k_B T} \left(\frac{a}{a_0}\right)^2 - \left(\frac{a}{a_0}\right)^{2-\gamma} \tag{5.52}$$

$$G = \frac{3 B_{y0}^2}{8\pi n_\Sigma k_B T \sqrt{2-F}} \left(\frac{a}{a_0}\right)^3 \tag{5.53}$$

Here, w, n_Σ, and B_{y0} are the prominence volume, combined density of ions, and neutral atoms, and the y-component of the magnetic field in the equilibrium state for $a = a_0$, respectively. The parameter D is specified by Eq. (5.35). For $W' > 0$, energy releases. For example, for $F = 0.9, a' = 0$, and the degree of compression a/a_0 that corresponds to the maximum of the energy release power, we obtain $W' = 1.2 \times 10^{23}$; 2.8×10^{24}; 5.7×10^{26} erg/s for $B_{y0} = 1$; 10; 500G, respectively. For triggering a flare, the energy release power should exceed the energy losses on optical radiation.

$$W' > W'_R = n_e n_\Sigma L(T) \tag{5.54}$$

where $L(T)$ is the radiation loss function. Equation (5.54) divides the phase plane $(a'/a_0; a/a_0)$ of the system (Eq. (5.42)) into the region in which radiative energy losses exceed the energy release, and the region in which the energy release exceeds the radiative losses. When the position of the system on the phase plane is in the region, for which $W' > W'_R$, the prominence is heated. In particular, when density oscillations in the system are developed, the dynamical mode is possible, for which the image point of the system in the course of its moving along a spiral phase trajectory (Figure 5.5) from time to time leaves the heating region. In this case, the prominence heating will be quasi periodic. In the Sun, this model may be confirmed by sometimes observed preflare long-period soft X-ray and radio pulsations, which may be related to oscillations of a prominence in the active region [10].

5.6 Stellar and Interstellar Prominences

Studying stellar counterparts of the solar activity offers many additional insights not available on the Sun. The stars that rotate much faster than the solar rate ($P_{rot} \geq 0.2$ day) possess convective regions extending almost to the core of the star. These result in more efficient magnetic field generation and hence solarlike phenomena are scaled by orders of magnitude. There are the evidences supporting the contention that clouds of neutral hydrogen exist in the coronae of active latetype stars, held above their photosphere in magnetic loops [2]. For example, Cameron and Robinson [28] suggested that the neural material in AB Dor (An K0V star at the distance of ~16 pc with $R = 0.9\ R_\odot$ and $P_{rot} = 0.514$ day) forms at the apexes of magnetic loops that extend beyond $r_{co\text{-}rot}$, where centrifugal acceleration leads to the condensation. Here $\omega^2 r_{co\text{-}rot} = g$. The material is fed from the lower, hotter portions of the loops by thermal expansion and then centrifugal acceleration. Clouds dissipate on the timescale of approximately 2–4 days, while new clouds form at the rate of ~1–2/day. Cloud masses can be as high as 10^{17}–10^{18} g [29], two to three orders of magnitude larger than typical solar prominences. Observations also indicate on the presence of prominences at HK Aqr (An M2Ve star at a distance ~20 pc with $P_{rot} \sim 10.34$ h and $R = 0.6 R_\odot$) and at the II Peg (a 6.72 day SB1 binary of spectral type K2IV). Such structures are believed to be much larger than typical solar prominences [2].

Transient dips during photometric flares, during spectroscopic emission-line flares, and during X-ray flares give us the evidences in favor of large-scale eruptive prominences or surges associated with the flare site [2, 30]. Moreover, as discussed in Section 5.5, the flare processes in stellar prominences are possible.

In order to discuss the origin of interstellar flares in RS CVn binaries, Haisch *et al.* [31] pointed out that the filament located near the midpoint of the region between stars can play an important role in the flare energy release. van den Oord [32] has shown that the total energy stored in such current-carrying filament is sufficient for the largest flares in RS CVn and Algol systems. When this energy is released, an intense "core" radio radiation should be generated between the two components. Filament oscillations can serve as a trigger for the recurrent flares [33]. However, van den Oord did not consider any energy release mechanism in the interstellar filament. We have discussed in Chapter 2 and Section 5.5 that because of nonfully ionized plasma and nonsteady-state conditions a powerful energy release in a current-carrying filament is possible. The nonlinear resistance from Eq. (2.58) can be written as

$$R_{\rm nl}(I) = \frac{2\pi F^2 I^2 \Delta l}{c^4 n m_i \nu'_{\rm ia} S^2} \approx 10^{12} \frac{F}{(2-F)} \frac{I^2 \Delta l}{c^4 m_i n^2 S^2 T^{1/2}} \ \text{cgs} \tag{5.55}$$

where F is the relative density of neutrals, n is the total number density, T is the filament temperature, and S and Δl are the cross-sectional area and length of the filament. Here, we consider the case when the electric current flows through the filament located between two stars of the binary system. The current value under this condition is equal to or above 10^{12} A [32]. Substituting, for example, into Eq. (5.55) $I = 3 \times 10^{12}$A, $F = 0.5$, $n = 10^{12}$ cm^{-3} $\Delta l = 2 \times 10^{10}$ cm, $S = 10^{17}$ cm^2, $T = 10^4$ K, we find the resistance of the current-carrying filament $R_{\rm nl} \approx 10^{-12}$ cgs $\approx$ 1 Ω. Thus, the energy release rate at the filament site $W = RI^2 \approx 10^{32}$ erg s^{-1} and the total flare energy $E = W\Delta t = 10^{36}$erg for $\Delta t = 10^4$ s. It should point out the very important peculiarity of the circuit flare model: an increase in the current I leads to an increase in energy release rate according to the formula $W \sim I^4$. Hence, for $I = 10^{13}$ A, the flare energy can be as high as 10^{38} erg, thus being sufficient to explain the energetics of the largest flares observed in close binaries.

The potential drop $U = RI$ over the length of the filament due to the nonlinear resistance $R_{\rm nl} \approx 1$ Ω and for the current $I = 3 \times 10^{12}$ A is $U = 3 \times 10^{12}$ V. With this potential drop, the high energetic charged particles appear in the energy release site and accelerated electrons can be the reason for the intense nonthermal radio radiation from an interstellar flare. Several mechanisms of the nonthermal radio emission are possible: gyrosynchrotron, synchrotron, electron cyclotron maser, and plasma radiation mechanisms are a few among them [34].

It is not excluded that for interstellar flares in the young binary system V773 Tau A observed on 11–17 March 2004 with Very Long Baseline Interferometry (VLBI) at 8.4 GHz [35], an interstellar filament was responsible. Using both Very Long Baseline Interferometry (VLBI) and single dish data [36] on the polarization, spectrum, and time behavior of the radio emission, one can determine the radiation mechanism of interstellar flares and, hence, can perform the diagnostics of the prominence as well as surrounding corona in the binary system. UX Ari, HR 1099, HR 5110, Algol, and YY Gem are good targets in this context.

References

1. Zirin, H. (1987) *Astrophysics of the Sun*, Cambridge University Press.
2. Byrne, P.B. (1996) in Prominence on Late Type Stars *Magnetodynamic Phenomena in the Solar Atmosphere – Prototypes of Stellar Magnetic Activity* (eds Y. Uchida *et al.*), Kluwer Academic Publishers, pp. 139–146.
3. Kippenhahn, R. and Schlüter, A. (1957) *Z. Astrophys.*, **43**, 36.
4. Kuperus, M. and Raadu, M.A. (1974) *Astron. Astrophys.*, **31**, 189.
5. Oliver, R. (2009) *Space Sci. Rev.*, **149**, 175.
6. Tripathi, D., Isobe, H., and Jain, R. (2009) *Space Sci. Rev.*, **149**, 283.
7. Hyder, C.L. (1966) *Z. Astrophys.*, **63**, 78.
8. Kleczek, J. and Kuperus, M. (1969) *Sol. Phys.*, **6**, 72.
9. Lindsay, R.B. (1960) *Mechanical Radiation*, McGraw-Hill, New York.
10. Zaitsev, V.V. and Stepanov, A.V. (1988) *Sov. Astron. Lett.*, **14**, 193.
11. Hershaw, J., Foullon, C., Nakariakov, V.M., and Verwichte, E. (2011) *Astron. Astrophys.*, **531**, 53.
12. Harvey, J.W. (1983) *Adv. Space Res.*, **2**, 31.
13. de Jager, C. and Svestka, Z. (1985) *Sol. Phys.*, **100**, 435.
14. Pustil'nik, L.A. (1974) *Sov. Astron.*, **17**, 763.
15. Bashkirtsev, V.S. and Mashnich, G.P. (1993) *Astron. Astrophys.*, **279**, 610.
16. Pinter, B., Jain, R., Tripathi, D., and Isobe, H. (2008) *Astrophys. J.*, **700**, L182.
17. Martin, S.F. (1973) *Sol. Phys.*, **31**, 3.
18. Zaitsev, V.V. and Khodachenko, M.L. (1992) *Sov. Astron.*, **36**, 81.
19. Malherbe, M., Schmieder, B., Ribes, E., and Mein, P. (1983) *Astron. Astrophys.*, **119**, 197.
20. Schmieder, B., Malherbe, M., Mein, P., and Tandberg-Hanssen, E. (1984) *Astron. Astrophys.*, **136**, 81.
21. Priest, E.R. (1982) *Solar Magnetohydrodynamics*, D. Reidel Publishing Company, Dordrecht, Boston.
22. Sakai, J., Colin, A., and Priest, E. (1987) *Sol. Phys.*, **114**, 253.
23. Kulikovsky, A.G. (1958) *Dokl. Acad. Nauk. SSSR*, **120**, 984.
24. Bulanov, S.V. and Olshanetskii, M.A. (1984) *Phys. Lett.*, **100A**, 35.
25. Bakhareva, N.M., Zaitsev, V.V., and Khodachenko, M.L. (1992) *Sol. Phys.*, **139**, 299.
26. Foullon, C., Verwichte, E., and Nakariakov, V.M. (2009) *Astrophys. J.*, **700**, 1658.
27. Foullon, C., Verwichte, E., and Nakariakov, V.M. (2004) *Astron. Astrophys.*, **427**, L5.
28. Cameron, A.C. and Robinson, R.D. (1989) *MNRAS*, **238**, 657.
29. Cameron, A.C., Duncan, D.K., Ehrenfreund, P. *et al.* (1990) *MNRAS*, **247**, 415.
30. Houdebine, E.R., Foing, B.H., Doyle, J.C., and Rodonó, M. (1993) *Astron. Astrophys.*, **278**, 109.
31. Haisch, B., Strong, K., and Rodonó, M. (1991) *Ann. Rev. Astron. Astrophys.*, **289**, 262.
32. van den Oord, G.H.J. (1988) *Astron. Astrophys.*, **205**, 167.
33. Doyle, J.G., van den Oord, G.H.J., Butler, C.J., and Kiang, T. (1990) A Model for the Observed Periodicity in the Flaring Rate on Yy-Geminorum, Kluwer Academic, Dordrecht, The Netherlands, Boston, MA, in *Flare Stars in Star Clusters* (eds L.V. Mirzoyan *et al.*), p. 325.
34. Ipatov, A.V. and Stepanov, A.V. (1997) *Vistas Astron.*, **41**, 203.
35. Massi, M., Ros, E., Menten, K.M. *et al.* (2008) *Astron. Astrophys.*, **480**, 489.
36. Trigilio, C., Umana, G., and Migenes, V. (1993) *MNRAS*, **260**, 903.

6
The Coronal Loop as a Magnetic Mirror Trap

Solar and stellar flares are accompanied by both plasma heating and charged particle acceleration. Some of the accelerated particles are confined in coronal magnetic loops. Superthermal electrons in the coronal part of a loop generate broadband radio emission, from meter to millimeter wavelengths. As for the energetic particles precipitating to the loop footpoints, electrons with the energy above 10 keV generate hard X-ray emission at the photosphere level, while protons above 1 MeV produce γ-ray emission. At the footpoints of a coronal loop, which are located in the photosphere, the magnetic field strength exceeds that at the top of the loop, so that the loop can form a magnetic trap (magnetic bottle) for charged particles.

6.1
Particle Distribution in a Coronal Loop

A flaring loop contains both collisional particles, the mean free path of which is less than the loop size $l_{\text{fp}} < l$ and collisionless particles with $l_{\text{fp}} = v/\nu_{\text{eff}} > l$, where $\nu_{\text{eff}} = \nu_{\text{ei}}(V_{\text{Te}}/v)^3$. The velocity of collisionless electrons exceeds the value of v_{c} determined by the following formula:

$$\frac{v_{\text{c}}}{V_{\text{Te}}} = \left(\frac{l\nu_{\text{ei}}}{V_{\text{Te}}}\right)^{1/4} \tag{6.1}$$

Here, v_{Te} is the velocity of thermal electrons, l the loop length, and the electron–ion collision frequency is [1]

$$\nu_{\text{ei}} \approx \frac{5.5n}{T^{3/2}}\Lambda, \quad \Lambda \approx \ln\left[10^4\frac{T^{2/3}}{n^{1/3}}\right] \tag{6.2}$$

For example, in a compact solar flaring loop with $n = 10^{11}$ cm^{-3}, $T = 10^7$ K ($V_{\text{Te}} \approx 1.26 \times 10^9$ cm s^{-1}), $l = 10^9$ cm, from Eqs. (6.1) and (6.2) we obtain $\Lambda \approx 11.5$, $\nu_{\text{ei}} \approx 2 \times 10^2$ s^{-1}, and $v_{\text{c}}/V_{\text{Te}} \approx 3.6$. Hence, electrons with the velocity $v > v_{\text{c}} \approx 4.5 \times 10^9$ cm s^{-1} are collisionless.

Since for the magnetic field with $B = 100$ G, the gyroradii of high-energy ($v \approx c$) electrons and ions $r_{\text{c}} = v/\omega_{\text{c}}$ are roughly equal to 10 and 10^3 cm, respectively,

Coronal Seismology: Waves and Oscillations in Stellar Coronae, First Edition.
A. V. Stepanov, V. V. Zaitsev, and V. M. Nakariakov.

which is substantially smaller than the typical loop scale, and the condition for the adiabatic approach in the inhomogeneous magnetic field is fulfilled. Therefore, for a charged particle with the mass m, the conservation equations for the energy and magnetic momentum are [2, 3]

$$\frac{1}{2}mv_{\|}^2 + \mu B(s) = \mu B_{\max}, \quad \mu = \frac{m{v_{\perp}}^2}{2B} = \text{constant} \tag{6.3}$$

Here, s is the distance along a magnetic line and $\mu B(s)$ the one-dimensional effective potential. A symmetric configuration suggests that there are two turning points in the motion of a particle, at s_1 and s_2, where $B = B_{\max}$. Such a configuration is often called a *magnetic bottle*; it determines motion of particles in planetary magnetospheres and radiation belts, as well as under laboratory conditions, in plasma containers, such as the Budker–Post mirror machine. Since the parallel kinetic energy at the turning points is equal to zero, $E = \mu B_{\max}$ is the total kinetic energy of the particle. For the parallel velocity, the energy conservation equation can be solved

$$v_{\|} = \frac{\mathrm{d}s}{\mathrm{d}t} = \pm\sqrt{\frac{2\mu}{m}\,(B_{\max} - B)} \tag{6.4}$$

which can be integrated to determine completely the motion of the particle along the field line. Let us define the pitch angle θ as the angle between the velocity vector and the magnetic field, that is, $v_{\perp} = v\sin\theta$. Since $E = \mu B_{\max}$ and $mv_{\perp}^2/2 = \mu B$, the equation for the pitch angle at any point along the magnetic line is

$$\sin^2\theta = \frac{B}{B_{\max}} \tag{6.5}$$

In a coronal magnetic loop, an upper limit to the magnetic field strength $B = B_{\max}$ (mirror point) is at the footpoints, and so if the minimum field strength at the loop top $B = B_{\min} = B_0$, the minimum pitch angle, below which a particle escapes from the magnetic trap (Figure 6.1a), is

$$\sin^2\theta_0 = \frac{B_0}{B_{\max}} = 1/\sigma \tag{6.6}$$

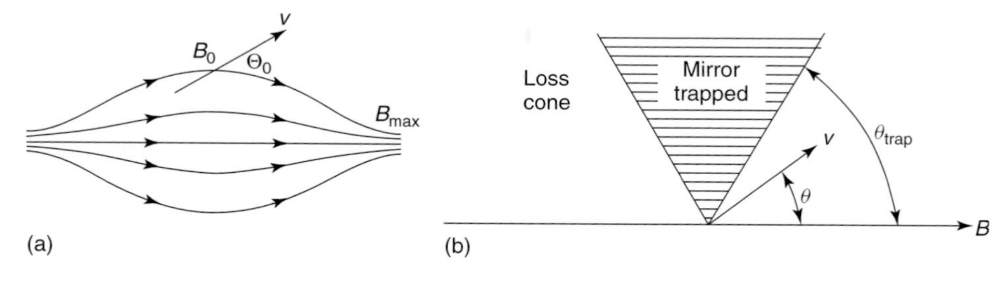

Figure 6.1 The magnetic field in an axially symmetrical magnetic trap. Particles with pitch angles smaller than θ_0 escape from the trap (a). A loss-cone velocity distribution. Particles with $\theta > \theta_0 = \theta_{\text{trap}}$ are mirror trapped, and the others are lost, that is, precipitate to the footpoints (b).

Here, σ is the mirror ratio of the magnetic trap. Any particle with the pitch angle smaller than θ_0 will escape from the trap. The cone of directions around the magnetic field with the angle θ_0 is called the *loss cone* (Figure 6.1b). We will show later that the existence of the loss cone results in various instabilities, which is important for both the emission from the loop and the detrimental effect on the confinement of collisionless energetic particles in coronal loops.

Coronal loops on the Sun and stars are sources of intense nonthermal radio emission. It originates from high-energy electrons confined by the magnetic field of the loop. The time of confinement for high-energy particles is determined not only by Coulomb collisions but also by the wave–particle interaction. The radio emission of solar and stellar flares is primarily caused by gyrosynchrotron [4] and nonlinear plasma mechanisms [5, 6]. They provide a relatively high intensity of emission related to particle population inversion, which results in instabilities in small-scale waves. These waves, in turn, can alter the distribution of the particles and affect their dynamics and propagation. If the pressure exerted by high-energy particles is sufficiently high, large-scale instabilities develop in the loops, which make it possible for high-energy particles to escape. The presence of accelerated particles in a loop forms the "thermal plasma + high-energy particles with a loss-cone" distribution, which is unstable with respect to the generation of waves of various types. The distribution of trapped energetic particles in the momentum space can be represented as

$$f_1(p) = \frac{n_1}{(2\pi)^{3/2} a^3 \cos\theta_0} \exp\left(-\frac{p^2}{2a^2}\right) \left[H(\theta - \theta_0) - H(\theta + \theta_0 - \pi)\right] \qquad (6.7)$$

where $H(x)$ is the Heaviside step function defined as

$$H(x) = \begin{cases} 1, x \geq 0 \\ 0, x < 0 \end{cases}$$

where θ is the pitch angle, the loss-cone angle θ_0 is determined by Eq. (6.6), a is the variance, and $n_1 \ll n_0$. Observations of X-ray emission from solar flares indicate that the energy distribution of accelerated electrons is represented by a power-law spectrum:

$$f_1 \propto n_1 E^{-\delta} \left[H(\theta - \theta_0) - H(\theta + \theta_0 - \pi)\right] \qquad (6.8)$$

It cannot be excluded that there are streams of high-energy particles in a loop with the distribution described as a *hollow beam*:

$$f_1 \propto n_1 \exp\left[-\frac{(p_{||} - p_{||0})^2}{a_{||}^2} - \frac{(p_\perp - p_{\perp 0})^2}{a_\perp^2}\right] \qquad (6.9)$$

In some cases, a convenient choice for the particle momentum distribution in a coronal loop is [7]

$$f_i \propto n_i p_\perp^{2\alpha} \exp\left(-p^2/a_i^2\right) \qquad (6.10)$$

where $\alpha = 0$ for the background equilibrium plasma ($i = 0$) and $\alpha = 1, 2, 3, \ldots$ for high-energy particles ($i = 1$, $n_1 \ll n_0$). Various models for high-energy particle distributions in magnetic traps can be found in the monograph by Mikhailovskii [8].

6.1.1
Gyrosynchrotron Emission from a Flaring Loop

The X-ray and radio emission radiated in solar and stellar flares provides evidence that the distribution of accelerated high-energy electrons formed in coronal loops is described by a power law (Eq. (6.8)). In some cases, however, the mirror ratio in stellar coronal loops can reach $\sigma \geq 50$, and the loss-cone effect becomes negligible. The most popular mechanism of microwave radiation emitted in solar and stellar flares is gyrosynchrotron emission of power-law electrons [4]:

$$f_1 \propto n_1 E^{-\delta} \tag{6.8a}$$

The spectral fluxes of the microwave emission for optically thin and thick sources are, respectively,

$$F_{\nu 1} = \eta_{\nu 1} \Omega d, \quad F_{\nu 2} = \frac{\eta_{\nu 2}}{\kappa_{\nu 2}} \Omega \tag{6.11}$$

Here, Ω is the solid angle of the source and d its characteristic thickness. The coefficients of emission η_ν and absorption κ_ν for the gyrosynchrotron radiation are expressed by approximate Dulk's formulae [4]:

$$\frac{\eta_\nu}{B n_1} \approx 3.3 \times 10^{-24} 10^{-0.52\delta} (\sin \vartheta)^{-0.43+0.65\delta} \left(\frac{\nu}{\nu_c}\right)^{1.22-0.9\delta} \tag{6.12}$$

$$\frac{\kappa_\nu B}{n_1} \approx 1.4 \times 10^{-9} 10^{-0.22\delta} (\sin \vartheta)^{-0.09+0.72\delta} \left(\frac{\nu}{\nu_c}\right)^{1.3-0.98\delta} \tag{6.13}$$

where $\vartheta \geq 20^\circ$ is the angle between the magnetic field direction and the line of sight, the spectral index for the accelerated electrons lies within the interval $2 < \delta < 7$, and $\nu_c = eB/(2\pi mc)$ is the electron gyrofrequency. The optically thin mode occurs at the frequencies exceeding the spectral peak of the gyrosynchrotron emission: $\nu > \nu_{peak}$, the while optically thick mode at $\nu < \nu_{peak}$, where

$$\nu_{peak} \approx 2.72 \times 10^3 10^{0.27\delta} (\sin \vartheta)^{0.41+0.03\delta} (n_1 l)^{0.32-0.03\delta} B^{0.68+0.03\delta} \tag{6.14}$$

and l is the length of the source along the line of sight. For a typical solar loop, $\nu_{peak} \approx 10$ GHz. Note that in the case of relatively dense plasma, the effect of Razin suppression becomes important at $\nu < \nu_R = \frac{2\nu_p^2}{3\nu_c} \approx \frac{20 n_0}{B}$. Razin effect can result in the appearance of an optically thin source even in the $\nu < \nu_{peak}$ domain. We use Eqs. (6.12) and (6.13) for diagnostics of the parameters of a flaring loop discussed in Chapter 8.

Since the magnetic field of a loop oscillates ($B = B_0 + B_\sim \sin \omega_0 t$) due to sausage oscillations or electric current oscillations in a loop as an RLC circuit, the gyrosynchrotron emission of accelerated electrons will be modulated with the same frequency. Taking into account the conservation law for the longitudinal magnetic flux ($d \propto B^{-1/2}$, $\Omega \propto B^{-1/2}$), from Eqs. (6.11)–(6.13) we obtain [9]

$$F_{\nu 1} \propto B^{0.90\delta - 1.22}, \quad F_{\nu 2} \propto B^{-1.02-0.08\delta} \tag{6.15}$$

It follows from Eq. (6.15) that, as the magnetic field strength B increases, the microwave radiation flux increases for an optically thin and decreases for an optically thick source, that is, corresponding oscillations are antiphased.

Defining the modulation depth of the spectral flux as $M_F = (F_{max} - F_{min})/F_{max}$, where F_{max} and F_{min} are the maximum and minimum spectral flux densities, respectively, we find from Eq. (6.15) that for optically thin and optically thick sources

$$M_{F1} = 2(0.90\delta - 1.22)\left(\frac{\Delta B}{B}\right), \quad M_{F2} = 2(0.08\delta + 1.02)\left(\frac{\Delta B}{B}\right) \tag{6.16}$$

Comparing these expressions, we conclude that $M_{F1} > M_{F2}$ for $\delta > 2.7$ and, in addition, magnetic field perturbations may drive the oscillations of gyrosynchrotron emission with the maximum modulation depth $\sim$30%. If the same population of accelerated electrons is responsible for the radiation, we can determine the particle spectral index from Eq. (6.16):

$$\delta = \frac{1.22 + 1.02(M_{F1}/M_{F2})}{0.90 - 0.08(M_{F1}/M_{F2})} \tag{6.17}$$

Using Eqs. (6.11)–(6.13), we express the optical depth in terms of the radiation fluxes $F_{\nu 1}$ and $F_{\nu 2}$ and the spectral index δ:

$$\tau_{\nu 2} = \kappa_{\nu 2} d = \frac{F_{\nu 1}}{F_{\nu 2}}\left(\frac{\nu_1}{\nu_2}\right)^{0.90\delta - 1.22} \tag{6.18}$$

From Eqs. (6.13) and (6.18), we can estimate the optical depth τ_ν at any frequency and determine the frequency ν_{peak}, at which the gyrosynchrotron emission flux reaches its maximum ($\tau_{peak} \approx 1$):

$$\tau_\nu = \tau_{\nu 2}\left(\frac{\nu_2}{\nu}\right)^{1.30+0.98\delta}, \quad \nu_{peak} = \nu_2(\tau_{\nu 2})^{1/(1.30+0.98\delta)} \tag{6.19}$$

Finally, eliminating the density of emitting electrons n_1 from Eq. (6.14) and using Eq. (6.13), we obtain

$$B = \left(\frac{\nu_{peak}}{D(\sin\vartheta)^{0.439-0.203\delta+0.022\delta^2}}\right)^{1/(0.584-0.237\delta+0.030\delta^2)} \tag{6.20}$$

where $D = 10^{3.583-1.695\delta+0/182\delta^2}\nu^{0.416+0.275\delta-0.030\delta^2}\tau_\nu^{0.320-0.030\delta}$.

Let us now analyze a particular event: 23 May 1990 solar radio flare [10]. The most interesting feature in the temporal structure of the microwave radiation at the pulse phase ($F_{max} \sim 10^3$ s.f.u. $= 10^{-19}$ W m^{-2} Hz^{-1}) was that the fast quasi-periodic ($P \approx 1.5$ s) pulsations at $\nu_1 = 15$ GHz and $\nu_2 = 9.375$ GHz were antiphased. Cross-correlation analysis of the time profiles at 9.375 and 15 GHz showed a relative time delay of 0.75 s between the corresponding peaks. The modulation depths were $M_{F1} \approx 5\%$ and $M_{F2} \approx 2.5\%$. The observed emission features may be ascribed to magnetic field variations driven by the sausage oscillation or the current oscillations of a loop as an RLC circuit. In this case, the gyrosynchrotron radiation of the trapped electrons was optically thin at $\nu_1 = 15$ GHz and optically thick at $\nu_2 = 9.375$ GHz. As it follows from Eq. (6.11), the oscillations should then be antiphased. Taking into account the ratio of the modulation depths $M_{F1}/M_{F2} \approx 2$, from Eq. (6.17), we determine the spectral index of the emitting electrons, $\delta \approx 4.4$. For the considered event, the flux ratio is $F_{\nu 1}/F_{\nu 2} \approx 0.5$ [10]; hence, using Eqs. (6.19) and (6.20), we obtain $\tau_{\nu 1} \approx 0.1$ and $\tau_{\nu 2} \approx 2$, which is consistent with our model. From Eq. (6.16), we can determine the frequency that corresponds to the maximum of gyrosynchrotron radiation, $\nu_{\text{peak}} \approx 10.6$ GHz. If we substitute $\vartheta = \pi/4$, $\nu = 9.375$ GHz, and $\tau_\nu \approx 2$ into Eq. (6.20), we obtain the magnetic field in the flaring loop $B \approx 190$ G.

On the other hand, the analysis of quasi-periodic radio and X-ray pulsations in the 15 June 2003 solar flare, carried out by Fleishman *et al.* [11], revealed in-phase pulsations in microwave emission for both optically thin and optically thick gyrosynchrotron radiation sources. It was concluded that in this event the role of magnetohydrodynamic (MHD) oscillations was negligible; instead, quasi-periodic acceleration and injection of fast electrons were shown to be more likely to cause the quasi-periodic pulsations (QPPs). Hence, along with imaging, radio and X-ray data should be used to determine whether QPPs are produced by quasi-periodic injections of fast electrons or by MHD oscillations of the loop (see also Chapter 7).

The loss-cone anisotropy effect on the gyrosynchrotron emission of power-law electrons in coronal loops was analyzed by Fleishman and Melnikov [12] by using the refined Ramaty's formulae [13]. This effect can be important mainly for the footpoint sources. Numerical calculations showed that anisotropy provides softer spectra of gyrosynchrotron radiation and is important for the observed variations of the spectral index of the microwave radiation along the loop with $\sigma \leq 10$ [12].

6.2 Kinetic Instabilities in a Loop

Numerous kinetic instabilities emerge in a coronal loop, since the system "background plasma + inverse populated superthermal particles" is formed. Some of the instabilities produce intense radio emission, while others play an important role in dynamics and propagation of fast particles.

6.2.1 A Loop as an Electron Cyclotron Maser

The instability of ordinary (o) and extraordinary (x) waves at the harmonics of the electron gyrofrequency, also known as electron cyclotron maser (ECM), was considered as a possible cause of the intense radio emission in solar and stellar flares [4,14–17], auroral emission from the Earth at kilometer wavelengths [18], and Jovian decameter radio emission [19]. The interaction between waves and energetic particles occurs under the cyclotron resonance condition:

$$m\omega - s m_0 \omega_c - k_{||} p_{||} = 0 \tag{6.21}$$

where $m = m_0\Gamma$, m_0 is the electron rest mass, p the particle impulse, and $\Gamma = (1 - v^2/c^2)^{-1/2}$ the Lorenz factor. The most favorable conditions for ECM emission exist in strong magnetic fields, where the electron cyclotron frequency exceeds the plasma frequency, $\omega_c > \omega_p$, that is, in active regions in the Sun and stars. The wave instability growth rate for the sth harmonic is [17]

$$\gamma_{o,x} = \int \mathrm{d}\vec{p}A_{o,x}(\vec{p},\vec{k})\delta\left(\omega - \frac{s\omega_c}{\Gamma} - k_{||}v_{||}\right)$$
$$\left(\frac{s\omega_c}{\Gamma v_\perp}\frac{\partial}{\partial p_\perp} + k_{||}\frac{\partial}{\partial p_{||}}\right) f_1\left(p_\perp, p_{||}\right) \tag{6.22}$$

where

$$A_{o,x}(\vec{p},\vec{k}) = \frac{4\pi^2 e^2 v_\perp^2}{\omega N_{o,x}\left[\partial\left(\omega N_{o,x}\right)/\partial\omega\right]\left(1 + T_{o,x}^2\right)}$$
$$\times \left|\frac{K_{o,x}\sin\vartheta + (\cos\vartheta - N_{o,x}v_{||}/c)T_{o,x}}{N_{o,x}\left(v_\perp/c\right)\sin\vartheta} J_s + J_s'\right|^2 \tag{6.23}$$

$J_s\left(\omega N_{o,x}v_\perp \sin\vartheta/\omega_c c\right)$ is the Bessel function, ϑ the angle between the wave vector and magnetic field vector, $T_{o,x}$ and $K_{o,x}$ the polarization coefficients [17], and $N_{o,x}$ the wave refractive index [1]

$$N_{o,x}^2 = 1 - \frac{2v(1-v)}{2(1-v) - u\sin^2\vartheta \pm \left[u^2\sin^4\vartheta + 4u(1-v)^2\cos^2\vartheta\right]^{1/2}} \tag{6.24}$$

Here, the subscript "o" (and the upper sign) corresponds to the ordinary wave, $v = \omega_p^2/\omega^2$, and $u = \omega_c^2/\omega^2$. The results of the studies of the linear phase of ECM instability, described by Eqs. (6.22) and (6.23), for various loss-cone distributions of fast electrons, for different ω_p/ω_c ratios, and at the quasi-linear phase of the ECM [20] can be summarized as follows. The extraordinary wave at the first harmonic of the gyrofrequency is predominantly generated in strong magnetic fields ($\omega_p/\omega_c < 0.24$–0.4). The ordinary wave generation prevails within the interval 0.24–$0.4 < \omega_p/\omega_c < 1$ at the harmonic $s = 1$. The maximum growth rates are of the order

$$\gamma_x \approx 10^{-2}\frac{n_1}{n_0}\omega_c, \quad \gamma_0 \approx 2\times 10^{-3}\frac{n_1}{n_0}\omega_c \tag{6.25}$$

and in both cases are realized for the wave propagating at $\vartheta_m \approx 70^\circ$ to the magnetic field vector. The angular width of the directivity pattern for the excited waves is $\Delta\vartheta \approx 3^\circ$. If $\omega_p/\omega_c > 1$, the maximum growth rate is found for potential waves with the upper hybrid frequency $\omega_{uh} = \left(\omega_p^2 + \omega_c^2\right)^{1/2}$, while the growth rate for ECM emission in the form of the x-mode at the frequency $\omega \approx 2\omega_c$ is by several orders of magnitude smaller.

6.2.2
The Plasma Mechanism of the Radio Emission from Coronal Loops

In the sufficiently dense plasma in coronal loops, with $\omega_p/\omega_c > 1$, the plasma mechanism for radio emission dominates. It was first suggested by Ginzburg and Zheleznyakov [21] to explain type III solar radio bursts. According to this mechanism, because of inverse population of energetic electrons, Langmuir waves are initially excited and are subsequently transformed to electromagnetic waves escaping the corona. The plasma density may reach $n_0 \approx 10^{11}-10^{12}$ cm^{-3} in flaring loops in the Sun and most late-type stars [22], so that the condition $\omega_p/\omega_c > 1$ remains valid even in magnetic fields $B \approx 10^3$ G. For solar flare loops, the plasma mechanism can explain even the microwave emission at 1–10 GHz [5]. In the case $\omega_p^2/\omega_c^2 \gg 1$ and $k_{||}v_{||} \gg \omega_c$, where the emission at the frequency ω includes several harmonics of the gyrofrequency. Under these conditions, the instability growth rate of plasma waves with $\omega \approx \omega_p$ and $k_\perp \gg k_{||}$ is [8]

$$\gamma = \frac{\pi}{n_0}\frac{{\omega_p}^4}{k^3}\int_{-\infty}^{\infty} dv_{||}\int_{\omega^2/k^2}^{\infty} d{v_\perp}^2 \frac{\partial f_1/\partial {v_\perp}^2}{\sqrt{v_\perp^2-\omega^2/k^2}} \tag{6.26}$$

Substituting the distribution function of the type (Eq. (6.7)) for high-energy electrons into Eq. (6.26), we can find the instability growth rate for the Langmuir waves. Its maximum value for a sufficiently large volume of the loss cone ($\sigma \approx 2$) is

$$\gamma_{max} \approx 0.1\frac{n_1}{n_0}\omega_p \tag{6.27}$$

The rate of plasma waves damping due to electron–ion collisions $\nu_{ei} \approx 60n_0/T^{3/2}$ imposes the lower threshold for the instability with respect to the density of high-energy particles:

$$\frac{n_1}{n_0} > 10\frac{\nu_{ei}}{\omega_p} \approx 5\times 10^{-7} \tag{6.28}$$

In addition to collisional damping, we should take into account Landau damping in hot coronal loops. Figure 6.2a presents the plasma wave instability versus the mirror ratio of a compact solar flare loop [5] (see also [17]). In order to describe more adequately the population of $\geq$10 keV electrons in a loop, a power-law velocity distribution with the loss cone less sharp than that in Eq. (6.7) and with the energy-depending boundary is used [23]:

$$f_1(v_{||}, v_\perp) \propto \frac{n_1}{(v_{||}^2+{v_\perp}^2)^\delta}\left[1-\exp(-y)\right], \quad y=(\sigma-1)\frac{{v_\perp}^2}{v_{||}^2} \tag{6.29}$$

Substituting the distribution Eq. (6.29) into Eq. (6.26), we find the instability region with respect to the density of high-energy electrons, taking into account both Landau damping (Figure 6.2b) and damping due to collisions (Figure 6.2c).

The excited plasma waves are transformed into electromagnetic waves through Rayleigh scattering on thermal plasma particles. The conservation law in this case is

$$\omega_t - \omega = (\vec{k}_t - \vec{k})\vec{v} \tag{6.30}$$

where ω_t and $\boldsymbol{k}_t$ are the frequency and wave vectors of electromagnetic waves and $\boldsymbol{v}$ the velocity of scattering particles. The Rayleigh scattering on plasma ions is the most efficient mechanism of radio emission at the fundamental tone $\omega_t \approx \omega_p$. The

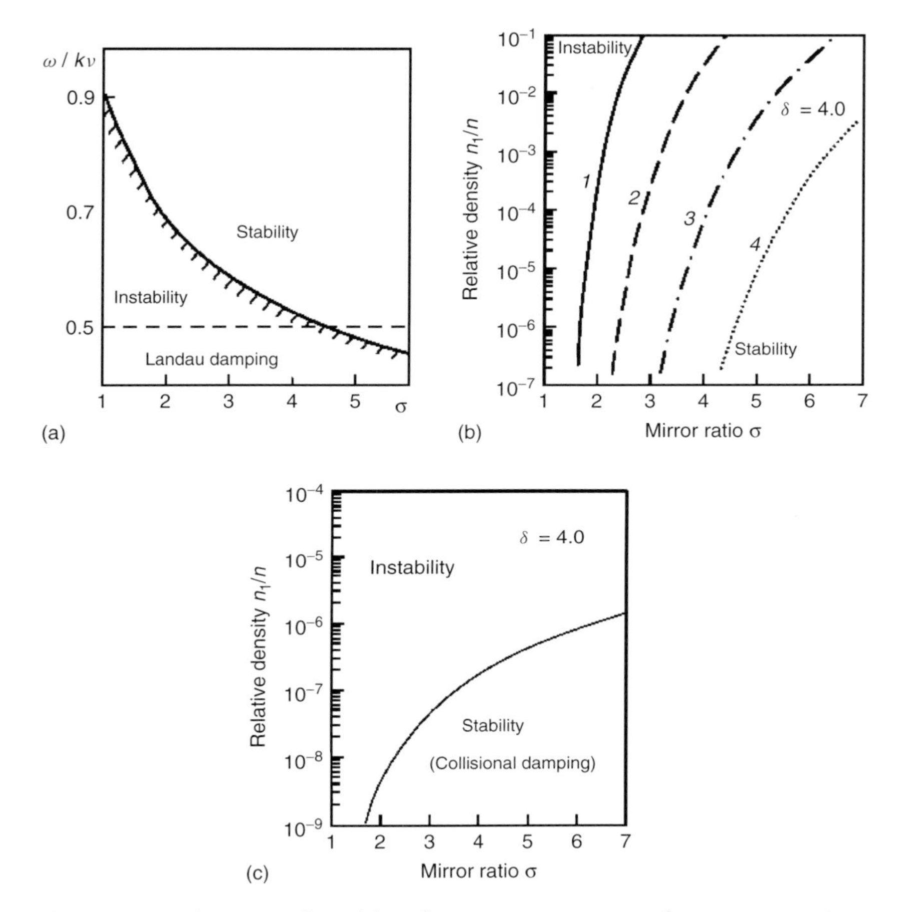

Figure 6.2 (a) The region of instability of plasma waves for the high-energy particles distribution (Eq. (6.7)) for $T = 10^7$ K and $V_e = a/m \approx 10^{10}$ cm s^{-1}. (b) The region of instability for the distribution (Eq. (6.19)) at $T = 3 \times 10^7$ K (curve 1), $T = 1.2 \times 10^7$ K (curve 2), $T = 7 \times 10^6$ K (curve 3), and $T = 4.5 \times 10^6$ K (curve 4). (c) The region of instability for the distribution (Eq. (6.19)) for $n_0 = 3 \times 10^{10}$ cm^{-3} and the plasma temperature $T < 3 \times 10^6$ K$/(\sigma - 1)$; the collision damping exceeds Landau damping.

dispersion relations for plasma and electromagnetic waves for $\omega_p^2/\omega_c^2 \gg 1$ take the following form:

$$\omega^2 = \omega_p^2 + 3k^2 {V_{Te}}^2, \quad \omega_t^2 = \omega_p^2 + k_t^2 c^2 \tag{6.31}$$

The dispersion relations (Eq. (6.31)) and the energy conservation law (Eq. (6.30)) specify the region of nonlinear interaction between the waves:

$$L_{nl} = 3L_n \frac{V_{Te}^2}{\omega_p^2} \left(k_{max}^2 - k_{min}^2\right) \approx 3L_n \frac{V_{Te}^2}{v^2} \tag{6.32}$$

where $L_n = n_0/|\nabla n_0|$ is the scale of particle density inhomogeneity for a source located in a stellar coronal loop; k_{max} and k_{min} restrict the spectrum of excited plasma waves.

The transfer equation for the brightness temperature of the emission is [1]

$$\frac{dT_b}{dl} = \eta - (\mu_{nl} + \mu_c)T_b \tag{6.33}$$

Here, η is the emissivity, μ_{nl} the coefficient of absorption (amplification) due to nonlinear processes, and μ_c the collisional absorption coefficient.

The solution of the transfer Eq. (6.33) is given by

$$T_b = \frac{\eta}{\mu_c + \mu_{nl}} \left\{ 1 - \exp\left[- \int_0^{L_{nl}} (\mu_c + \mu_{nl})\, dl \right] \right\} \tag{6.34}$$

where L_{nl} is expressed by Eq. (6.32). For Rayleigh scattering, the coefficients of emission and absorption are [1, 5]

$$\eta_1 \approx \frac{\pi}{36} \frac{\omega_p}{v_g} m v^2 w, \; \mu_c = \frac{\omega_p{}^2}{\omega_t{}^2} \frac{\nu_{ei}}{v_g} \approx \frac{\nu_{ei}}{v_g}, \; \mu_{nl1} \approx -\frac{\pi}{108} \frac{m}{m_i} \frac{\omega_p}{v_g} \frac{v^2}{V_{Te}{}^2} w \tag{6.35}$$

where v_g is the group velocity of electromagnetic waves and $w = W/n_0 \kappa_B T$ the plasma turbulence level. Since μ_{nl1} is negative, for $|\mu_{nl1}| > \mu_c$, that is, when the plasma turbulence level is sufficiently high, the emission brightness temperature may grow exponentially, which results in the *maser effect*.

The radio emission at the doubled plasma frequency is generated as a result of Raman scattering of plasma waves, $l_1 + l_2 \to t$. The energy and momentum conservation laws take the following form:

$$\omega_1 + \omega_2 = \omega_t, \quad \vec{k}_1 + \vec{k}_2 = \vec{k}_t \tag{6.36}$$

where ω_1, $\boldsymbol{k}_1$, and ω_2, $\boldsymbol{k}_2$ are the frequency and wave vectors of interacting plasma waves, respectively. The emissivity and absorption coefficients are [1, 5]

$$\eta_2 \approx \frac{(2\pi)^5}{15\sqrt{3}} \frac{\omega_p^4 n_0 T}{v c^3 \Delta^2} w^2, \quad \mu_{c2} \approx \frac{\nu_{ei}}{2\sqrt{3}c}, \quad \mu_{nl2} \approx \frac{(2\pi)^2}{5\sqrt{3}} \frac{\omega_p^4}{v c^3 \Delta} w \tag{6.37}$$

Here, $\Delta = (4\pi/3)\left(k_{max}^3 - k_{min}^3\right)$ is the phase volume of plasma waves. Figure 6.3 shows an example for the behavior of the radiation brightness temperature at the fundamental tone and the second harmonic as a function of the plasma turbulence level w, for the red dwarf AD Leo [24]. The radio emission at the fundamental tone dominates when the plasma wave turbulence level is relatively high, $w \geq 10^{-5}$ to 10^{-3} (Figure 6.3). For lower levels of plasma turbulence, the emission at the double plasma frequency dominates, reaching values of $T_b \approx 10^{13}-10^{14}$ K for $w \approx 10^{-5}$, whereas the brightness temperature at the fundamental tone is $T_b \approx 10^{11}-10^{12}$ K. The value of w that marks the exponential amplification of the fundamental tone emission (*maser effect*) depends on flare loop parameters and belongs to the interval $w \approx 10^{-5}$ to 3×10^{-4}.

According to Eqs. (6.32) and (6.34), in hot ($T \geq 10^7$ K) stellar coronae, the efficiency of the plasma radio emission mechanism exceeds that in the solar

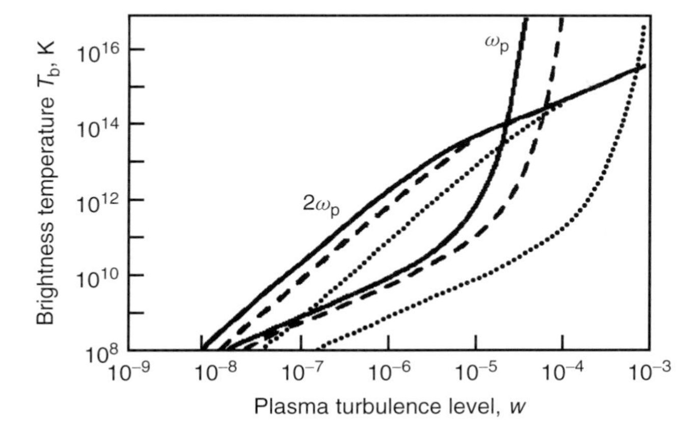

Figure 6.3 The brightness temperature of radio emission as a function of the plasma turbulence level $w = W/n_0\kappa_B T$ for AD Leo flare loop at $T = 10^7$ K and the energy of accelerated electrons $E = 30$ keV; the fundamental tone corresponds to the frequency 1.4 GHz. The solid, dashed, and dotted lines correspond to $L_n = 3 \times 10^9$, 10^9, and 10^8 cm, respectively [24].

corona, owing to the essentially larger transformation length $L_{nl} = 3L_n(T/E)$. Moreover, the high temperature of stellar coronal loops leads to reduction in the phase volume Δ of plasma waves and thus favors the increase in the radio emission brightness temperature of flaring stars up to $T_b \approx 10^{16}$ K [14]. This fact also answers the question set by Abada-Simon *et al.* [25] as to why the plasma mechanism of radio emission is more efficient in stars than that in the Sun.

The flux density of the radio emission from a star is

$$F_\nu = \frac{2\kappa_B T_b \nu^2}{c^2} \frac{S}{d^2} \tag{6.38}$$

where S is the source area and d the distance to the star. Assuming the brightness temperature $T_b \approx 10^{14}$ K and $L_n = S^{1/2} \approx 3 \times 10^9$ cm, for AD Leo ($d = 4.85$ pc $\approx 1.55 \times 10^{19}$ cm), we find $F \approx 3$ Jy at the frequency $\nu = 4.85$ GHz (here 1 Jy $= 10^{-26}$ W m^{-2} Hz^{-1}). Such flux densities are inherent in energetic flares in red dwarves. For close binaries, exemplified by AR Lac, for $T_b = 10^{14}$ K, $S = 9 \times 10^{20}$ cm^2, $d = 50$ pc $\approx 1.5 \times 10^{20}$ cm, and $\nu = 1$ GHz, it follows from Eq. (6.38) that $F_\nu \approx 0.12$ Jy, which is also consistent with the observed radio fluxes from close binaries.

6.2.3
Instabilities of Whistlers and Small-Scale Alfvén Waves

In addition to the wave instabilities described earlier, coronal magnetic loops are also unstable with reference to other modes, in particular, to whistlers and small-scale Alfvén waves. Both the whistlers and Alfvén wave turbulence play an essential role in the dynamics and propagation of energetic particles in space plasma as well as in the formation of the particle spectrum and the emission spectra due to wave–particle interaction (see [3] and Chapter 7).

From the dispersion relation for whistlers

$$\omega = \omega_c \frac{k^2 c^2 |\cos\vartheta|}{\omega_p^2 + k^2 c^2} \tag{6.39}$$

and the cyclotron resonance condition (Eq. (6.21)), the velocity of high-energy electrons that are at resonance with the whistler waves is determined as follows:

$$|v_{||}| = c \frac{\omega_c - \omega}{\omega_p} \sqrt{\frac{\omega_c |\cos\vartheta| - \omega}{\omega}} \tag{6.40}$$

For the distribution of high-energy electrons (Eq. (6.8)), the instability growth rate is [26]

$$\gamma_w = 2\pi^2 \omega_c \cos\theta_0 C \frac{n_1}{n_0} E_r \left(\tan^2\theta_0 - \frac{2}{2\delta - 1} \frac{\omega}{\omega_c - \omega} \right) \tag{6.41}$$

where $C = (\delta - 1)\frac{E_{\min}^{\delta-1}}{4\pi\cos\theta_0}$, $E_r = \frac{m_e}{2}\left(\frac{\omega_c - \omega}{k\cos\vartheta\cos\theta_0}\right)^2$, and $E_{\min}$ is the lower energy boundary for high-energy particles. Equation (6.41) makes it possible to obtain the necessary condition for the instability, $\omega < \omega_c(\delta - 0.5)/(\sigma + \delta - 1.5)$, which, for instance, at $\delta = 3$, $\sigma = 10$ yields $\omega < 0.22\omega_c$. Combining the latter inequality with Eq. (6.40), we obtain the threshold energy for fast electrons [27]:

$$E > E_{cr} = 5.26\xi \left(\frac{c_A}{10^8}\right)^2 \text{keV} \tag{6.42}$$

where $\xi = (\sigma - 1)/\left[(\delta - 0.5)(\sigma + \delta - 1.5)^2\right]$. For $\delta = 3$, $\sigma = 10$, Eq. (6.42) yields $E_{cr} = 11.6$ keV. Equation (6.42) is different from that obtained by Melrose and Brown [28] in the presence of the coefficient ξ. The growth rate may be estimated with the use of [29]

$$\frac{\gamma_w}{\omega_c} \approx A\delta \frac{\pi}{2} \frac{n_1}{n_0} \tag{6.43}$$

where $A \approx (\sigma - 1)^{-1}$ is the degree of anisotropy. Landau damping is the most significant for whistlers in solar and stellar coronal loops. Note that whistlers propagating along the magnetic field display no Landau damping, since the longitudinal component of the wave electric field is zero. However, this component occurs in whistlers propagating in the curved magnetic field of a loop, so that Landau damping arises with the relative decrement [30]

$$\frac{\nu_L}{\omega_c} = \frac{\sqrt{\pi}}{4} \frac{\sin^2\vartheta}{\cos\vartheta} \frac{\omega}{\omega_c} \Phi(x), \quad x = \frac{\omega}{\sqrt{2} k V_{Te} \cos\vartheta} \tag{6.44}$$

The function $\Phi(x)$ was tabulated in [30]: $\Phi(x \ll 1) \approx 1/x$ and $\Phi(1) \approx 1$. Landau damping restricts the energy increase in excited waves, especially in stellar flare loops with the plasma temperature $T \geq 10^7$ K. However, in whistlers that propagate in a loop composed of multiple thin filaments or in a plasma condensation along the loop axis (the duct), the amplification of their amplitude may be significant. In such cases, the threshold for the instability with respect to the density of high-energy electrons is specified by Coulomb collisions with $\nu_{coll} \approx \nu_{ei}(\omega/\omega_c)$, which according to Eq. (6.44) and for $\delta = 3$, $\sigma = 10$, $n_0 = 10^{11}$ cm^{-3}, $B = 100$ G, and $T = 10^7$ K, yields a fairly low threshold $n_1/n_0 > 10^{-7}$.

In contrast to accelerated electrons, which predominantly excite high-frequency waves (electromagnetic waves at the harmonics of electron cyclotron frequency, Langmuir waves, whistlers, and Bernstein modes), anisotropic high-energy ions in coronal loops are responsible for the generation of low-frequency ($\omega \ll \omega_i$) Alfvén waves, where ω_i is the ion cyclotron frequency. The Alfvén waves generated in the Galaxy by cosmic rays with the energy of 1–100 GeV can significantly affect propagation of cosmic rays in the interstellar gas [31]. Expressions for the instability growth rates of small-scale ($\lambda \approx 2\pi V_i/\omega_i \leq 10^5$ cm for $v_i \approx c$, $B \geq 100$ G) Alfvén waves for various types of anisotropy of high-energy ions are presented in [32]. For example, for the momentum distribution of type (Eq. (6.10)), the instability growth rate at the ion cyclotron resonance is

$$\gamma_A = \sqrt{\frac{\pi}{8}\frac{n_1}{n_0}}\omega_i \left[\frac{\omega_i}{\omega} \frac{\alpha}{\sqrt{1 + k_\perp^2/k_0^2}} - (\alpha - 1) \right] y \exp\left(-\frac{y^2}{2}\right) \tag{6.45}$$

where $k_0^2 = 2(\omega\omega_i)k_{||}^2$ and $y = \omega_i/k_{||}V_{Ti}$. In the case of propagation along the magnetic field ($k_\perp = 0$), Eq. (8.35) coincides with the expression obtained by Kennel and Petscheck [33]. The maximum growth rate (Eq. (6.45)) can be represented as

$$\gamma_A^{max} \approx 0.4\alpha\omega\beta_p \tag{6.46}$$

where $\beta_p = 8\pi p_p/B^2$ and p_p is the pressure exerted by high-energy protons.

Small-scale Alfvén waves refract when propagating in the curved magnetic field of a loop. It may seem that, similar to whistlers, they may subsequently quit the quasi-parallel regime, $k_\perp > k_{||}\sqrt{\omega/\omega_i}$, and undergo Landau damping, which will restrict the increase in their amplitudes. However, because of the dispersion caused by plasma gyrotropy, $\omega/\omega_i \ll 1$, Alfvén waves maintain quasi-longitudinal propagation, since the loop forms a waveguide [32].

6.3 The Fine Structure of Radio Emission from Coronal Loops

In the earliest days of radioastronomy, the emission from solar coronal loops was attributed to broadband bursts of types IV and V at meter and decimeter wavelengths. As the frequency and time resolution of radio telescopes increased, the true complexity of the fine structure of the radio emission from the Sun and stars became visible. One element of the fine structure, QPPs, was mentioned in Chapter 3. The pulsations do not exhaust, however, the total diversity of the fine structure in the radiation of solar and stellar flares, which also includes sudden reductions, zebra patterns, fiber bursts, and spike bursts [14, 34]. It is vitally important to clarify the mechanisms responsible for the formation of the fine structure in order to diagnose flare plasma and understand the nature of stellar flares.

6.3.1 Sudden Reductions

They were observed against the continuum radio emission background in the Sun [34] and red dwarves AD Leo and YZ CMi [14]. Zaitsev and Stepanov [35] suggested that rapidly drifting sudden reductions originate from beams of high-energy electrons entering a loop and filling the loss cone – the reason of continuum emission. This leads to the quenching of the loss-cone instability, which results in the reduction in continuum radio emission. In this case, the relative density of a new portion of energetic particles entering the loss cone remains relatively low, $n_2/n_1 \geq 0.2$. This mechanism naturally explains the sudden reductions (Figure 6.4) and clearly confirms the existence of stellar coronal loops – magnetic traps for high-energy particles.

6.3.2 Zebra Pattern

Against the background of broadband radio emission of solar flares at meter and decimeter wavelengths, quasi-equidistant (with respect to frequency) emission strips are often observed (Figure 6.5ab). Their number can exceed 10 (Figure 6.5b). Currently, the most elaborated is the model of a distributed source of the zebra pattern [1, 36]. This model assumes that the emission bands arise when the condition of double plasma resonance (DPR) in a coronal loop is fulfilled, $s\omega_c = \omega_{uh} = \left(\omega_p^2 + \omega_c^2\right)^{1/2}$, which provides a maximum instability growth rate for plasma

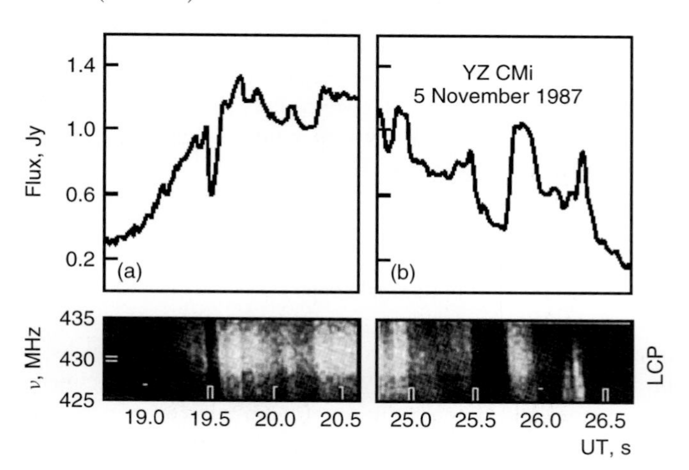

Figure 6.4 The light curve (at 430 MHz) and dynamic spectrum (intensity as a function of frequency and time) of two fragments of flaring radio emission of YZ CMi, with rapidly drifting sudden reductions, observed with the Arecibo 305 m radio telescope [14]. The first fragment starts at 9:48:19 UT; left circular polarization.

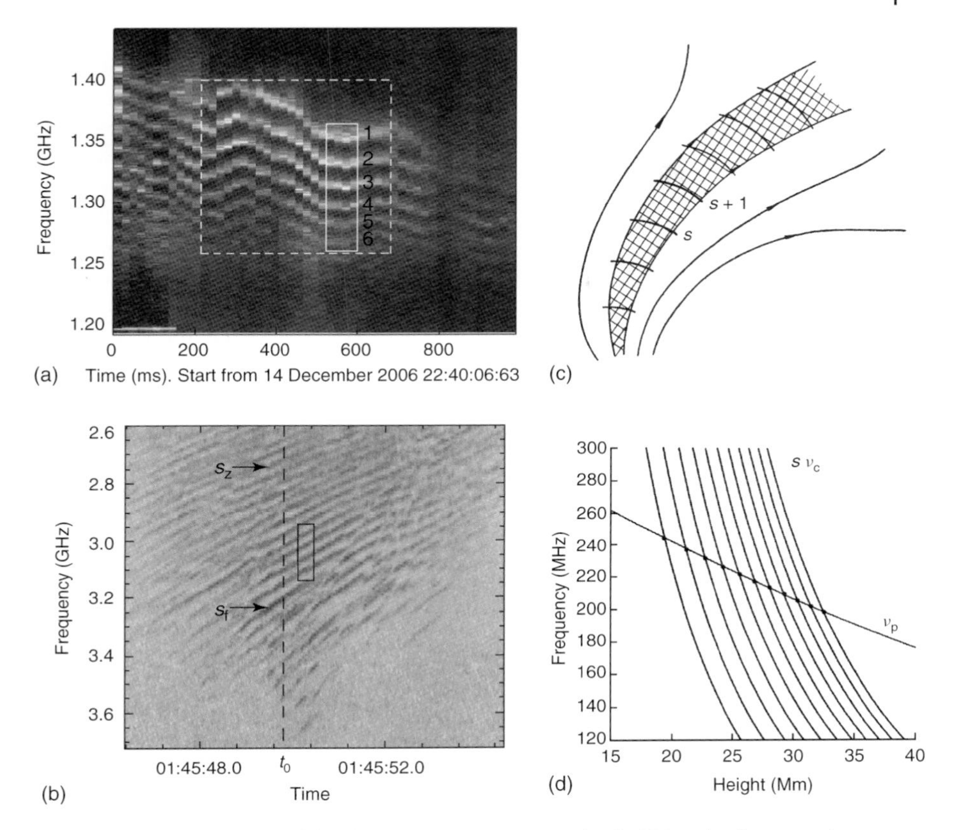

Figure 6.5 (a,b) Examples of the dynamic spectra of the zebra pattern in solar radio emission: (a) The flare event on 14 December 2006 with six successive strong stripes observed with FST and OVSA in the frequency interval 1.25–1.4 GHz [37]. (b) The event on 21 April 2002 in the frequency interval 2.6–3.6 GHz [38]. (c) The distributed source model. (d) An example of the levels of the double plasma resonance at meter wavelengths [36].

waves. Figure 6.5c,d presents an example for the along-the-loop distribution of the upper hybrid frequency $\omega_{\rm uh} = \left(\omega_{\rm p}^2 + \omega_{\rm c}^2\right)^{1/2}$ and the harmonics of the gyrofrequency $s\omega_{\rm c}$. The points of intersection of the curves specify the regions of DPR, at which the enhanced amplification of plasma waves occurs. The frequency interval between the zebra stripes, in which the radio emission displays the double upper hybrid frequency, is

$$\Delta\omega \approx \frac{2(s-1)\omega_{\rm c} L_{\rm B}}{|(s-1)L_{\rm B} - sL_{\rm n}|} \tag{6.47}$$

Depending on the ratio $L_{\rm B}/L_{\rm n}$, this interval varies from $2\omega_{\rm c}$ (if $L_{\rm B}/L_{\rm n} \gg 1$) to $2\omega_{\rm c} L_{\rm B}/L_{\rm n}$ (if $L_{\rm B}/L_{\rm n} \ll 1$). Here, $L_{\rm B}$ is the inhomogeneity scale of the magnetic field in the loop. The interval (Eq. (6.47)) decreases by the factor of two for the fundamental tone emission.

It should be noted that zebra pattern appearance and the stripes number strongly depend on the shape of particle distribution function. A study of the instability of high-energy electrons at the DPR [39] demonstrates that the observed uncommonly large number of stripes ($\sim$40) at the centimeter wavelengths [38] (Figure 6.5b) results from the specific fast-electron distribution. For a power-law distribution with the loss cone (Eq. (6.10)), the number of emission bands in the zebra pattern increases with the spectral index α, reaching 40–50 for $\alpha = 5$. Additionally, this study estimates the plasma parameters in the source of the zebra pattern: $n_0 \approx 10^{11}$ cm^{-3}, $B \approx 40$ G, $T \approx (1-5) \times 10^6$ K, and plasma $\beta \approx 0.2-1.0$ for the event on 21 April 2002 [39].

Except DPR model, there is an explanation of zebra pattern based on the nonlinear interaction of propagated whistler wave packets with Langmuir waves, $l + w \rightarrow t$ ("whistler model") [34, 40]. Recently, Bin Chen *et al.* [37] presented the first interferometric observation of a zebra-pattern radio burst in the X1.5 flare on 14 December 2006 by using the Frequency-Agile Solar Radiotelescope Subsystem Testbed (FST) and the Owens Valley Solar Array (OVSA). After calibrating FST against OVSA, they obtain the absolute locations of radio fine structure and study their spatial and spectral structures. It was found in [37] that radiation of each of six zebra stripes originates from different spatial locations. The conclusion was that the zebra-pattern burst is consistent with the DPR model [36, 39] in which the radio emission occurs in resonance layers where the upper hybrid frequency is harmonically related to the electron cyclotron frequency in a coronal magnetic loop.

Theoretical indication on the possibility of the generation of zebra pattern in the flaring stars due to the DPR was given in Ref. [6].

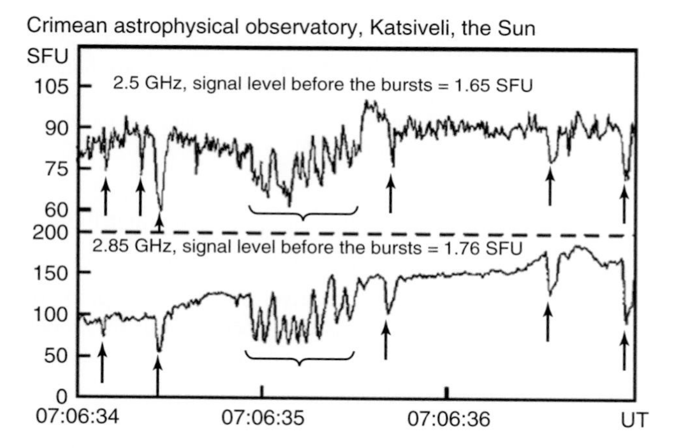

Figure 6.6 An example of millisecond pulsing structure with separate sudden reductions (arrows) at 2.5 and 2.85 GHz [41]. The event features the presence of pulsations at a lower emission level, which originate from the reduction in the anisotropy of high-energy electrons on an injection of a new portion of particles into the coronal loop.

6.3.3
Diagnostics of Coronal Plasma Using the Fine Structure of Radio Emission

A study of the solar event on 17 November 1991 provides an example of diagnosing parameters of loops and energetic particles using the fine structure of radio emission, on the basis of the plasma mechanism of radio emission with the frequency $\omega_t = 2\omega_{uh}$ [41]. Drifting sudden reductions at 2.5 and 2.85 GHz were observed against the background of the pulsing structure with a typical period $P_{puls} \approx 60$ ms (Figure 6.6), which is difficult to explain by MHD loop oscillations.

Pulsations with such a small period are interpreted as a manifestation of nonlinear transfer of plasma waves through the scattering on background plasma ions into the nonresonance domain (the decay domain) and are described in terms of the Lotka–Volterra equations for the "predator-pray" problem [42]:

$$\frac{dw}{dt} = \gamma w - \zeta w w^*, \quad \frac{dw^*}{dt} = -\nu w + \zeta w w^* \tag{6.48}$$

where w and w^* are the relative energy density of plasma waves in the resonance (unstable) and nonresonance domains, $\zeta \approx \omega_p/40$ is the coefficient of induced scattering, and $\nu = \nu_{ei}$ is the decrement. Equation (6.48) allows periodic solutions corresponding to closed trajectories around a center type point $w_0 = \nu/\zeta$ and $w_0^* = \gamma/\zeta$. The period of pulsations around this point is given by

$$P_{puls} = 2\pi\,(\gamma\nu)^{-1/2} \tag{6.49}$$

From the frequency of radiation for the event on 17 November 1991, the plasma density was found to be $n_0 = 2.5 \times 10^{10}$ cm^{-3}, and the energy of electrons injected into the loss cone 50–400 keV, from the rate of the sudden reduction frequency drift. The growth rate of the plasma waves $\gamma \approx 10^{-2}(n_1/n_0)\omega_p$ was used to derive the density of high-energy electrons trapped by the loop and providing the generation of plasma waves, $n_1 = 2.5 \times 10^4$ cm^{-3}, and the density of particles causing the disruption of loss-cone instability, $n_2 > 5 \times 10^3$ cm^{-3}. Close to instability threshold, $\gamma \approx \nu_{ei}$; hence, the temperature of the plasma in the loop is $T \approx 10^7$ K. The condition $\beta = 8\pi n_0 \kappa_B T/B^2 < 1$ yields the estimate for the magnetic field in the emission source $B > 30$ G.

References

1. Zheleznyakov, V.V. (1996) *Radiation in Astrophysical Plasmas*, vol. 204, Kluwer Academic Publishers, 462 p.
2. Alfvén, H. and Fälthammar, C.-G. (1963) *Cosmical Electrodynamics: Fundamental Principles*, Clarendon Press, Oxford, 228 p.
3. Gurnett, D.A. and Bhattacharjee, A. (2004) *Introduction to Plasma Physics. With Space and Laboratory Applications*, Cambridge University Press, 452 p.
4. Dulk, G.A. (1985) *Ann. Rev. Astron. Astrophys.*, **23**, 169.
5. Zaitsev, V.V. and Stepanov, A.V. (1983) *Sol. Phys.*, **88**, 297.
6. Stepanov, A.V., Kliem, B., Krüger, A. *et al.* (1999) *Astrophys. J.*, **524**, 961.
7. Dory, R.A., Guest, G.E., and Harris, E.G. (1965) *Phys. Rev. Lett.*, **14**, 131.

8. Mikhailovsky, A.B. (1974) *Theory of Plasma Instabilities*, Instabilities of a Homogeneous Plasma, Vol. 1, Consultant Bureau, New York.
9. Kopylova, Y.G., Stepanov, A.V., and Tsap, Y.T. (2002) *Astron. Lett.*, **28**, 783.
10. Qin, Z., Li, C., Fu, Q., and Gao, Z. (1996) *Sol. Phys.*, **163**, 383.
11. Fleishman, G.D., Bastian, T.S., and Gary, D.E. (2008) *Astrophys. J.*, **684**, 1433.
12. Fleishman, G.D. and Melnikov, V.F. (2003) *Astrophys. J.*, **587**, 823.
13. Ramaty, R. (1969) *Astrophys. J.*, **158**, 753.
14. Bastian, T.S., Bookbinder, J., Dulk, G.A., and Davis, M. (1990) *Astrophys. J.*, **353**, 265.
15. Kukes, A.F. and Sudan, R.N. (1971) *Sol. Phys.*, **17**, 194.
16. Stepanov, A.V. (1978) *Sov. Astron. Lett.*, **4**, 103.
17. Melrose, D.B. (1980) *Plasma Astrophysics: Nonthermal Processes in Diffuse Magnetized Plasma*, vol. 2, Gordon and Breach, New York.
18. Wu, C.S. and Lee, L.C. (1979) *Astrophys. J.*, **230**, 621.
19. Hewitt, R.G., Melrose, D.B., and Rönnmark, K.G. (1982) *Aust. J. Phys.*, **35**, 447.
20. Aschwanden, M.J. (1990) *Astron. Astrophys. Suppl.*, **85**, 1141.
21. Ginzburg, V.L. and Zheleznyakov, V.V. (1958) *Sov. Astron.*, **2**, 653.
22. Mullan, D.J., Mathioudakis, M., Bloomfield, D.S., and Christian, D.J. (2006) *Astrophys. J. Suppl.*, **164**, 173.
23. Zaitsev, V.V., Kruger, A., Hildebrandt, J., and Kliem, B. (1997) *Astron. Astrophys.*, **328**, 390.
24. Zaitsev, V.V., Koupriyanova, E.G., and Stepanov, A.V. (2000) *Astron. Lett.*, **26**, 736.
25. Abada-Simon, M., Lecacheux, A., Aubier, M., and Bookbinder, J.A. (1997) *Astron. Astrophys.*, **321**, 841.
26. Kawamura, K., Omodaka, T., and Suzuki, I. (1981) *Sol. Phys.*, **71**, 55.
27. Stepanov, A.V. and Tsap, Y.T. (2002) *Sol. Phys.*, **211**, 135.
28. Melrose, D.B. and Brown, J.C. (1976) *Monthly Notices Roy. Astron. Soc.*, **176**, 15.
29. Wentzel, D.G. (1976) *Astrophys. J.*, **208**, 595.
30. Akhiezer, A.I., Akhiezer, I.A., Polovin, R.V., Sitenko, A.G., and Stepanov, K.N. (1975) *Plasma Electrodynamics, Linear Theory*, vol. 1, Pergamon Press, New York.
31. Wentzel, D.G. (1974) *Ann. Rev. Astron. Astrophys.*, **12**, 71.
32. Mazur, V.A. and Stepanov, A.V. (1984) *Astron. Astrophys.*, **139**, 467.
33. Kennel, C.F. and Petscheck, H.E. (1966) *J. Geophys. Res.*, **71**, 1.
34. Chernov, G.P. (2006) *Space Sci. Rev.*, **127**, 195.
35. Zaitsev, V.V. and Stepanov, A.V. (1975) *Astron. Astrophys.*, **45**, 135.
36. Zlotnik, E.Y., Zaitsev, V.V., Aurass, H., and Mann, G. (2009) *Sol. Phys.*, **255**, 273.
37. Chen, B., Bastian, T.S., Gary, D.E., and Jing, J. (2011) *Astrophys. J.*, **736**, 64.
38. Fu, Q., Ji, H., Qin, Z., Xu, Z., and 20 co-authors (2004) *Sol. Phys.*, **222**, 167.
39. Kuznetsov, A.A. and Tsap, Y.T. (2007) *Sol. Phys.*, **241**, 127.
40. Chernov, G.P. (1990) *Sol. Phys.*, **130**, 75.
41. Fleishman, G.D., Stepanov, A.V., and Yurovsky, Y.F. (1994) *Space Sci. Rev.*, **68**, 205.
42. Zaitsev, V.V. (1971) *Sol. Phys.*, **20**, 95.

7
Flaring Events in Stellar Coronal Loops

7.1
Particle Acceleration and Explosive Joule Heating in Current-Carrying Loops

In solar and stellar flares, which frequently occur in coronal magnetic loops, a substantial fraction of their energy is released in the form of high-energy particles. The basic part of the electrons and ions in impulsive solar flares is accelerated up to energies of 100 keV and 100 MeV, respectively [1], thereby producing hard X-ray and γ-line radiation. Moreover, continuum γ-radiation and the radiation from neutral pions, observed in some cases, provide evidence that the energy of electrons and ions during the flares may reach 10 MeV and 1 GeV, respectively. If we assume that hard X-ray radiation of flares originates from bremsstrahlung of fast electrons in the chromosphere (e.g., by the nonthermal "thick target" model) [2–4], then an impulsive solar flare should produce about 10^{37} electrons with the energy $>20\,\text{keV s}^{-1}$ in 10–100 s. This implies that the rate of the energy release in the form of accelerated electrons is $\dot{E}_e \approx 3 \times 10^{29}\,\text{erg s}^{-1}$ for 100 s, which corresponds to the total energy of the electrons $E_e(>20\,\text{keV}) \approx 3 \times 10^{31}$ erg, with the total number of the accelerated electrons $N_e(>20\,\text{keV}) \approx 10^{39}$. The requirements on the acceleration rate are less stringent if we assume that the spectrum of hard X-ray radiation with the energy $<30\,\text{keV}$ is associated with the radiation of a hot ($\sim 3 \times 10^7$ K) plasma, while the radiation with a higher energy is generated by nonthermal electrons with a power-law energy spectrum. This assumption underlies the hybrid thermal/nonthermal (T/NT) model [5]. In this case, the necessary rate of acceleration of electrons with the energy $>20\,\text{keV}$ decreases to $\dot{N}_e \approx 2 \times 10^{35}$electrons s^{-1} for the duration of the process of about 100 s, which gives $N_e(>20\,\text{keV}) \approx 2 \times 10^{37}$ electrons, $\text{d}E(>20\,\text{kev})/\text{d}t \approx 6 \times 10^{27}\,\text{erg s}^{-1}$, and $E_e(>20\,\text{keV}) \approx 6 \times 10^{29}$ erg.

In order to explain the generation of fast particles during flares, several acceleration mechanisms were suggested, which may be divided into three basic classes.

1) **Acceleration by waves in the course of interaction between waves and particles at the cyclotron and Cherenkov resonance.** In the random phase approximation, this interaction is described by quasi-linear equations for the wave energy density and the particle distribution, and is similar to the second-type Fermi

Coronal Seismology: Waves and Oscillations in Stellar Coronae, First Edition.
A. V. Stepanov, V. V. Zaitsev, and V. M. Nakariakov.

acceleration. Different versions of the self-consistent spectrum of waves and particles applied to solar flares and other astrophysical objects are presented, for example, in the monograph by Kaplan and Tsytovich [6]. When applied to flares, this mechanism may play only a secondary role, since wave excitation itself requires the presence of fast particles.

2) **Acceleration by shock waves.** Shocks may emerge, for example, in coronal magnetic loops in the course of evaporation of dense chromospheric plasma from loop footpoints. In this case, mutually approaching shock fronts will form magnetic mirrors for a certain group of charged particles. Repeatedly reflected, these particles will increase their energy (similar to the first-type Fermi acceleration). This mechanism was used by Bai *et al.* [7] to analyze acceleration of protons and to explain the time delay between the appearance of hard X-ray and γ-radiation of a flare, and also by Somov and Kosugi [8] to interpret particle acceleration in the region of the high-temperature turbulent current sheet at the top of a flare magnetic loop. If the plasma is sufficiently rarefied, collisionless shock waves may be formed. In this case, due to the relative motion of electrons and ions, current instabilities may develop inside the shock front, which leads to plasma heating and acceleration of particles [9, 10].
3) **Quasi-stationary electric (DC) field acceleration**. Large-scale electric fields in coronal magnetic loops are formed, for example, when loop footpoints are located in the nodes of several supergranulation cells. In this case, converging convective flows of plasma, interacting with the magnetic field in the loop footpoints, generate the electric field of charge separation due to different magnetization of electrons and ions. Under certain conditions, this field may effectively accelerate particles. Such accelerating electric fields are called *direct current* (*DC*).

In addition to the above-mentioned mechanisms, betatron acceleration or magnetic pumping are sometimes used [11, 12], for example, in the case of a collapsing magnetic arch [8, 13]. Particle acceleration in solar flares is discussed in detail in monographs [14, 15].

Taking into account the good correlation between impulsive flares and coronal magnetic loops, in this section, we consider the acceleration of electrons with electric fields of charge separation in the footpoints of a coronal magnetic loop, and also acceleration by induction electric fields originating in a coronal magnetic loop due to the variations of electric current in it. We assume, of course, that in stellar corona other acceleration mechanisms also occur. The loop may contain sufficient energy, up to $10^{32}-10^{33}$ erg, and may provide energy release observed in an intensive flare. This energy is stored in the nonpotential part of the magnetic field that emerges due to the large (up to $\sim 3 \times 10^{12}$ A) electric current in the loop.

7.1.1
Where Is the Acceleration Region Located?

In order to provide beams of fast electrons observed in impulsive solar flares, a sufficient number of particles should be injected for acceleration. What serves as the reservoir for these particles, if the acceleration occurs in a flaring magnetic loop? The total number of particles in a flaring loop with plasma density 10^{10} cm^{-3}, cross section 10^{18} cm^2, and length $(1-5) \times 10^9$ cm is $(1-5) \times 10^{37}$ electrons. If we take into account that in any reasonable mechanism only an insignificant fraction of particles is accelerated, we should conclude that the obtained total number of electrons in the coronal part of a flare loop is obviously insufficient to provide the acceleration even in the most favorable case of the hybrid model ($\sim 2 \times 10^{37}$ electrons).

In a magnetic loop, two sources exist, which essentially may provide the necessary number of particles. First, these are parts of the loop near its footpoints located in the chromosphere. In the chromospheric part of the loop, the column from the temperature minimum to the transition region between the chromosphere and the corona contains $\sim 5 \times 10^{40}$ particles, if we assume the cross-sectional area of the loop in this region to be $\sim 10^{18}$ cm^2. If the acceleration occurs in the chromospheric part of the loop, the above number of particles is sufficient to provide the injection of the required number of electrons to the acceleration mode.

The other opportunity to enrich a flaring magnetic loop by particles appears when the loop interacts with a solar prominence [16, 17]. In this case, the flare may be initiated by flute instability, which develops at the top of the loop and results in penetration of the dense plasma of the prominence into the current channel of the magnetic loop. The number of particles supplied by the prominence during the flare $t_f \sim 100$ s may be estimated as $N \approx 2\pi r r_f n_p V_p t_f$, where $r_f \approx 4 \times 10^9$ cm is the thickness of the prominence "tongue" that penetrates into the current channel, $r \approx 2 \times 10^8$ cm is the half-thickness of the loop, $n_p \approx 10^{12}$ cm^{-3} is the plasma number density in the prominence, and $V_p \approx V_{Ti} \approx 2 \times 10^6$ cm s^{-1} is the characteristic rate of plasma inflow into the current channel of the loop, approximately equal to the thermal velocity of ions at the temperature of the prominence matter $\sim 5 \times 10^4$ K. For these parameters, we obtain $N \approx 10^{39}$ electrons, which exceeds the value required for the hybrid (T/NT) model by more than an order of magnitude, and which is close to the value required for the nonthermal "thick target" model. Hence, it may be concluded that, to cover the needs of the acceleration mechanisms for particles in the most powerful flares, the acceleration region is preferably located in the chromospheric part of the coronal arch. For flares with moderate energy release, the acceleration region may be located in the vicinity of the top of the loop, where the necessary storage of particles is provided, for example, by a plasma flow from the prominence (if the latter exists).

7.1.2
Large-Scale Electric Fields in Flare Loops

The most efficient way to accelerate a particle in the energy release region of a flare is direct acceleration by the large-scale electric field $\boldsymbol{E}$ of the flare magnetic loop. It is essential that if magnetic field $\boldsymbol{B}$ presented in a plasma, so that $|\boldsymbol{B}| > |\boldsymbol{E}|$, then particles will be accelerated only by the component of the electric field projected on the magnetic field $E_{||} = \boldsymbol{EB}/B$. If $E_{||}$ is smaller than the Dreicer field

$E_D = e\Lambda\omega_p^2/V_{Te}^2$, then the acceleration process (runaway) involves electrons with the speed $V > (E_D/E_{||})^{1/2} V_{Te}$, where V_{Te} is the thermal speed of the electrons, Λ is the Coulomb logarithm, and ω_p is the Langmuir frequency. The kinetic theory gives the following formula for the generation rate of the runaway electrons [18]:

$$\dot{N}_e = 0.35 n \nu_{ei} V_a x^{3/8} \exp\left[-\sqrt{2x} - \frac{x}{4}\right] \tag{7.1}$$

where $x = E_D/E_{||}$, V_a is the volume of the acceleration region, $\upsilon_{ei} = (5.5n\Lambda)/T^{3/2}$ is the effective frequency of electron collisions, n is the electron concentration in the plasma, T is the temperature, and Λ is the Coulomb logarithm. In the coronal part of the magnetic loop, even for a sufficiently large electric current $I = 10^{12}$A and for typical values of the parameters $n = 10^{10}$ cm^{-3}, half-thickness $r_2 = 5 \times 10^8$ cm, and $T = 10^6 - 10^7$ K, the electric field caused by the finite conductivity of the plasma is too small to result in any noticeable acceleration ($x = E_D/E_{||} > 200$).

7.1.2.1 The Charge-Separation Electric Field

The largest electric fields are generated in the dynamo region in the footpoints of a magnetic loop, where charges are effectively separated due to a convective flow of photospheric matter inward of the flux tube and to different magnetization of electrons and ions. In a vertical cylindrical flux tube with a radial converging flow of plasma (this is our approximation for the part of a realistic loop, located in the vicinity of the photospheric footpoints), only the radial component of the charge-separation electric field exists. This component appears to be perpendicular to the magnetic field of the stationary flux tube (B_φ, B_z); therefore, in stationary conditions, no acceleration by the charge-separation field occurs. The acceleration occurs when the magnetic field of the flux tube is deformed so that, for example, the radial component of the magnetic field B_r originates. In this case, the projection of the electric field on the direction of the magnetic field is [19]:

$$E_{||} \approx \frac{1-F}{2-F} \frac{\sigma V_r B^2}{enc^2(1+\alpha B^2)} \frac{B_r}{B} \tag{7.2}$$

Here, the radial component of the magnetic field $B_r \ll B$; $F = n_a m_a/(n_a m_a + n m_i)$ is the relative density of neutrals, $\sigma = e^2 n/m_e(\nu'_{ei} + \nu'_{ea})$ is the Coulomb conductivity; $\alpha = \sigma F^2/(2-F)c^2\ n m_i \nu'_{ia}$, ν'_{jk} is the effective frequency of collisions between a j-type and a k-type particles; and V_r is the radial component of the rate of the convective motion in the loop footpoint. Particle acceleration due to the charge-separation field may appear, for example, in the course of development of flute instability in a footpoint of a magnetic tube (see Chapter 2), when a plasma "tongue" penetrating the current channel with velocity V_r is inhomogeneous along the height. In this case, it may be shown that the radial component of the magnetic field is generated

$$B_r = B\frac{\partial}{\partial z}\int_0^t V_r(t')\mathrm{d}t' \tag{7.3}$$

and the electric field $E_{||}$ appears, which results in the acceleration. In general, in the region of generation of the charge-separation electric field, that is, in chromospheric footpoints of a coronal magnetic loop, the condition $\alpha B^2 \gg 1$ is fulfilled. Then Eq. (7.2) takes the form:

$$E_{||} \approx \frac{1-F}{F^2}\frac{m_i V_r \nu'_{ia}}{e}\frac{B_r}{B}, \quad \alpha B^2 \gg 1 \tag{7.4}$$

Note that for $V_r < 0$ (the convective flow is directed inward of the tube) the $E_{||}$ component is directed downward and accelerates electrons toward the corona, while ions toward the photosphere, that is, high-energy ions and electrons move to different footpoints. This may result in a situation, in which one footpoint of an arch will emit γ-radiation caused by high-energy ions, while the other the hard X-ray radiation related to fast electrons. This was observed, for example, with Reuven Ramaty High-Energy Solar Spectroscopic Imager (*RHESSI*) space observatory during the 23 July 2003 flare [20]. Figure 7.1 shows that the centroid of the 2.223 MeV image was found to be displaced by $20'' \pm 6''$ from that of the 0.3–0.5 MeV image. In the vicinity of the 2.223 MeV centroid, no significant X-ray emission was detected. This displacement implies that there are significant differences in location between the electron- and ion-associated sources.

The heating of footpoints of the magnetic loop results in an increase in $E_{||}$, since the heating decreases the relative density of neutrals F. Under the conditions in the lower chromosphere, the ionization equation is controlled by spontaneous

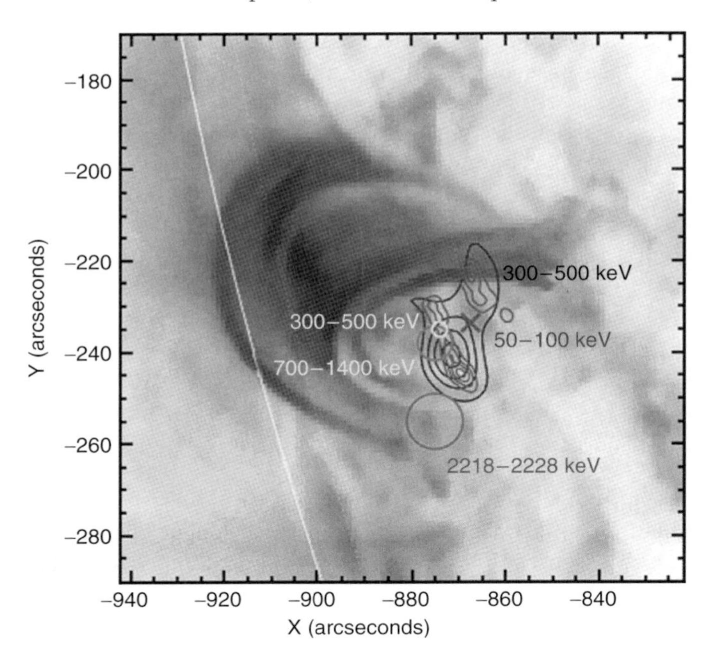

Figure 7.1 Centroid of γ-ray emission (2218–2228 keV) that indicates the energetic ion interaction region, and hard X-ray sources in the solar flare of 23 July 2002 obtained with *RHESSI*, superposed on a *TRACE* image [20].

recombination to the lower level of the hydrogen atom and by photoionization from this level by the optical radiation of the photosphere. Therefore, when the characteristic time of recombination appears to be shorter than the characteristic scale of flare heating, the ionization is specified by the modified Saha equation, with the instant electron temperature being obtained from energy equation [21]. For hydrogen atoms

$$\frac{(n+n_a)x^2}{1-x} = 7.2 \times 10^{18} T^{1/2} \exp\left(-6.583 - \frac{1.185 \times 10^5}{T}\right) \tag{7.5}$$

where n, n_{a} are the concentrations of electrons and neutral atoms, respectively, and $x = n/(n + n_{\mathrm{a}})$ is the degree of ionization. In the temperature interval, $10^4\ \mathrm{K} < T < 10^5\ \mathrm{K}$, for the effective frequency of collisions between ions and neutral atoms, $\nu'_{\mathrm{ia}} \approx 2.12 \times 10^{-11}\ F(n + n_{\mathrm{a}})T^{1/2}$. Taking into account that $F = 1 - x$, from Eq. (7.4) we obtain

$$E_{||} \approx 0.7 \times 10^{-9} \frac{(1-F)TV_r}{x^2} \left(\frac{B_r}{B}\right) \exp\left(-\frac{1.185 \times 10^5}{T}\right) \text{statvolt cm}^{-1} \tag{7.6}$$

Plasma number density is included in Eq. (7.6) only implicitly, through the coefficients F and x. In the case of relatively high degree of ionization, it may be assumed that $F \ll 1, x \approx 1$. In this case, the charge-separation electric field does not depend on the plasma number density and is specified only by the temperature and the convection rate. The Dreiser field $E_{\mathrm{D}} \approx 6 \times 10^{-8} n/T$ volt cm^{-1} is proportional to the plasma density; therefore, for a given temperature, the maximum contribution to flows of escaping electrons comes from regions of the flare with lower concentration. Assuming, for example, the plasma number density in the acceleration region in the chromosphere to be $n + n_{\mathrm{a}} \approx n \approx 10^{12}\mathrm{cm}^{-3}$, the volume of the acceleration region $V_{\mathrm{a}} \approx 10^{24}\ \mathrm{cm}^3$, the temperature in the optical flare $T \approx 2 \times 10^4$ K, the speed of the convective flow $V_{\mathrm{r}} \approx 2 \times 10^4\ \mathrm{cm\,s}^{-1}$, and the radial component of the magnetic field in the chromospheric base of the tube $B_{\mathrm{r}}/B \approx 0.2$, from Eq. (7.1) we obtain the flow of the accelerated electrons $\dot{N}_{\mathrm{e}} \approx 10^{35}\ \mathrm{el\,s}^{-1}$. This is sufficient to explain the hard X-ray flux of the flare within the framework of the hybrid T/NT model. In this case, the maximum energy acquired by runaway electrons on the scale of the acceleration region $\approx 10^8$ cm may reach 6 MeV, while the ratio of the parallel component of the electric field to the Dreiser field is $E_{||}/E_{\mathrm{D}} \approx 2.4 \times 10^{-2}$.

7.1.2.2 **Inductive Electric Field**

In a coronal magnetic loop with a current, induction electric field may be generated at the impulse phase of a flare, when the electric current along the loop varies. This field is directed along the loop axis, and its circulation is specified by the equation

$$\oint E_z \mathrm{d}z = -\frac{L}{c^2}\frac{\partial I_z}{\partial t} \tag{7.7}$$

Here, z is the coordinate along the loop axis, L is the inductivity of the loop, and I_z is the electric current through the cross section of the loop. The integration is carried out over the coronal part of the loop and over the region of the current closure under the photosphere (see Chapter 2). If the current variations are sufficiently slow (for the time exceeding the period of intrinsic oscillations of the coronal magnetic loop as an equivalent electric circuit), the equation for the current may be written in the form

$$\frac{L}{c^2}\frac{\partial I_z}{\partial t} + R(I_z)I_z = \Xi(I_z) \tag{7.8}$$

where the effective resistance R and the electromotive force Ξ caused by photospheric convection are determined by Eqs. (2.55) and (2.56) in Chapter 2. In the steady-state situation, the photospheric electromotive force compensates Ohmic losses, and a stationary current $I_z = I_0$ is established in the circuit. Assume that the velocity of the flow that generates the electromotive force increases from the steady-state value V_r to $V_r + \Delta V_r$ for the characteristic time τ_V. The velocity may increase, for example, due to 5 min oscillations of the rate of the photospheric convection or due to the flute instability in the magnetic loop footpoints. If the characteristic time of the velocity variation is substantially smaller than the time of establishing of the current in the circuit, that is, $\tau_V \ll L/c^2 R$, then the variation of the current in the course of the velocity variation may be neglected. Substituting the derivative from Eq. (7.8) to Eq. (7.7), we obtain

$$\oint E_z \mathrm{d}z = -\frac{|\Delta V_r|\, l_1 I_0}{c^2 r_1} \tag{7.9}$$

Here, l_1, r_1 are the height of the region of action of the photospheric electromotive force and the radius of the magnetic tube in this region. Denoting the length of the integration contour by Z, from Eq. (7.9) we obtain the average induction electric field emerging in a coronal magnetic loop due to a variation of the electromotive force:

$$E_z = -\frac{|\Delta V_r|\, l_1 I_0}{c^2 r_1 Z}\ \text{statvolt cm}^{-1} \tag{7.10}$$

The characteristic time of the existence of the induction electric field is of the order of τ_V. For typical values of the parameters $\Delta V_r \approx 10^5$ cm s^{-1}, $l_1 \approx 5 \times 10^7$ cm, $r_1 \approx 10^7$ cm, $I_0 \approx 10^{10}$ A $= 3 \times 10^{19}$ statampere, and $Z \approx 10^{10}$ cm, the induction electric field is $E_z \approx 1.7 \times 10^{-6}$ statvolt cm^{-1} $\approx 5 \times 10^{-4}$ V cm^{-1}.

The Dreicer field $E_D = 6 \times 10^{-8} n/T$ V cm^{-1} in coronal magnetic loops reaches its minimum at the top of the loop, since at this point the lowest plasma concentration and the highest temperature occur. Therefore, at the top of a loop, the conditions for electrons to run away are more favorable than those in other parts of the loop. This implies that the induction electric field forms the largest flows of accelerated electrons from the top of the loop. For sufficiently low plasma density at the top of the loop ($n < 10^{10}$ cm^{-3}), the electric field becomes super-Dreicerian. In this case, the bulk of electrons at the top of the loop will be running away and accelerated by the induction electric field. The maximum energy of accelerated electrons may reach several mega electronvolts.

7.1.3
The Impulsive and Pulsating Modes of Acceleration

As noted above, the appearance of the radial component of the magnetic field at a loop footpoint, where the powerful charge-separation electric field is located, may be caused by the development of flute instability. In this case, due to aperiodicity of the flute instability, the acceleration is apparently impulsive. It follows from Eq. (2.84) of Chapter 2 that the time of development of the flute instability with the perturbation wavelength of the order of the height of inhomogeneous atmosphere is 10–30 s for the temperature of the plasma in the footpoint of a magnetic loop $T \approx 10^4 - 10^5$ K. For this time, the plasma that surrounds the flux tube penetrates inside. It generates the radial component of the magnetic field, due to which the charge-separation electric field is projected onto the magnetic field, thus resulting in particle acceleration. In the first approximation, the time of developing of the flute instability may be taken as the characteristic duration of the impulsive phase of the acceleration. The acceleration may cease due to a decrease in the radial component of the convection rate, which results from an increase in the gas pressure inside the tube in the course of the flare process. Another reason for the termination of the acceleration may be a slowdown of an outer plasma flute penetrating into the magnetic loop. With the above parameters of the plasma in the region of acceleration caused by the charge-separation electric field, the total number of electrons accelerated in a footpoint of a current-carrying magnetic loop for the time of developing of the flute instability is $\dot{N}_e \tau_b \approx 3.5 \times 10^{36}$ electrons, which is close to the productivity of the acceleration mechanism within the framework of the T/NT model of generation of hard X-ray radiation of impulse solar flares [22].

The presence of broadband quasi-periodical pulsations in some type IV solar radio bursts provides evidence that sometimes a flare is directly followed by a pulsating mode of electron acceleration, which displays a substantially smaller acceleration rate of fast electrons, but usually lasts appreciably longer than the main flare. An example of such an event was studied in [23], where the spatial structure of a source of radio pulses was reconstructed from combined analysis of spectral (Figure 7.2) and heliographic radio, optical, and X-ray data, and the source appeared to be a coronal magnetic loop. Electrons were periodically injected at the speed of the order of $0.3c$ from the loop footpoint with a stronger magnetic field; further, they moved along the trap axis, generating a sequence of type III bursts with fast frequency drift. The pulse period was on average 1.33 s, while the pulse phase duration was around 3.5 min, that is, pulses displayed a high-quality factor. Another good example of quasi-periodic particle acceleration was that which provided the flare on 15 June 2002. Microwave and hard X-ray observations demonstrated that quasi-periodic acceleration and injection of fast electrons is a more likely cause of quasi-periodic oscillations than a magnetohydrodynamic (MHD) oscillation of the emission source [24].

The analysis of the efficiency of the acceleration mechanism for this event showed [25] that the average electron acceleration rate in the pulsations was

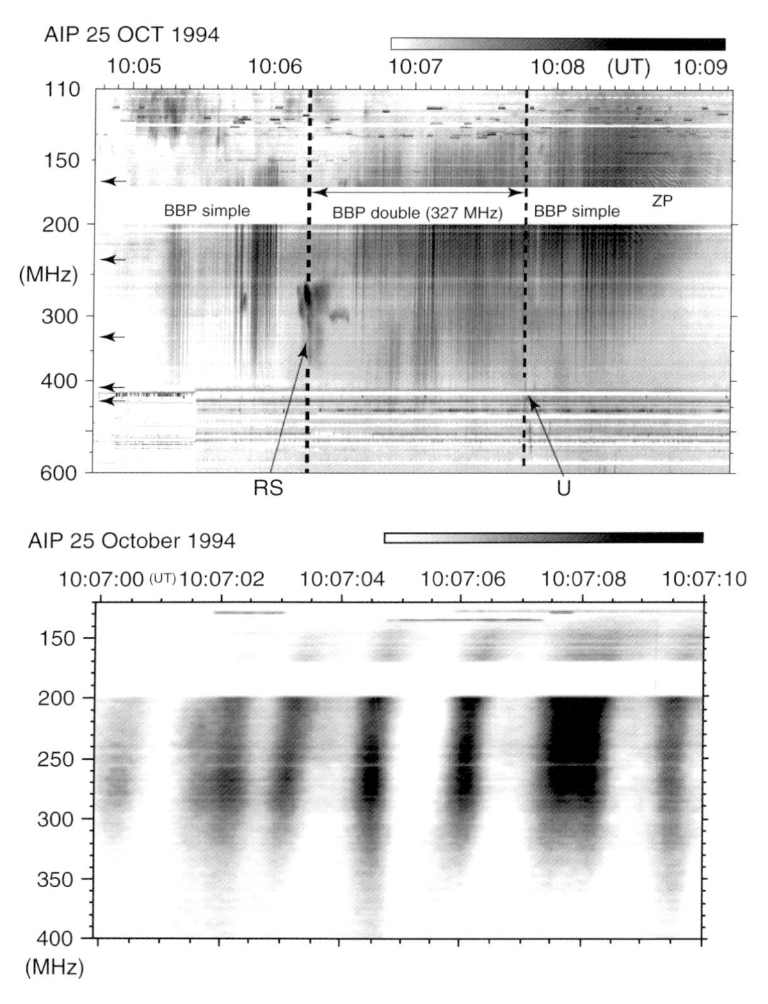

Figure 7.2 Broadband radio pulsations (BBPs) in solar flares of 25 October 1994. Zoomed spectrum of neighboring pulses is shown on the bottom block [23] (AIP – Astrophysikalisches Institut Potsdam, Germany).

$3 \times 10^{32}\ \mathrm{s}^{-1}$, approximately smaller by three orders of magnitude than that during a medium-power solar flare. Pulsations are also detected in radio emission of flares in late-type stars [26–30]. The rapid negative frequency drift of radio pulsations provides evidence of periodic acceleration of particles in lower layers of the atmosphere and their subsequent injection into the corona. In this case, the radio emission may be generated in a current-carrying magnetic loop, while the pulsations may be caused by the excitation of intrinsic oscillations of the loop as an equivalent electric circuit with the frequency specified by Eq. (2.67) from Chapter 2.

7.1.4
The Current of Accelerated Electrons (Colgate's Paradox)

Let us consider an important issue discussed in the context of a large electric current, which is associated with the accelerated electrons [5, 31]. If the efficiency of the accelerating mechanism is $dN_e/dt \geq 10^{35}$ electrons s^{-1}, then an electric current $I = e\dot{N}_e \gtrsim 1.6 \times 10^{15}$ A should arise. If this current flows in a magnetic loop with the cross section $\sim 10^{18}$ cm^2, it should induce a magnetic field $B \geq 6 \times 10^6$ G, which indeed is not observed in coronal structures. This long-standing paradox was first mentioned by Colgate [32]. Generally, two ways to eliminate this contradiction are considered.

The first is related to the assumption of filamentation of the current of accelerated electrons, that is, of subdivision of the current channel into a number of thin current filaments with opposite directions of the currents in the adjacent filaments. As a result, the total magnetic field of the current channel does not exceed the observed value [33, 34]. However, it remains unclear how a system of "threads" with oppositely directed currents may be formed in an expanding electron beam. It was shown in the study [35] that the current filamentation is indeed possible under certain conditions; however, the current direction in adjacent filaments does not undergo inversion, and the problem of generation of strong magnetic fields at the periphery of the current channel remains unsolved.

The other explanation is related to the formation of reversed current in the plasma [36–39]. Assume, for example, that an electron beam with radius r_0 is injected into the plasma along the z-axis parallel to the outer magnetic field. Then in each point of the plasma the field B_ϕ will vary when passing the front of the flow. The variation of B_ϕ will result in the formation of the electric field E_z on the front; the action of this field on plasma electrons will result in the origination of a current directed opposite to that injected. Consequently, the total current will decrease, and eventually will be totally compensated. If the radius of the beam of fast electrons exceeds the screening length ($r_0 > c/\omega_p$), then no magnetic field exists for $r > r_0$. The beam current is compensated by the inverse current in the plasma, which almost totally flows inside the beam. The condition for total neutralization is $c/\omega_p \ll r_0$, $\nu_{ei}t < 1$, where t is the time after the injection. For the times $\nu_{ei}t \gg 1$, the reverse current decays and neutralization gradually disappears. However, the characteristic decay time for the reverse current is specified by the time of magnetic diffusion $t_D = \pi\sigma r_0^2/c^2$, which for r_0 of the order of the loop thickness substantially exceeds all characteristic timescales for flare processes. Therefore, it may be suggested that the injection of accelerated electrons does not result in a variation of the outer magnetic field.

7.1.5
Explosive Joule Energy Release. The Role of Flute Instability and Cowling Conductivity

The analysis of dissipation of the electric current in a coronal magnetic loop shows that neither classical (Spitzer) resistivity nor the largest microinstabilities of plasma, for example, the Buneman instability, can provide the necessary resistance, which would yield the observed rate of plasma heating during a flare. In the studies [16, 17], a possible increase in the current dissipation in a coronal loop was considered taking into account the neutral component of the plasma and the instability of the process. In this case, the Cowling conductivity [40] is essential, which is lower than the classical Spitzer conductivity by many orders of magnitude. The effect of a considerable increase in Joule dissipation in partly ionized gas was for the first time found in 1950 by Schlüter and Biermann [41]. This effect is caused by large energy losses undergone by ions moving through the gas of neutral particles under the action of the ampere force $\boldsymbol{j} \times \boldsymbol{B}$. For example, in interstellar clouds, the Cowling conductivity is smaller than the Spitzer conductivity by 10–12 orders of magnitude [42].

In Ref. [43], it was shown that the generalized Ohm's law

$$\boldsymbol{E}^* = \frac{\boldsymbol{j}}{\sigma} + \frac{\boldsymbol{j} \times \boldsymbol{B}}{enc} - \frac{\nabla p_e}{en} + \frac{F}{cnm_i\nu_{ia}}\left[\left(n_a m_a \boldsymbol{g} - \nabla p_a\right) \times \boldsymbol{B}\right] - \frac{F^2}{cnm_i\nu_{ia}}\rho\frac{\mathrm{d}\boldsymbol{V}}{\mathrm{d}t} \times \boldsymbol{B} \tag{7.11}$$

along with the Maxwell equations, the equation of the bulk motion of the plasma,

$$\rho\frac{\mathrm{d}\mathbf{V}}{\mathrm{d}t} = \rho\mathbf{g} - \nabla p + \frac{1}{c}\mathbf{j} \times \mathbf{B} \tag{7.12}$$

and the continuity equation

$$\frac{\partial \rho}{\partial t} + \mathrm{div}(\rho\vec{V}) = 0 \tag{7.13}$$

self-consistently describe the behavior of plasma and electromagnetic fields. Here, $\mathbf{E}^* = \mathbf{E} + \mathbf{V} \times \mathbf{B}/c$ is the electric field in the coordinate frame that moves together with the plasma, $\rho = n_a m_a + n_e m_e + n_i m_i$ is the density of partly ionized plasma, $p = p_a + p_e + p_i$ is the pressure, $\sigma = \frac{ne^2}{m_e(\nu'_{ei}+\nu'_{ea})}$ is the Spitzer conductivity, and $F = \rho_a/\rho$ is the relative density of neutrals.

Joule current dissipation is described by the parameter $q = \mathbf{E}^*\mathbf{j}$, which, taking into account relations (Eqs. (7.11) and (7.12)), may be presented in the form:

$$q = \frac{j^2}{\sigma} + \frac{F^2}{c^2 nm_i\nu'_{ia}}\left(\mathbf{j} \times \mathbf{B}\right)^2 \tag{7.14}$$

It is seen from Eq. (7.14) that in the force-free field ($\boldsymbol{j} \parallel \mathbf{B}$) the second term on the right-hand side is insignificant and the current dissipation is specified by Spitzer conductivity. The dissipation reaches the highest efficiency for $\boldsymbol{j} \perp \boldsymbol{B}$. The increase in the current dissipation in a coronal arch and triggering of a flare may be due to ballooning instability in chromospheric footpoints of a magnetic loop or at its top, where a prominence may be localized. When the ballooning mode of the flute instability develops, a "tongue" of partly ionized plasma penetrates into the current channel and distorts the magnetic field in accordance with the equation

$$\frac{\partial \boldsymbol{B}}{\partial t} = \nabla \times \left[\left(\boldsymbol{V} + \frac{F^2}{cnm_i\nu'_{ia}}\rho\frac{\mathrm{d}\boldsymbol{V}}{\mathrm{d}t}\right) \times \boldsymbol{B}\right] \tag{7.15}$$

As a result, the Ampere force appears, which provides the increased current dissipation. Integrating Eq. (7.14) over the arch volume, we find the energy release power:

$$\frac{dW}{dt} = \left[\frac{m_e(\nu'_{ei} + \nu'_{ea})d}{e^2 nS} + \frac{2\pi F^2 I^2 d}{c^4 nm_i \nu'_{ia} S^2}\right] I^2 = \left[R_c + R_{nl}(I)\right] I^2 \tag{7.16}$$

where d is the characteristic size of the "tongue" of partly ionized plasma penetrating into the current channel. Assuming $d \approx 5 \times 10^7$ cm, $S \approx 3 \times 10^{16}$ cm^2, $I \approx 3 \times 10^{11}$ A, $n \approx 10^{11}$ cm^{-3}, $T \approx 10^4$ K, and $F \approx 0.5$, from Eq. (7.16) we find the Cowling resistance $R_{nl} \approx 10^{-3}$ Ω, which provides the flare energy release power 10^{27} erg s^{-1}. For the current $I \approx 10^{12}$ A, the power increases to 10^{29} erg s^{-1}.

Wheatland and Melrose [44] considered the problem of current dissipation in a force-free magnetic tube, suggesting that the energy release power (Eq. (7.16)) may be overestimated. The study [43] presents a detailed analysis of the penetration of the "tongues" of partly ionized plasma into the current channel of a force-free magnetic tube. The flute instability appears to be developing as a series of oscillations accompanied by plasma heating inside the tube. As a result, the tube ceases to be force free, and the explosive phase of Joule heating begins.

The above may be illustrated by the example of a magnetic tube with a force-free magnetic field ($\boldsymbol{j} \times \boldsymbol{B} = 0$) specified as [45].

$$B_{\varphi 0} = \frac{r}{r_0}\frac{B_0}{1 + r^2/r_0^2}, \quad B_{z0} = \frac{B_0}{1 + r^2/r_0^2} \tag{7.17}$$

Let us assume, for definiteness, the following approximation for the velocity of plasma inflowing to the current channel [46]: $V_r(r,t) = V_0(t) r/r_0, r \leqslant r_0$. In this case, the magnetic field components B_φ and B_z will vary as follows:

$$B_\varphi(r,t) = e^\gamma B_{\varphi 0}(re^\gamma),\ B_z(r,t) = e^{2\gamma} B_{z0}\left(re^\gamma\right)$$

$$\gamma = -\frac{1}{r_0}\int_0^t \left(V_0(t) + \frac{F^2 \rho}{nm_i \nu'_{ia}}\frac{dV_0}{dt}\right) dt' \tag{7.18}$$

Equations (7.14), (7.15), and (7.18) make it possible to study the time and coordinate dependence of the energy release rate in a magnetic flux tube in the course of development of flute instability. In the simplest form, it may be done for the near-the-axis region in the tube, where the pressure depends on the radius quadratically, $p(r,t) = p_{00}(t) + p_0(t) r^2/r_0^2$, and the function γ may be considered to be minor compared to unity. Physically, this means that the amplitude of oscillations of a plasma tongue penetrating into the tube is small compared to the tube radius. In this case, taking into account (Eq. (7.12)), we obtain the system of equations for V_0, p_0, γ:

$$\frac{\partial V_0}{\partial t} = -\frac{2p_0}{r_0} + \frac{B_0^2}{\pi r_0}\gamma \tag{7.19}$$

$$\frac{\partial p_0}{\partial t} = \frac{(\gamma - 1)F^2 B_0^4}{nm_i \nu'_{ia} \pi^2 r_0^2}\gamma^2 \tag{7.20}$$

$$\frac{\partial y}{\partial t} = -\frac{V_0}{r_0} - \frac{F^2 \rho}{n m_i \nu'_{ia} r_0} \frac{\partial V_0}{\partial t} \tag{7.21}$$

Combining Eqs. (7.19)–(7.21), we obtain the equation:

$$\frac{\partial^3 y}{\partial \tau^3} + \varepsilon_1 \frac{\partial^2 y}{\partial \tau^2} + \frac{\partial y}{\partial \tau} - 2\varepsilon\varepsilon_1 y \frac{\partial y}{\partial \tau} = \varepsilon y^2 \tag{7.22}$$

Here, the following notation has been introduced:

$$\tau = \frac{t}{t_A}, \quad t_A = \frac{r_0}{2c_A}, \quad c_A = \frac{B_0}{\sqrt{4\pi\rho_0}}, \quad \varepsilon_1 = \frac{\varepsilon}{2(\gamma - 1)},$$
$$\varepsilon = \frac{t_c}{t_A}, \quad t_c = \frac{2(\gamma - 1) F^2 (n + n_a)}{n \nu'_{ia}} \tag{7.23}$$

In the general case, Eq. (7.22) describes pulsating modes of energy release, when a flute of the outer partially ionized plasma penetrates into a loop via a series of oscillations. In addition to this, the explosive mode occurs, which appears when plasma pressure increases from the loop axis toward its periphery, that is, $p_0 > 0$, and partly balances the Ampere force that appears when the flute instability is developing. In this case,

$$y = \frac{y_0}{1 - \frac{t}{t_0}}, \quad t_0 = \frac{n m_i \nu'_{ia} \pi r_0^2}{2(\gamma - 1) F^2 B_0^2 y_0} \tag{7.24}$$

the ratio between the gas pressure and the pressure of magnetic field in the loop increases with the increase in y explosively, up to the stage at which effects restricting the energy release and not having been taken into account in this model come into play. For example, the increase in the degree of ionization in the course of plasma heating results in a decrease in the energy release due to ion–atom collisions. Figure 7.3 illustrates two time profiles of solar microwave bursts, which can be understood in terms of dense, partly ionized plasma penetration into a current-carrying loop. In Figure 7.3a, the burst starts with several ~10 s pulses at the preflash phase and then evolves into explosive energy release. This time profile can occur if the lifetime of the pulsating phase $t_P \approx r_0/\varepsilon_1 c_A$ is shorter than the timescale of the explosive phase $t_E \approx r_0/y_0 \varepsilon c_A$. In the second case, $t_P > t_E$ and the pulsations are followed by the explosive phase (Figure 7.3b). Detailed analysis of Eq. (7.22) is given in Ref. [43].

7.2
The Kinematics of Energetic Particles in a Loop and the Consequent Radiation

Once charged particles have been accelerated in a coronal magnetic loop, for example, during a flare, their further development may be influenced by the Coulomb scattering, or by wave–particle interaction. In the absence of scattering, the particles with the pitch angle $\theta > \theta_0$ are trapped in a magnetic bottle. When scattered, accelerated particles can enter the loss cone and leave the magnetic

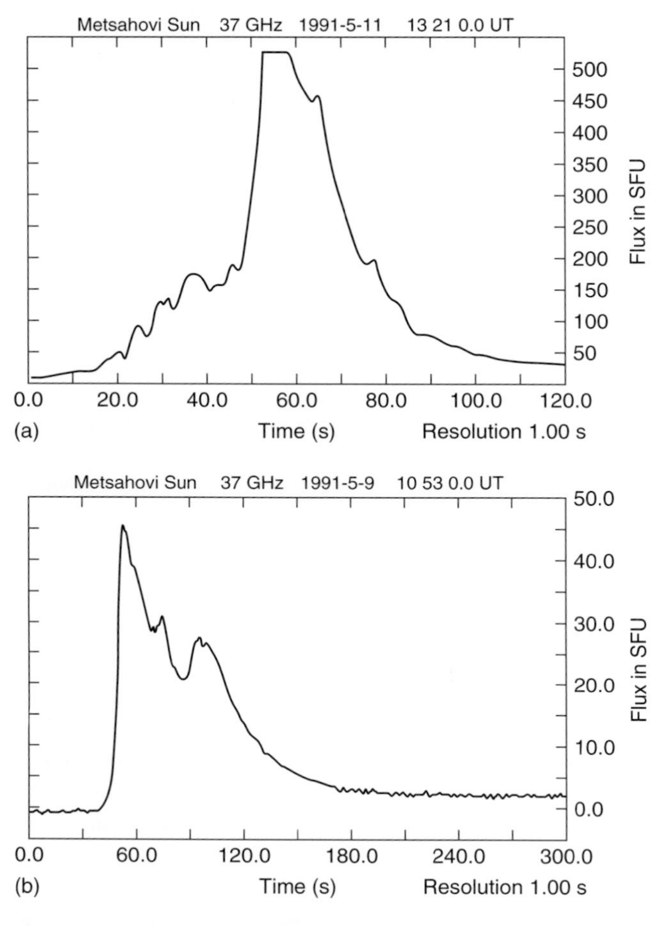

Figure 7.3 (a,b) Examples of a light curve from solar flares at a 37 GHz observer with Metsähovi radio telescope [43], which illustrates two regimes of penetration of dense plasma tongue into flaring loop described by Eq. (7.22).

bottle due to the pitch-angle diffusion. Melrose and Brown [47] introduced the *trap-plus-precipitation* model, taking into account the consequences of scattering in a coronal magnetic trap, either by Coulomb collisions or by some anomalous scattering mechanism. MacKinnon [48] considered the generalized scattering problem, and Fletcher [49] carried out a numerical simulation of coronal particle trapping. An extended analysis of particle acceleration and kinematics in solar flares was presented by Aschwanden [15, 50].

Nevertheless, the kinematics of energetic particles in coronal magnetic loops is still intensively discussed. Two extreme approaches are commonly used here: (i) free propagation (the "time-of-flight" approach), which suggests that there are no particle–particle or wave–particle collisions and (ii) collision-dominated propagation [47–49]. The first approach was used by Aschwanden and Schwartz [51] in order to explain the energy-dependent time delays of hard X rays in solar flares. This section is devoted to the second approach with the emphasis on the wave–particle interaction.

7.2.1
Diffusion Regimes of Accelerated Particles in Coronal Loops

There are three types of diffusion of particles in a magnetic trap depending on scattering peculiarities and particle injection rate: weak, intermediate, and strong [52, 53]. Each case is associated with the typical mean lifetime of a magnetically trapped particle, defined in the following way [54, 55]:

$$\tau_l = \frac{N}{J} \tag{7.25}$$

where $N = \iint (B_{\max}/B) f d^3 v ds$ is the number of energetic particles stored in a flux tube with the unity cross section at the footpoint, where $B = B_{\max}$, and J is the rate of injection (particle per second) into the trap. For a symmetric coronal magnetic trap, $J = 2S$, where S is the flux of particles toward a loop footpoint. A particle moves in the adiabatic invariant space undergoing particle–particle or wave–particle collisions, in a way similar to the Brownian motion. Therefore, the variations of its pitch angle can be described in terms of diffusion. For physical reasons, it is clear that the peculiarity of particle diffusion in a coronal loop depends on the following timescales: the diffusion time τ_{D}, which corresponds to the mean time of the pitch-angle variation by about $\pi/2$, the time of filling of the loss cone τ_{D}/σ, and $\tau_0 = l/2v$ that is the time of a particle transit to escape the loss cone.

In the case of weak diffusion

$$\sigma \tau_0 < \tau_{\mathrm{D}} \tag{7.26}$$

The loss cone is empty, since the particle diffusion is too slow. The number of trapped particles does not depend on the injection power, $N = \text{const}$, and the mean lifetime is inversely proportional to the source power $\tau_l \propto J^{-1}$.

In the regime of intermediate diffusion, the following inequality is satisfied:

$$\tau_0 < \tau_{\mathrm{D}} < \sigma \tau_0 \tag{7.27}$$

In this case, the loss cone is partially filled, and particle distribution function is close to isotropic. Therefore, the number of particles stored in the trap is $N \approx \sigma n l$ because the particle density depends weakly on the coordinates. The flux of precipitating particles is $S \approx nv$, and the mean lifetime is $\tau_l \approx \sigma \tau_0$.

Strong diffusion means that the mean time of pitch-angle scattering by about $\pi/2$ is smaller than the free travel time of particles escaping the loss cone

$$\tau_{\mathrm{D}} < \tau_0 = \frac{l}{2v} \tag{7.28}$$

Inequality (Eq. (7.28)) implies that during one transit time of the magnetic trap the particle alters the direction of its motion a number of times. This led Budker *et al.* [56] to the idea of trapping hot collisional plasma. According to this idea, in the case $\tau_D < \tau_0$, instead of free propagation of plasma along the trap axis, slow diffuse extension occurs. Strong diffusion suggests that the particle distribution function is nearly isotropic, and hence, $N \approx \sigma nl$, as it was previously. Nevertheless, even under the steady-state condition near the magnetic trap footpoints, a deficit of particles reflected from dense plasma exists in this region. This anisotropy of the distribution function causes the excitation of electromagnetic waves, which scatter a part of precipitating flux so that $S < nv$. Therefore, the lifetime of particles in the trap increases from the minimum value of $\sigma\tau_0$. It appears as if the particles are "locked" by the turbulence that they excite.

Weak and intermediate diffusion modes were suggested by Kennel and Petschek [57], while strong pitch-angle diffusion was first introduced by Bespalov and Trakhtengertz [54, 55]. The different modes of pitch-angle diffusion may be explained by extending the "leaky pail model" suggested by O'Brien [58], illustrated in Figure 7.4. A faucet is used as the analog for a particle source; two holes in the side of a pail represent the drain of particles into the loss cone; the water level corresponds to the number of trapped particles. If the number of particles is sufficient to induce instability, a weak source leads to weak precipitation, and the number of trapped particles does not depend on the source power. A moderate source increases the precipitation rate, and the number of trapped particles increases. A strong source makes the outflowing stream turbulent and increases its viscosity, thus leading to a greater increase in the fluid level in the pail.

The wave particle interaction is described by the equation of the quasi-linear theory for the distribution function of energetic particles f and for the spectral density of the energy of electromagnetic turbulence W_k [54, 55]:

$$\frac{df}{dt} = \frac{(2\pi)^3}{m^2} \sum_{s=-\infty}^{\infty} \int \left(\hat{\hat{\Lambda}} G_{ks} W_k \hat{\Lambda} f\right) d^3k + j \tag{7.29}$$

$$\frac{dW_k}{dt} = \frac{(2\pi)^3}{m} W_k \sum_{s=-\infty}^{\infty} \left(G_{ks} \hat{\Lambda} f\right) d^3v - \nu W_k, \tag{7.30}$$

where G_{ks} is the power emitted by a particle at the s-harmonics of the cyclotron frequency in the range of wave vectors $(\vec{k}, \vec{k} + d\vec{k})$ under the cyclotron resonance

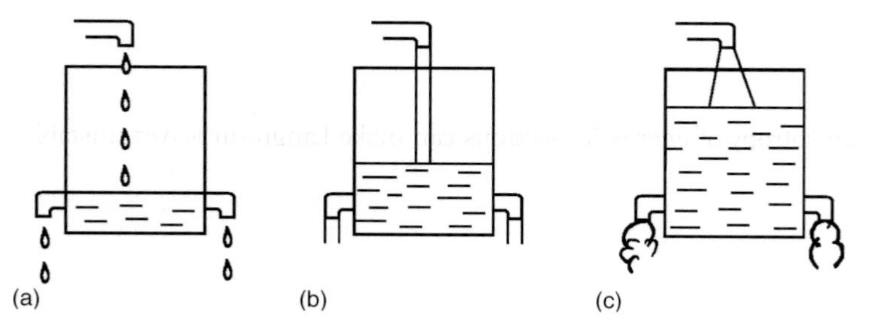

Figure 7.4 Weak (a), intermediate (b), and strong (c) diffusion represented by a hydrodynamical analogy.

condition $\omega - k_{||}v_{||} = s\omega_c$, j is the particle source power, and ν is the wave decrement. Equations (7.29) and (7.30) include differential operators

$$\hat{\hat{\Lambda}} = \frac{1}{v_\perp}\frac{\partial}{\partial v_\perp}\frac{\omega - k_{||}v_{||}}{\omega} + \frac{\partial}{\partial v_{||}}\frac{k_{||}}{\omega}, \quad \hat{\Lambda} = \frac{\omega - k_{||}v_{||}}{\omega}\frac{1}{v_\perp}\frac{\partial}{\partial v_\perp} + \frac{k_{||}}{\omega}\frac{\partial}{\partial v_{||}}$$

The strong diffusion is the most interesting case for the interpretation of peculiarities of dynamics, propagation, and emission of accelerated particles, observed in stellar coronal loops. The inequality $\tau_D < \tau_0$ is necessary but insufficient for the strong diffusion. The complete criterion was derived by Bespalov and Trakhtengertz [55] from the analysis of the equations of quasi-linear theory. The solution for Eqs. (7.29) and (7.30) indicates that strong diffusion is established if the particle source integral $J = \iint (B_{\max}/B) j d^3 v ds$ exceeds the value

$$J > J_* \approx \frac{\omega_c n_0 c^2 \sigma}{\omega_p^2 l} \cong \frac{cB\sigma}{4\pi el} \tag{7.31}$$

which does not depend on the particle mass. $J_* \approx 5 \times 10^{13}\ \text{cm}^{-2}\ \text{s}^{-1}$ for typical parameters of stellar flaring loops ($l \approx 10^9$ cm, $B \approx 10^3$ G). Under *steady-state* conditions, the number of particles injected into a magnetic loop per second is equal to the particle flux precipitating to the footpoints, $J = 2S$. Hard X-ray data for solar flares suggest that the flux of electrons with the energy >10 keV toward footpoints is about $S_e \approx 10^{15}-10^{17}\ \text{cm}^{-2}\ \text{s}^{-1}$. The flux of ≥ 10 MeV ions is usually 10^2-10^3 times smaller. Thus, during flare events, electrons are accelerated so efficiently (see also Section 7.1) that strong pitch-angle diffusion is highly possible. For energetic ions, strong, weak, and intermediate diffusion modes are possible.

Calculations presented in [55] for the *steady-state* case of strong pitch-angle diffusion of energetic electrons due to whistler waves with dispersion Eq. (6.29) yield the following expression for the mean lifetime of fast electrons within a magnetic trap:

$$\tau_l^e \approx \frac{m_e}{m_i}\left(\frac{v}{c_A}\right)^2 \sigma\tau_0 \tag{7.32}$$

The *nonsteady-state case* of strong pitch-angle diffusion or turbulent propagation apparently matches the conditions of the impulse phase of a flare. It may be shown that the typical time of turbulent propagation of particles $\tau_{pr} = N|dN/dt|_{j=0}^{-1} \approx N/2S$ is also determined by Eq. (7.25). This is because the time of formation of "turbulent mirrors" is of the order of the inverse instability growth rate and is much smaller than the mean lifetime τ_l.

Note that anisotropy of energetic electrons can make Langmuir waves unstable. Still, the excitation conditions for Langmuir waves in flare loops are more stringent than those for whistlers. For example, no Langmuir wave instability occurs for the temperature anisotropy of energetic particles, $T_\perp > T_{||}$, while such a population of particles is unstable against whistler waves. Therefore, we restrict our consideration with the pitch-angle diffusion of electrons due to whistlers.

If the energetic electrons in a flare loop generate high-frequency waves, the maximum growth rate for energetic ions corresponds to the cyclotron instability of

Alfvén waves $\omega = k_{||}c_A$, where $c_A \ll v$. Taking into account the latter inequality, we can express the operator $\hat{\Lambda}$ as

$$\hat{\Lambda} = \frac{1}{v}\frac{\partial}{\partial v} + \frac{1}{v c_A}\frac{\partial}{\partial x}, \quad x = \frac{v_{||}}{v} \tag{7.33}$$

Under strong diffusion conditions, the particle distribution function adjusts itself (on the time scale of turbulent propagation) to a steady-state level, and the growth rate $\gamma \leqslant 0$ for all wave vectors, due to quasi-linear relaxation. It follows from Eqs. (7.29), (7.30), and (7.33) that the condition $\gamma = 0$ in the region with a high level of Alfvénic turbulence may be satisfied if $\hat{\Lambda} f = 0$. The solution for such a differential equation is [59]

$$f = \Phi(v - x c_A) \approx \Phi(v) - x c_A \frac{\partial \Phi}{\partial v} \tag{7.34}$$

where Φ is an arbitrary function of velocity, which depends only on the initial conditions. For the given angular dependence of the distribution function, the particle flux toward the footpoint is

$$S = 2\pi \int_0^\infty \int_{-1}^1 f v^3 \mathrm{d}v x \mathrm{d}x \approx n c_A \tag{7.35}$$

where $n = 2\pi \int_0^\infty \int_{-1}^1 f v^2 \mathrm{d}v \mathrm{d}x$ is the density of energetic particles. Equation (7.35) is valid for minor anisotropy of the distribution function ($\sim c_A/v \ll 1$). Dividing the distribution function in Eq. (7.29) into the isotropic part and a small anisotropic part, we can show that the energetic particle density in the magnetic trap satisfies the equation [55, 59]

$$\frac{\partial n}{\partial t} = B\frac{\partial}{\partial z}\left(\frac{n c_A}{B}\frac{\partial n/\partial z}{|\partial n/\partial z|}\right) + \int j_0 d^3 v \tag{7.36}$$

Here, the sign of $\partial n/\partial z$ determines the direction of the propagation of the cloud of energetic particles, while j_0 is the isotropic part of the particle source. The particle distribution function is isotropic at the center of the trap; therefore, there is no instability ($\gamma < 0$) and $\partial n/\partial z = 0$ (Figure 7.5). The expansion of the cloud is accompanied by a decrease in the density at the center of the trap.

In a homogeneous magnetic field, for example, from Eq. (7.36) one gets

$$n(t) = \frac{N}{2 c_A t} \tag{7.37}$$

For small anisotropy of the distribution function, the number of energetic particles within the magnetic trap is about $N \approx \sigma n l$. Taking into account Eq. (7.35), we obtain the mean lifetime of energetic protons in the magnetic trap during turbulent propagation under a strong pitch-angle diffusion condition:

$$\tau_l^i = \frac{N}{\mathrm{d}N/\mathrm{d}t} = \frac{N}{2S} \approx \frac{\sigma l}{2 c_A} \tag{7.38}$$

Hence, the trapping time of fast protons is very close to the wave propagation time $t_w \approx l/V_{gr}$, where V_{gr} is the wave group velocity, which is equal to the phase speed

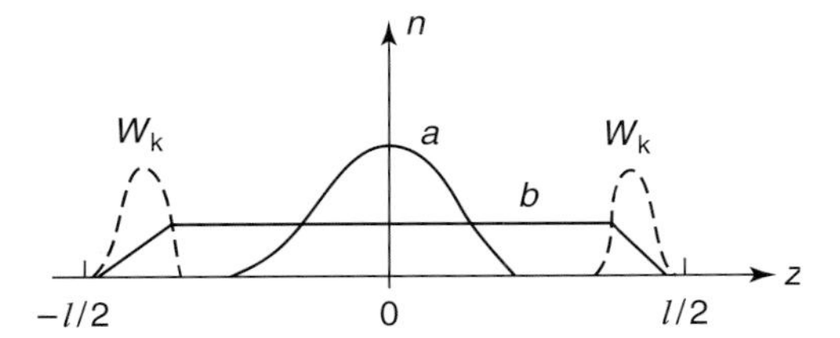

Figure 7.5 The density profile of energetic particles along the axis of a magnetic trap at different times. Dashed curves indicate two regions at the leading edges of the particle cloud with a high level of small-scale turbulence. Curve "a" corresponds to the initial stage, while curve "b" to the expansion stage; l is the length of the trap.

c_A for Alfvén waves with the dispersion equation $\omega = k_{||} c_A$. Indeed, let us assume that a sufficiently large fraction of energetic particles is injected into a magnetic loop. The "turbulent mirrors" are formed due to the instability of fast particles. If the wave damping and wave escape is not essential, the slowest particles not only undergo scattering but also absorb the waves. As a result, the mean impulse of the entire particle flux does not vary, and only a small part of particles precipitate to the loop footpoints due to the presence of magnetic mirrors.

It is easy to show that the mean lifetime of fast electrons, determined with Eq. (7.32), is of the order of the propagation time of whistler waves, since the whistler group velocity $\partial\omega_w/\partial k = 2\left(c^2/v\right)\left(\omega_c/\omega_p\right)^2 = 2\omega_w/k$ is twice as large as the phase velocity.

Three points should be emphasized with reference to strong diffusion:

1) The mean lifetime of energetic particles is derived from the wave phase speed (of whistlers for electrons and Alfvén waves for ions), which is substantially smaller than the speed of accelerated particles v.
2) The proton-to-electron lifetime ratio within the magnetic trap is $\tau_l^i/\tau_l^e \approx m_i c_A/m_e v \geq 10$ for $c_A \approx 10^8$ cm s^{-1}, $v \leq c$.
3) The proton lifetime (Eq. (7.38)) depends only on the parameters of the magnetic loop and is independent of the energy of protons.

7.2.2 Consequences of the Strong Diffusion of Energetic Particles

The wave–particle interaction in the strong diffusion mode is essential not only for the storage conditions of energetic particles in coronal magnetic loops but also for the propagation and radiation of accelerated particles.

7.2.2.1 Turbulent Propagation of Fast Electrons in a Loop

Strong diffusion is well exemplified with the 28 August 1999 solar flare observed by the Nobeyama Radioheliograph (NoRH) at 17 and 34 GHz. This two-loop event

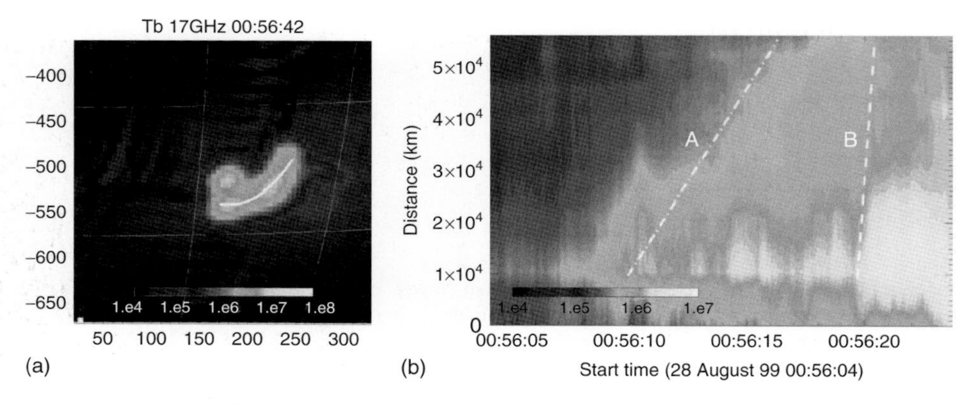

Figure 7.6 The time variation of the 17 GHz intensity distribution along the microwave loop observed by Nobeyama Radioheliograph (NoRH). The solid white line in (a) indicates the propagation trajectory of the emission front. (b) The dash-and-dotted line A corresponds to the velocity of $\approx 10^4$ km s^{-1} and the dashed line B to the speed of light [60].

revealed unusual behavior of the microwave source after an injection of $\sim$1 MeV electrons responsible for the gyrosynchrotron emission at 17 GHz (Figure 7.6). The microwave structure of the event consists of two sources. The point source was unresolved by the NoRH (as it is smaller than its spatial resolution in this band, 10 arcseconds). The well-resolved source with a propagating emission front has length $l \approx 7 \times 10^9$ cm. The observations indicated two injections of energetic electrons into the larger loop. After the first injection (A), the velocity of propagation of the emission front along the loop was about $V_A \approx 10^4$ km s^{-1}, which is smaller by a factor of 30 than the velocity of relativistic electrons generating the gyrosynchrotron emission at 17 and 34 GHz. After the second injection (B), the emission front propagated with the speed of light.

It is generally accepted that the most important small-scale turbulence that determines the behavior of high-energy electrons in solar coronal loops is the whistler-wave turbulence [47, 57–59, 61–63]. Stepanov *et al.* [60] interpreted this anomalous propagation (A) in terms of strong diffusion of relativistic electrons interacting with plasma turbulence. A cloud of highly energetic electrons responsible for the microwave emission generates low-frequency whistler waves, and a turbulent "wall" is formed in the loop. The electrons undergo strong resonant scattering due to wave–particle interaction, and the emission front propagates with the wave phase velocity, which is substantially lower than the particle velocity.

To confirm this model, the parameters of the flaring loop and high-energy electrons should be determined. As it follows from spatial co-alignment of the radio map with *SOHO/MDI* (Solar and Heliospheric Observatory)/(Michelson Doppler Imager) magnetograms, the magnetic field along the loop varies from 255 to 120 G. This means that the mirror ratio of the extended coronal loop is $\sigma \approx 2$. Because the plasma temperature in the extended loop before the first injection (00:56:10 UT) was close to the coronal temperature ($T \approx 2 \times 10^6$ K), we can estimate the loop plasma density, taking into account that the bremsstrahlung radio emission flux is proportional to $n_0^2 T^{-1/2}$ and that this flux was before the NoRH detection limit, $n_0 \approx 7 \times 10^9$ cm^{-3}. The corresponding plasma frequency was $\omega_p \approx 5 \times 10^9$ s^{-1}, and the electron cyclotron frequency for the mean magnetic field of the loop $B = 150$ G was about $\omega_c \approx 2.6 \times 10^9$ s$^{-1} \approx 0.5\omega_p$.

To construct the model of strong diffusion, we also need to know the power flux of the particle source $J \approx n_1 V_A$, where n_1 is the number density of accelerated electrons and $V_A \approx 10^4$ km s^{-1}. From the NoRH observations at 17 and 34 GHz, we can estimate the density of radio-emitting electrons assuming that the electron population has a power-law spectrum $\propto E^{-\delta}$. From two-frequency radio data, $\delta = 5.0$ was obtained and, with the use of Ramaty formalism [64] for gyrosynchrotron emission, the density of emitting electrons was estimated as $n_1(>0.5\text{MeV}) \approx 3 \times 10^3$ cm^{-3} [60]. Hence, the injection rate is $J \approx n_1 V_A \approx 3 \times 10^{12}$ cm^{-2} s^{-1}.

On the other hand, the strong diffusion condition (Eq. (7.31)) requires $J > J_* = cB\sigma/4\pi el$. For the event on 28 August 1999 with $B = 150$ G, $l \approx 7 \times 10^9$ cm, and $\sigma = 2$, we obtain $J_* \approx 2 \times 10^{11}$ cm^{-2} s^{-1}. This is more than an order of magnitude lower than the source power estimated previously from the observations.

Since the range of whistler frequencies is rather wide ($\omega_i < \omega < \omega_c$, where $\omega_i = (m_e/m_i)\omega_c$ is the ion gyrofrequency), another problem consists in the condition of phase synchronism between waves and high-energy particles to form "turbulent propagation." The whistler phase speed ω/k for wave propagation along the magnetic field ($k = k_{||}$) in cold plasma approximation is determined from the dispersion relations [65].

$$N_w^2 = \left(\frac{kc}{\omega}\right)^2 = 1 - \frac{y^2}{\chi(\chi - 1)} - \mu\frac{y^2}{\chi(\chi + \mu)} \tag{7.39}$$

Here, $y = \omega_p/\omega_c$, $\chi = \omega/\omega_c$, and $\mu = m_e/m_i$. Figure 7.7 presents the inverse whistler phase speed versus χ for different y. It displays the minimum $N_w \approx 2.3$–8.8 at $\chi \approx 0.5$ and grows for $\chi \to 1$ and for $\chi \to \mu$.

In our case, after the first injection, the emission front moves at the speed $V_A \approx 3 \times 10^{-2}\, c$. Therefore, the conditions must be determined under which the turbulence level of whistlers with the phase speed $\omega/k = 3 \times 10^{-2}\, c$ ($N_w \approx 30$) should be sufficiently high. It follows from Figure 7.7 that, for $\omega/k = 3 \times 10^{-2}\, c$, high ($\omega \to \omega_c$)- and low-frequency whistlers ($\omega \to \omega_i$) are possible.

For the distribution function of accelerated electrons $f_1 \propto n_1 E^{-\delta}\varphi(\theta)$, where $E = m_e c^2(\Gamma - 1)$, Γ is the Lorenz factor and $\varphi(\theta)$ describes the pitch-angle anisotropy in a manner similar, for example, to Eq. (6.8). The whistler growth rate under the cyclotron resonance condition

$$\omega - k_{||}v_{||} - \omega_c/\Gamma = 0 \tag{7.40}$$

has the form [61]

$$\frac{\gamma_w}{\omega_c} \approx \frac{\pi}{2}\frac{\delta}{\Gamma}\frac{n_1}{n_0}\left(A - \frac{\omega}{kc}\right) \tag{7.41}$$

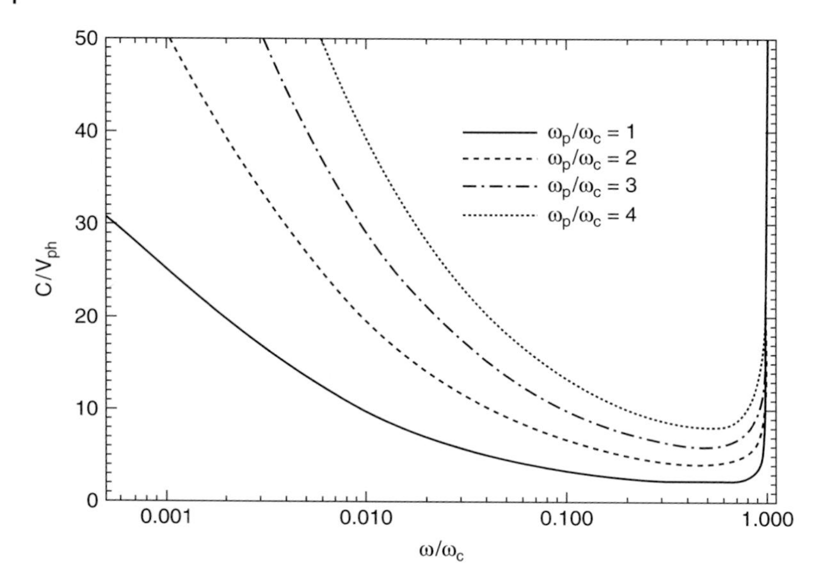

Figure 7.7 Inverse phase velocity of whistlers as a function of χ for $y = 1,2,3,$ and 4.

In contrast to Eq. (6.43), Eq. (7.41) takes into account the Lorenz factor and the instability threshold ($\gamma_\mathrm{w} > 0$ at $A > \omega/kc$). For $\delta = 5$, with $\omega/kc \ll A$ and $\Gamma = 3$, we have

$$\gamma_\mathrm{w}/\omega_\mathrm{c} \approx 2.6A(n_1/n_0) \tag{7.42}$$

From the phase speed (Figure 7.7), we conclude that, in the event on 28 August 1999, flare whistlers with relative phase speed $\omega/kc \approx 3 \times 10^{-2}$ display the frequency $\chi >$ 0.95. These high-frequency whistlers, however, undergostrong cyclotron damping on electrons of thermal background plasma. Indeed, the damping rate is [66]

$$\nu_\mathrm{c}/\omega_\mathrm{c} = (\pi/2)^{1/2}\left[\frac{(1-\chi)(c/V_\mathrm{Te})}{N_\mathrm{w}}\right]\exp\left[\frac{-(1-\chi)^2(c/V_\mathrm{Te})^2}{2(N_\mathrm{w}\chi)^2}\right] \tag{7.43}$$

For the frequency $\chi = 0.95$ and plasma temperature $T = 2 \times 10^6$ K, it follows from Eq. (7.43) that $\nu_c/\omega_c \approx 0.1$. Comparing this value with the growth rate specified by Eq. (7.43), we find the instability threshold $n_1/n_0 > 0.1/2.6A$. Since radio observations give $n_1/n_0 \approx 4.3 \times 10^{-7}$ and the anisotropy measure $A \leq 1$, we can see that no instability for high-frequency whistlers occurs. Moreover, the cyclotron resonance condition suggests that high-frequency whistlers $\chi > 0.9$ interact with comparatively low-energy ($\approx$10 keV) electrons. These electrons, however, do not contribute to the gyrosynchrotron emission at 17 GHz. Hence, we have to consider low-frequency whistlers.

The strong diffusion regime means that, due to pitch-angle scattering, only weak anisotropy occurs in the electron cloud [54]. Indeed, immediately after injection, for the time of about $10/\gamma_\mathrm{w} \approx 3.8n_0/A_0\omega_\mathrm{c}n_1 \approx 3 \times 10^{-3}$ s, which is substantially shorter than the free travel time of fast electrons $\tau_0 \approx l/c \approx 0.23$ s, the quasi-linear pitch-angle diffusion of particles on small-scale turbulence makes the electron distribution nearly isotropic. Here it was suggested that the initial anisotropy measure $A_0 = (\sigma - 1)^{-1} \approx 1.0$, $n_0/n_1 \approx 2.3 \times 10^6$, and $\omega_c \approx 2.6 \times 10^9\ s^{-1}$. The pitch-angle diffusion rate is known to appreciably exceed the momentum diffusion rate $(\Delta\theta/\theta)/(\Delta p/p) = (k_{||}p_{||}/m\omega - 1) \gg 1$, since $k_{||}p_{||}/m \approx \omega_c \gg \omega$ [67]. In this case, the anisotropy of fast electrons propagating along a loop is very small. For the minor anisotropy $A << 1$, the frequency of excited whistlers is $\omega < A\omega_c \ll \omega_c$. Taking, for example, waves with the phase speed $\omega/kc \approx 0.03$ (the refractive index $N_w \approx 30$) from Eqs. (7.39) and (7.40) for 1 MeV electrons ($\Gamma = 3$), we obtain $\chi \approx (\Gamma N_w)^{-1} \approx 10^{-2}$, for example, $\omega \approx 20\omega_i$, and $\gamma \approx N_w\chi^{1/2} \approx 3$. For 0.5 and 2.0 MeV electrons, we obtain, respectively, $\chi \approx 1.7 \times 10^{-2}$ ($\omega \approx 30\omega_i$), $\gamma \approx 3.9$, and $\chi \approx 6.7 \times 10^{-3}$ ($\omega \approx 12\omega_i$), $\gamma \approx 2.4$. Note that accelerated electrons can display substantial dispersion $\Delta E \approx E_{res} \approx 1$ MeV.

The above approximation of homogeneous plasma is a rough idealization of a real magnetic loop. Since a real loop is inhomogeneous throughout its radius and length, the whistler-wave spectrum can be rather broad. In our example for (0.5–2.0) MeV electrons, the wave–particle interaction occurs in the frequency band $\omega \approx (12 - 30)\omega_i$. The wavelength band of whistlers interacting at the cyclotron resonance with (0.5–2.0) MeV electrons is $\lambda = 2\pi/k \approx 1.4 - 3.6$m.

From Eq. (7.43), it follows that for $\omega \ll \omega_c$ gyroabsorption of whistlers on the electrons of thermal plasma is negligible and the minimum instability threshold is determined by the Coulomb-collision damping with the decrement

$$\nu_{coll} = \chi\nu_{ei}, \quad \nu_{ei} \approx 60 n_0 T^{-3/2} \tag{7.44}$$

Comparing Eqs. (7.42) and (7.44), we obtain, for $A \approx 0.1$, $n_0 \approx 7 \times 10^9\ cm^{-3}$, and $T = 2 \times 10^6$ K, the instability threshold for the density of fast particles $(n_1/n_0)_{min} > \nu_{ei}\chi/2.6A\omega_c \approx 2 \times 10^{-9}$, which is far below the relative density of $\geq$0.5 MeV electrons, $n_1/n_0 \approx 4.3 \times 10^{-7}$.

In addition to cyclotron and Coulomb-collision damping of the whistlers, there is Landau damping, which can break off the strong diffusion regime in a loop [59]. Indeed, whistlers propagating parallel to the magnetic field do not undergo Landau damping, since the longitudinal component of the electric field of the waves is equal to zero. However, propagating in the curved magnetic field of the loop, whistlers generate a small electric field parallel to $\boldsymbol{B}$, and therefore Landau damping can occur, with the rate determined by Eq. (6.34). In our case, $x = 1.3$, thus $\Phi(x) \approx 0.3$ (see Chapter 6 [30]). Comparing Eqs. (7.42) and (6.34) with $A \approx 0.1$, $x \approx 10^{-2}$, we obtain the instability threshold against Landau damping:

$$n_1/n_0 \approx 5 \times 10^{-5} \sin^2\vartheta \tag{7.45}$$

Substituting the previous estimate for $n_1/n_0 \approx 4.3 \times 10^{-7}$ into Eq. (7.45), we concluded that low-frequency whistlers do not undergo Landau damping within the domain $\vartheta \approx 5.3^\circ$. With the loop length $l \approx 7 \times 10^9$ cm, from simple geometrical considerations, we can find the minimum loop thickness required to avoid Landau damping of whistlers: $\approx 3 \times 10^8$ cm. The extended source in the flare on 28 August 1999, with the thickness of about 10^9 cm, does not indicate high consistency with this condition. Thus, the whistlers undergo strong Landau damping outside the narrow domain $\vartheta < 5^\circ$, that is, their amplitude may increase when they propagate only almost parallel to the magnetic field.

What is the way out? There are indications for the fine (subarcsecond) structure of solar coronal loops along the loop radius. In order to explain the close-to-isothermal appearance of a loop in TRACE observations, Reale and Peres [68] proposed the multithread model. The multithread concept was applied by Aschwanden *et al.* [69] to the study of nonuniform heating of coronal loops, based on *TRACE* data; the authors suggested that the loop filaments displayed the typical cross section of about 0.1 arcseconds. Seismological evidence of subresolution, field-aligned structuring of the corona was found in [70, 71]. The mechanism for the loop plasma heating proposed by Zaitsev and Shibasaki [72] (see Chapter 9) also requires filamentation of the loop with the cross-sectional scale $\sim 10^6$ cm. Filaments with the temperature and density stratification are in fact ducts (wave guides) for whistler waves. The whistlers propagate in the ducts along the magnetic field without substantial Landau damping, and the lower instability threshold is determined by Coulomb collisions: $n_1/n_0 > 2 \times 10^{-9}$. Previously, we found that whistlers with a phase speed of $c/30$ are consistent both with the resonance condition and with the dispersion equation if $y \approx 3$. This means that the plasma density in the loop should be about 2×10^{10} cm^{-3}, which exceeds the preflare density by the factor of more than 3. Loop filamentation may provide these conditions.

The total energy of the injected electron cloud in the case (A) is about $n_1(\pi r^2 l)E \approx 3 \times 10^{25}$ erg. We see that this population of electrons contains a small part of the thermal plasma energy in the loop, $n_1 E/n_0 \kappa_B T \approx 2 \times 10^{-3}$. To check the strong diffusion model, recall that the lifetime of trapped electrons in a loop $\tau_{\rm strong} \approx \beta^* \tau_0$ must exceed the particle lifetime in the weak diffusion regime $\tau_{\rm weak} \approx \sigma \tau_0$ [54]. Here, $\beta^* = 8\pi n_0 E/B^2$. With $n_0 = 10^{10}$ cm^{-3}, $E = 1$ MeV, and $B = 150$ G, the result is $\beta^* \approx 20 \gg \sigma$. Thus, the strong diffusion suggestion is satisfied.

Note that the high propagation speed ($\approx c$) of the microwave source, which was detected at 00:56:19.5 UT (line "B" in Figure 7.6), indicates that physical conditions in the flare loop differ from those in the case "A." The conditions in the case "B" were unfavorable for whistler excitation. Indeed, the second injection started when the loss cone was filled by precipitating electrons, and therefore the inequality $A > \omega/kc$ in Eq. (7.41) might not have been satisfied, and the turbulent "wall" in the case "B" might not have been formed. Moreover, the second injection (B) might have been weaker, $J < J_*$, and electrons propagated with the speed of light.

The physical reasons for the considered "turbulent expansion" are as follows:

1) Electrons in a loop must be relativistic to produce gyrosynchrotron emission at 17 and 34 GHz.
2) The same population of electrons generates low-frequency whistlers due to anisotropy.
3) The strong diffusion conditions are satisfied: the injection power is sufficient, $J > J^* = 2 \times 10^{11}$ cm^{-2} s^{-1}, and $\beta^* \gg \sigma$.
4) To avoid Landau damping of the whistlers, the loop has to consist of filaments (ducts).

We conclude that the solar event of 28 August 1999 illustrates collective effects of interaction of high-energy particles with small-scale turbulence in astrophysical plasma.

7.2.2.2 Turbulent Propagation of Superthermal Ions: the Absence of Linear Polarization in Hα Emission of a Solar Flare

In Chapter 6, it was shown that anisotropic superthermal ions in coronal loops are responsible for the generation of low-frequency ($\omega \ll \omega_i$) Alfvén waves. Similar to the case of electron–whistler interaction, superthermal population of ions interacting with Alfvén waves can explain the peculiarities in particle propagation and in flare emission. Here we consider the Alfvén wave effect on the polarization of Hα emission from solar flares, which is produced by accelerated ions after their precipitation in the solar chromosphere.

Linear polarization up to 20% in Hα emission of solar flares has been reported in numerous papers. The linear impact polarization can be caused by a beam of low-energy ($\leq$100 keV) protons precipitating in the upper chromosphere. In high-sensitivity observations of Hα polarization of 30 flares with *Zürich IMaging POLarimeter* (*ZIMPOL*), Bianda *et al.* [73] did not detect any indications on polarization exceeding 0.07%. The absence of linear polarization in Hα emission means that the distribution of precipitating particles should be isotropic.

The degree of linear impact polarization of atoms excited by collisions with a beam of particles that forms an angle Θ with the line of sight is [74]

$$P(\Theta) = \frac{I_{||} - I_{\perp}}{I_{||} + I_{\perp}} \tag{7.46}$$

where $I_{||}$ and $I_{\perp}$ are the "parallel" and "perpendicular" emission intensities with respect to the plane specified by the beam axis and the line of sight. For a symmetric (relative to the beam axis) distribution function $f_1(v, \cos\theta)$, where θ is the pitch angle,

$$P(\Theta) = P(90^{\circ})\frac{\sin^2\Theta}{1 - P(90^{\circ})\cos^2\Theta}, \qquad P(90^{\circ}) \approx \frac{\int_{v_m}^{\infty} P(90^{\circ}, E)(3J_2 - J_0)v\sigma(v)dv}{\int_{v_m}^{\infty} 2J_0 v\sigma(v)dv} \tag{7.47}$$

Here, v_m is the minimum velocity, $P(90^{\circ}, E)$ is the maximum degree of polarization corresponding to $\Theta = 90^{\circ}$, $\sigma(v)$ is the excitation cross section, E is the particle energy, and

$$J_l(v) = -\int_0^{\pi} f_1(v, \cos\theta)\cos^l\theta d\cos\theta \tag{7.48}$$

In particular, it follows from Eqs. (7.47) and (7.48) that the observed degree of polarization is $P(\Theta) = 0$ for an isotropic particle distribution, $\partial f_1/\partial(\cos\theta) = 0$.

Bianda *et al.* [73] suggested the following reasons for the absence of linear polarization: (i) instability of Alfvén waves excited by energetic protons and isotropization of the protons on the waves; (ii) isotropization due to proton–neutral collisions; and (iii) defocusing by a converging magnetic field. The authors of [73] suggested that factors (ii) and (iii) are the most important. However, Tsap and Stepanov [75] concluded that under solar chromosphere condition, factor (i) may be the most significant.

Hénoux *et al.* [76] were the first to notice that collisions of accelerated protons with background loop plasma make the distribution function for the protons more anisotropic with a maximum along the magnetic field $\boldsymbol{B}$. This nontrivial phenomenon can be explained both by the negligible proton scattering in the "cold-target" approximation and by an increase in their collisional losses with the increase in the pitch angle θ.

As for the magnetic isotropization, in [71], it was mentioned that, since the magnetic moment of accelerated protons $\mu = m_i v_\perp^2/2B$ is conserved, an increase in the magnetic field strength with the depth does not sufficiently affect the pitch-angle distribution of the beam in the chromosphere. Indeed, from Eq. (6.6), it follows that the pitch angle θ is related to the local magnetic field B by

$$\theta = \arcsin\left(\sin\theta^*\sqrt{\frac{B}{B^*}}\right)$$

where index * refers to the initial value at the release from the injection or acceleration region. For example, an increase by a factor of 4 in the magnetic field from the acceleration site to the Hα-emitting region would approximately double a small pitch angle. The proton beam widens the velocity space by the same factor. Thus, the magnetic isotropization is insufficient to explain the absence of linear polarization.

Here, we consider the interaction of Alfvén waves with 10–100 keV protons (i.e., with the velocity $(1.4\text{–}4.4) \times 10^8\ \mathrm{cm\,s^{-1}} \gg V_{\mathrm{Ti}} \approx 10^6\ \mathrm{cm\,s^{-1}}$), since at the corresponding energies the hydrogen levels are excited most effectively [77]. It is suggested (see, i.e., [78]) that Alfvén waves cannot be excited by energetic protons in the chromosphere due to strong damping of waves in the dense partially ionized plasma.

Let us consider this problem in more detail. If the frequency ω of Alfvén waves substantially exceeds the effective frequencies of electron (ν_{ea}) and ion (ν_{ia}) collisions with neutrals, then the damping rate is [79]

$$\nu_{\mathrm{d}} = \frac{1}{2}\left(\frac{m_{\mathrm{e}}}{m_{\mathrm{i}}}\nu_{\mathrm{ea}} + \nu_{\mathrm{ia}}\right) \tag{7.49}$$

Under solar chromosphere conditions, ν_{ea} and ν_{ia} are determined by elastic collisions, and therefore [80]

$$\nu_{\mathrm{ea}} \approx 5 \times 10^{-10} n_{\mathrm{a}}\sqrt{T}\mathrm{s}^{-1}, \quad \nu_{\mathrm{ia}} \approx 10^{-10} n_{\mathrm{a}}\sqrt{T}\mathrm{s}^{-1} \tag{7.50}$$

From Eq. (6.45) for the instability growth rate, we can obtain the formula, similar to that presented in [57]:

$$\gamma_{\mathrm{A}} \approx \omega_{\mathrm{i}}\frac{\omega_{\mathrm{i}}}{\omega}\frac{n_1}{n_0} \tag{7.51}$$

From Eqs. (7.50) and (7.51), it follows that the condition for the excitation of Alfvén waves by protons in the upper chromospheres is

$$\frac{n_1}{n_0} > \frac{v_{\rm d}}{\omega_{\rm i}}\frac{\omega}{\omega_{\rm i}} \approx 5\times 10^{-15}\frac{n_0\sqrt{T}}{B}\frac{\omega}{\omega_{\rm i}} \tag{7.52}$$

For $T = 10^4$ K, $n_0 = 10^{11}{-}10^{12}$ cm^{-3}, $B = 500$ G, and $\omega/\omega_{\rm i} = 0.1$, from Eq. (7.52), we obtain $n_1/n_0 > 10^{-5}$ to 10^{-4}. This suggests that Alfvén waves can be effectively excited in the upper solar chromosphere even with moderate fluxes of accelerated protons.

It follows from the resonance conditions and from the dispersion relation for Alfvén waves that the effective proton pitch-angle scattering will occur only if the proton velocity $v > c_{\rm A}$ [57], that is, $E_{\rm keV} > E^* \approx 2.5\times 10^7 B^2/n_0$. Since the threshold energy $E^* = 2.3{-}62.5$ keV at $n_0 = 10^{11}{-}10^{12}$ cm^{-3}, $B = 300{-}500$ G, Alfvén waves can be effectively excited by ≤ 100 keV protons.

Alfvén turbulence is capable of governing the pitch-angle distribution of the accelerated protons only if the timescale of particle diffusion on waves $\tau_{\rm D} \approx (B/\delta B)^2\,\omega_{\rm i}^{-1}$ [57], where δB, which is the amplitude of Alfvén waves, is shorter than the collision time $\tau_{\rm coll}$. Thereby, the following inequality holds:

$$\left(\frac{\delta B}{B}\right)^2 > \frac{1}{\omega_{\rm i}\tau_{\rm coll}} = \frac{l_{\rm f}}{v\omega_{\rm i}} \tag{7.53}$$

The mean free path for protons with the energy $E = 10{-}100$ keV, that is, with the velocity $(1.4{-}4.4)\times 10^8$ cm s^{-1} is [79]

$$l_{\rm f} \approx 4.2\times 10^{15}\frac{E^2}{n_0\ln\Lambda + n_{\rm a}\ln(0.145E)}\,{\rm cm} \tag{7.54}$$

Here, n_0 and $n_{\rm a}$ are the number densities of electrons and atoms in cm^{-3}, $\ln\Lambda = \ln\left[5.2\times 10^7 (T/n_0)^{1/2}E\right]$ is the Coulomb logarithm, and the energy E and the temperature T are given in kiloelectronvolts and kelvin, respectively. For example, taking in Eq. (7.54) $E = 50$ keV, $n_0 = 10^{11}{-}10^{12}$ cm^{-3}, $n_0 \approx n_{\rm a}$, we obtain $l_{\rm f} = 10^6{-}10^7$ cm. Setting in Eq. (7.53) $v = 4\times 10^8$ cm s^{-1}, $B = 500$ G, and $l_{\rm f} = 10^6{-}10^7$ cm, we obtain $(\delta B/B)^2 > 6\times(10^{-6}{-}10^{-5})$, and the minimum wave energy density $W_{\rm min} = \delta B^2/8\pi \approx 6\times(10^{-2}{-}10^{-1})$ erg cm^{-3}. On the other hand, for the average energy of $\sim$100 keV protons and the number density $n_1 = 10^7{-}10^8$ cm^{-3}, the total energy of protons is $E_{\rm tot} \approx n_1 E \approx 2{-}20$ erg cm^{-3}. Since $E_{\rm tot} \gg W_{\rm min}$, the necessary level of Alfvén turbulence can be easily achieved in the upper chromosphere.

Nevertheless, the necessary condition for strong diffusion (Eq. (7.28)) requires $\tau_{\rm D} < \tau_0 = l/v$, where l is the distance from the acceleration site to the Hα emission region. Assuming $(\delta B/B)^2 \approx 10^{-5}$, $v = 4\times 10^8$ cm s^{-1}, $B = 500$ G, and $l = 4\times 10^8$ cm, we obtain $\tau_{\rm D} \approx 0.02s \ll \tau_0 \approx 1$s.

This is the necessary condition for strong diffusion. The sufficient condition (Eq. (7.31)) requires $J > J_* \approx 10^{13}$ pr cm^{-2} s^{-1}. The acceleration rate for <1 MeV protons in solar flares ranges from 10^{32} to 10^{34} pr s^{-1} [1]. For the loop cross section near the footpoints $\approx 10^{17}$ cm^2, it yields $J \approx (10^{15}{-}10^{17})$ pr cm^{-2} s$^{-1} \gg J^*$. Hence, Alfvén wave turbulence is the most probable reason for proton isotropization.

Note that the linear polarization is absent not only in the strong diffusion case but also in the regime of intermediate diffusion, since the particle distribution function is close to isotropic. In the case of a weak particle source $J < J_*$ and a low level of Alfvén wave turbulence $(\delta B/B)^2 < 6 \times 10^{-6}$, strong diffusion does not occur, and Hα emission is linearly polarized.

7.2.2.3 Time Delays in Hard X-Ray and γ-Ray Emission

If the flare energy releases in the coronal part of a magnetic loop, then >10 keV electrons move from the acceleration region down to the loop footpoints, where a thick target is located, and hard X-ray emission is generated. Interaction of protons and heavy ions accelerated to energies ≥10 MeV per nuclear with dense layers of the solar atmosphere leads to γ-ray line emission. Therefore, observations of hard X-ray and γ-ray line emission provide important information on processes of electrons and acceleration of ions and on their dynamics and propagation in the flare.

In our opinion, the most challenging and puzzling data are those concerning the dependence of the delays in the time profiles of electromagnetic emission on the photon energy. The data obtained with Solar Maximum Mission (SMM) and Yohkoh space observatories indicate that in some flares the peaks of γ-ray line emission in the 4.1–4.6 MeV band are delayed with respect to the corresponding X-ray peaks in 300 keV flux by 1–45 s (see, i.e., [81, 82]). Incidentally, impulse flares show short peak delays, while gradual flares display long delays. Figure 7.8 presents an example of such delays in an impulse flare.

Such delays are usually explained in terms of the two-step acceleration model: in the first step, primary nonrelativistic electrons are accelerated, and in the second step, ions and relativistic electrons are accelerated [83]. This interpretation, however, contradicts Chupp's idea that electrons and ions are accelerated simultaneously [81]. Whether ions and electrons are really accelerated simultaneously is crucial for understanding the physics of flare energy release. This raises the question whether it is theoretically possible for the time of propagation of energetic particles from the acceleration site (the coronal part of a loop) to the region of X-ray and γ-ray emission (footpoints) to increase significantly.

Bespalov *et al.* [84] proposed the "trap plus turbulent propagation" model, which to a certain extent explains this X-ray/γ-ray delay effect (Figure 7.9). The essential point here is that the study of energetic particle dynamics in stellar magnetic loops cannot be limited to Coulomb collisions but should consider particle interaction with electromagnetic turbulence. Under solar conditions, in this context, the most important factor for fast electrons is the whistler turbulence, and for energetic ions the small-scale Alfvén wave turbulence.

From Eqs. (7.32) and (7.38) for the nonsteady-state case, the strong diffusion suggests that the particle propagation velocity is close to the phase velocity (whistler and Alfvén waves). The proton-to-electron propagation time ratio within the magnetic loop is $\tau^{\mathrm{i}}_{\mathrm{pr}}/\tau^{\mathrm{e}}_{\mathrm{pr}} \approx m_{\mathrm{i}} c_{\mathrm{A}}/m_{\mathrm{e}} v \gg 1$. For example, with the travel distance $l \approx 10^9$ cm, $c_{\mathrm{A}} \approx 10^8\ \mathrm{cm\,s^{-1}}$, $\sigma \approx 4$, $v \approx 10^{10}\ \mathrm{cm\,s^{-1}}$, from Eqs. (7.32) and (7.38), we obtain the delay times $\Delta t_{\mathrm{X}} \approx \tau^{\mathrm{e}}_{\mathrm{pr}} \approx 2$ s and $\Delta t_{\gamma} \approx \tau^{\mathrm{i}}_{\mathrm{pr}} \approx 20$ s.

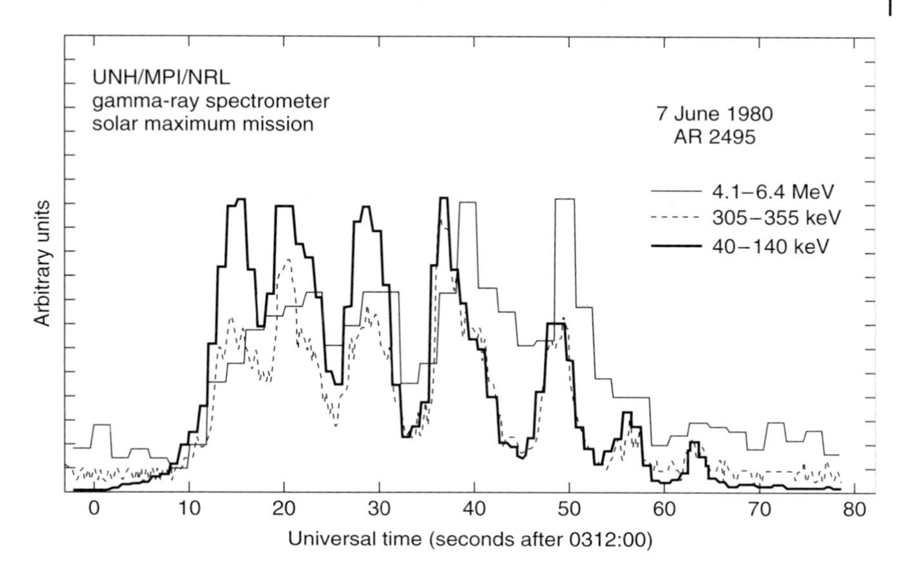

Figure 7.8 Time profiles of hard X-ray (40–140 and 305–355 ke V) emission and γ-ray line emission (4.1–4.6 Me V) for the solar flare on 7 June 1980 [81]. Two- to four-second delays of γ-ray peaks with respect to X-ray peaks are most distinctly seen for the fourth and fifth pulses.

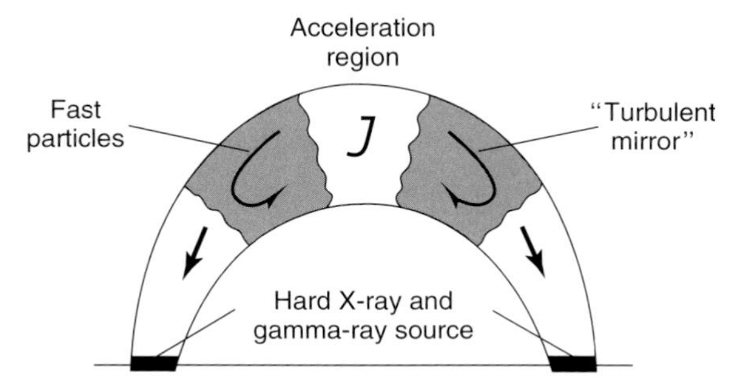

Figure 7.9 The trap plus turbulent propagation model. Waves scatter energetic particles and decelerate their propagation. The large arrows indicate the motion of "turbulent mirrors" at the speed compared with the wave phase velocity. J is the source of accelerated particles. Sources of X-ray and γ-ray emission are located at the footpoints of the flare loop [84].

Another problem of the theory of solar flares is associated with the interpretation of the lack of an abundant population of 10–100 keV electrons in particle fluxes observed at the Earth's orbit, while the hard X-ray emission provides evidence that a considerable number of such electrons are accelerated [85]. It is not excluded that such behavior of the particles can also be understood in terms of strong pitch-angle diffusion. It is known that energetic particles can escape from the coronal magnetic trap and move away into the interplanetary space, provided the gas pressure of the background plasma and superthermal particles p_Σ is sufficiently high. When $\beta = 8\pi p_\Sigma / B^2 \geqslant 0.3$, a flute instability develops, and energetic particles with the plasma and frozen-in magnetic field leave the flare loop on to open magnetic field lines. The flute instability timescale is $t_{fl} \approx \sqrt{l r_{loop}} / V_{Ti} \approx 10\,s$ for the radius of curvature of the magnetic field in the loop $r_{loop} \approx 10^8\,cm$, $l \approx 10^9\,cm$, and $V_{Ti}(T = 10^7\,K) \approx 3 \times 10^7 cm\,s^{-1}$. It was shown above that "turbulent mirrors" for fast electrons are less effective than those for protons. It seems plausible that energetic electrons precipitate – within a time shorter than t_{fl} – into the footpoints, but energetic protons will be locked by "turbulent mirrors" and escape after $t > t_{fl}$ into the interplanetary space. Figure 7.10 illustrates this situation. This might be the reason for the deficit (0.1–1%) in the number of particles accelerated by a flare of 10–100 keV electrons in proton-reach events in the interplanetary space [86]. Note that Figure 7.10 can be applied to impulse flares. For gradual γ-ray/proton flares, protons can additionally accelerated by coronal shock waves [87].

7.2.2.4 **Transformation of Energetic Particle Spectra in a Coronal Loop**

Owing to wave–particle interaction, the spectrum of energetic particles that generate hard X-ray and γ-ray emission can be different from the original spectrum of particles in the accelerated domain. Bespalov and Zaitsev [88] showed that energetic particle spectra do not depend on the spectrum of the particle source $j(E)$ under the condition of weak diffusion. These spectra are universal and depend only on peculiarities of whistler and Alfvén wave damping $\nu \propto k^q$, where k is the wave number and the value of q is determined by the type of turbulence. This is also true in the case of intermediate diffusion for the high-energy part of the accelerated particles. However, the particle energy spectrum in low energetic channels can be very close to the source spectrum. Thus, the formation of a "knee" in the particle spectrum is possible.

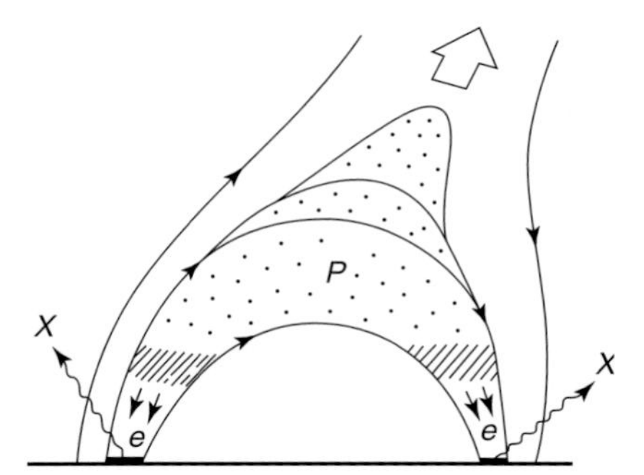

Figure 7.10 A flare loop undergoes flute instability at time $t_{fl} < t < \tau^{i}_{pr}$. The dashed region indicates the "turbulent mirrors." Thin arrows denote fluxes of precipitating fast electrons. Energetic protons are locked in the flare loop due to Alfvén turbulent mirrors and can escape into the interplanetary space. Whistler turbulent mirrors are less effective ($t_{pr}{}^{e} < t_{fl}$); hence, the energetic electrons precipitate into the footpoints and cause hard X-ray emission.

The situation is different in the case of strong pitch-angle diffusion. For steady-state conditions, we can substitute $-f/\tau_l$ instead of the diffusion term in Eq. (7.29), and the particle distribution can be written in the form:

$$f(E) \approx \tau_l j(E) \tag{7.55}$$

Equation (7.55) shows that the ion spectrum within a flare loop coincides with the source spectrum, since the mean lifetime of ions (Eq. (7.38)) does not depend on the particle energy. From Eq. (7.32), we obtain for electrons $\tau_l^e \propto E^{1/2}$. This means that the electron energy spectrum is harder than the source spectrum. The similarity between the ion spectrum and source spectrum is very important for the diagnostics of ion characteristics in the acceleration site, as well as for understanding the mechanism by which the corona and interplanetary medium affect the shape of the proton spectrum near the Earth.

References

1. Miller, J.A., Cargill, P.J., Emslie, A.G. *et al.* (1997) *J. Geophys. Res.*, **102**, 14631.
2. Emslie, A.G., Brown, J.C., and Machado, M.E. (1981) *Astrophys. J.*, **246**, 337.
3. McClymont, A.N. and Canfield, R.C. (1986) *Astrophys. J.*, **305**, 936.
4. Mariska, J.T., Emslie, A.G., and Li, P. (1989) *Astrophys. J.*, **341**, 1067.
5. Holman, G.D. (1985) *Astrophys. J.*, **293**, 584.
6. Kaplan, S.A. and Tsytovich, V.N. (1973) *Plasma Astrophysics*, Pergamon Press, Oxford.
7. Bai, T., Hudson, H.S., Pelling, R.M. *et al.* (1983) *Astrophys. J.*, **267**, 433.
8. Somov, B.V. and Kosugi, T. (1997) *Astrophys. J.*, **485**, 859.
9. Gubchenko, V.M. and Zaitsev, V.V. (1979) *Sol. Phys.*, **63**, 337.
10. Zaitsev, V.V. and Ledenev, V.G. (1976) *Sov. Astron. Lett.*, **2**, 443.
11. Alfvén, H. and Fälthammar, C.-G. (1963) *Cosmical Electrodynamics: Fundamental Principles*, Clarendon Press, Oxford, 228 p.
12. Kujipers, J., Fletcher, L., Abada-Simon, M. *et al.* (1997) *Astron. Astrophys.*, **322**, 242.
13. Karlicky, M. and Kosugi, T. (2004) *Astron. Astrophys.*, **419**, 1159.
14. Benz, A. (1993) *Plasma Astrophysics. Kinetic Processes in Solar and Stellar Coronae*, Kluwer Academic Publishers, Dordrecht, Boston, London.
15. Aschwanden, M.J. (2005) *Physics of the Solar Corona. An Introduction with Problems and Solutions*, Springer.
16. Zaitsev, V.V. and Stepanov, A.V. (1991) *Sov. Astron.*, **35**, 189.
17. Zaitsev, V.V. and Stepanov, A.V. (1992) *Sol. Phys.*, **139**, 343.
18. Knoepfel, H. and Spong, D.A. (1979) *Nucl. Fusion*, **19**, 785.
19. Zaitsev, V.V. and Khodachenko, M.L. (1997) *Radiophys. Quantum Electron.*, **40**, 114.
20. Lin, R.P., Krucker, S., Hurford, G.J., Smith, D.M., Hudson, H.S. *et al.* (2003) *Astrophys. J.*, **595**, L69.
21. Brown, J.C. (1973) *Sol. Phys.*, **29**, 421.
22. Holman, G.D. and Benka, S.G. (1992) *Atrophys. J.*, **400**, 79.
23. Aurass, H., Klein, K.-L., Zlotnik, E.Y., and Zaitsev, V.V. (2003) *Astron. Astrophys.*, **410**, 1001.
24. Fleishman, G.D., Bastian, T.S., and Gary, D.E. (2008) *Astrophys. J.*, **684**, 1433.
25. Zaitsev, V.V., Zlotnik, E.Y., and Aurass, H. (2005) *Astron. Lett.*, **31**, 285.
26. Bastian, T.S., Bookbinder, J.A., Dulk, G.A., and Davis, M. (1990) *Astrophys. J.*, **353**, 265.
27. Stepanov, A.V., Kliem, B., Zaitsev, V.V. *et al.* (2001) *Astron. Astrophys.*, **374**, 1072.

28. Güdel, M., Benz, A.O., Bastian, T.S., and Furst, E. (1989) *Astron. Astrophys.*, **220**, L5.
29. Abada-Simon, M., Lecacheux, A., Aubier, M., and Bookbinder, J.A. (1997) *Astron. Astrophys.*, **321**, 841.
30. Zaitsev, V.V., Kislyakov, A.G., Stepanov, A.V., Kliem, B., and Furst, E. (2004) *Astron. Lett.*, **30**, 319.
31. Melrose, D.B. (1991) *Astrophys. J.*, **381**, 306.
32. Colgate, S.A. (1978) *Astrophys. J.*, **221**, 1068.
33. Melrose, D.B. (1995) *Astrophys. J.*, **451**, 391.
34. Van der Oord, G.H.J. (1990) *Astron. Astrophys.*, **234**, 496.
35. Fadeev, A.A., Kvartskhava, I.F., and Komarov, N.N. (1965) *Nucl. Fusion*, **5**, 2002.
36. Hammer, D.A. and Rostoker, N. (1970) *Phys. Fluids*, **13**, 1831.
37. Cox, J.D. and Bennet, W.N. (1970) *Phys. Fluids*, **13**, 182.
38. Lee, R. and Sudan, R.N. (1971) *Phys. Fluids*, **14**, 1213.
39. Lovelace, R.V. and Sudan, R.N. (1971) *Phys. Rev. Lett.*, **16**, 1472.
40. Cowling, T.G. (1957) *Magnetohyrrodynamics*, Interscience Publication, London.
41. Schlüter, A. and Biermann, L. (1950) *Z. Naturforsch.*, **5A**, 237.
42. Pikelner, S.B. (1964) *Fundamentals of Cosmic Electrodynamics*, NASA, Washington, DC.
43. Zaitsev, V.V., Urpo, S., and Stepanov, A.V. (2000) *Astron. Astrophys.*, **357**, 1105.
44. Wheatland, M.S. and Melrose, D.B. (1995) *Sol. Phys.*, **159**, 137.
45. Priest, E.R. (1982) *Solar Magnetohydrodynamics*, D. Reidel Publishing Company, Dordrecht, Holland.
46. Henoux, J.C. and Somov, B.V. (1991) *Astron. Astrophys.*, **241**, 613.
47. Melrose, D.B. and Brown, J.C. (1976) *Mon. Not. R. Astron. Soc.*, **176**, 15.
48. MacKinnon, A.L. (1991) *Astron. Astrophys.*, **242**, 256.
49. Fletcher, L. (1997) *Astron. Astrophys.*, **326**, 1259.
50. Aschwanden, M.J. (2002) *Space Sci. Rev.*, **101**, 1.
51. Aschwanden, M.J. and Schwartz, R.A. (1996) *Astrophys. J.*, **464**, 974.
52. Kennel, C.F. (1969) *Rev. Geophys.*, **7**, 379.
53. Melrose, D.B. (1986) *Instabilities in Space and Laboratory Plasmas*, Cambridge University Press, 237 p.
54. Trakhtengertz, V.Y. (1984) Relaxation of Plasma with Anisotropic Velocity Distribution in *Basic Plasma Physics: Selected Chapters, Handbook of Plasma Physics*, vol. **1** (eds A.A. Galeev and R.N. Sudan), North-Holland Physics Publishing, 519 p.
55. Bespalov, P.A. and Trakhtengertz, V.Y. (1986) *Rev. Plasma Phys.*, **10**, 155.
56. Budker, G.I., Mirnov, V.I., and Rytov, D.D. (1971) *Pis'ma ZhETF*, **14**, 320.
57. Kennel, C.F. and Petschek, H.E. (1966) *J. Geophys. Res.*, **71**, 1.
58. O'Brien, B.J. (1962) *J. Geophys. Res.*, **67**, 3687.
59. Bespalov, P.A., Zaitsev, V.V., and Stepanov, A.V. (1991) *Astrophys. J.*, **375**, 369.
60. Stepanov, A.V., Yokoyama, T., Shibasaki, K., and Melnikov, V.F. (2007) *Astron. Astrophys.*, **465**, 613.
61. Wentzel, D.G. (1976) *Astrophys. J.*, **208**, 595.
62. Stepanov, A.V. and Tsap, Y.T. (2002) *Sol. Phys.*, **211**, 135.
63. Krucker, S., Battaglia, M., Cargill, P.J. *et al.* (2008) *Astron. Astrophys. Rev.*, **16**, 155.
64. Ramaty, R. (1969) *Astrophys. J.*, **158**, 753.
65. Brambilla, M. (1998) *Kinetic Theory of Plasma Waves: Homogenous Plasma*, Clarendon Press, Oxford.
66. Alpert, Y.L. (1982) *The Near-Earth and Iinterplanetary Plasma*, vol. **1**, Cambridge University Press.
67. Melrose, D.B. (1980) *Plasma Astrophysics*, vol. **2**, Gordon & Breach.
68. Reale, F. and Peres, G. (2000) *Astrophys. J.*, **528**, 753.
69. Aschwanden, M.J., Nightingel, R.W., and Alexander, D. (2000) *Astrophys. J.*, **541**, 1059.
70. King, D.B., Nakariakov, V.M., Deluca, E.E., Golub, L., and McClements, K.G. (2003) *Astron. Astrophys.*, **404**, L1.

71. Van Doorsselaere, T., Brady, C.S., Verwichte, E., and Nakariakov, V.M. (2008) *Astron. Astrophys.*, **491**, L9.
72. Zaitsev, V.V. and Shibasaki, K. (2005) *Astron. Rep.*, **49**, 1009.
73. Bianda, M., Benz, A.O., Stenflo, J.O. *et al.* (2005) *Astron. Astrophys.*, **434**, 1183.
74. Vogt, E. and Hénoux, J.-C. (1999) *Astron. Astrophys.*, **349**, 283.
75. Tsap, Y.T. and Stepanov, A.V. (2008) *Astron. Lett.*, **34**, 52.
76. Hénoux, J.-C., Chambe, G., Smith, D. *et al.* (1990) *Astrophys. J. Suppl.*, **73**, 303.
77. Fletcher, L. and Brown, J.C. (1995) *Astron. Astrophys.*, **294**, 260.
78. Hua, X.-M., Ramaty, R., and Lingenfelter, R.E. (1989) *Astrophys. J.*, **341**, 516.
79. Ginzburg, V.L. (1970) *The Propagation of Electromagnetic Waves in Plasmas*, Pergamon Press, Oxford.
80. Leake, J.E., Arber, T.D., and Kchodachenko, M.L. (2005) *Astron. Astrophys.*, **442**, 1091.
81. Chupp, E.L. (1983) *Sol. Phys.*, **86**, 383.
82. Yoshimori, M. (1984) *J. Phys. Soc. Jpn.*, **53**, 4499.
83. Bai, T. and Ramaty, R. (1979) *Astrophys. J.*, **227**, 1072.
84. Bespalov, P.A., Zaitsev, V.V., and Stepanov, A.V. (1987) *Sol. Phys.*, **114**, 127.
85. Lin, R.P. and Hudson, H.S. (1976) *Sol. Phys.*, **50**, 153.
86. Lin, R.P. (1974) *Space Sci. Rev.*, **16**, 189.
87. Bai, T. (1986) *Astrophys. J.*, **308**, 912.
88. Bespalov, P.A. and Zaitsev, V.V. (1987) Formation of Energetic Proton Spectra in Solar Flares in *Solar Maximum Analysis* (eds V.E., Stepanov and V.N., Obridko), VNU, Dordrecht, p. 131.

Further Reading

Orrall, F.Q. and Zirker, J.B. (1976) *Astrophys. J.*, **208**, 618.

8 Stellar Coronal Seismology as a Diagnostic Tool for Flare Plasma

As it was mentioned in the introduction (Chapter 1), the coronal seismology is the quite powerful diagnostic tool for the parameters of stellar flaring loops. Two illustrations of diagnostics of solar coronal loop were described in Chapters 3 and 6. The acoustic damping of sausage modes was used for the pulsing radio flare event on 16 May 1973. The plasma density in the loop $n_0 \approx 6.6 \times 10^8$ cm^{-3}, the density ratio inside and outside loop were obtained $n_0/n_e \approx 50$, and a height above photosphere of the oscillating source was founded, $h \approx 3.6 \times 10^{10}$ cm($\approx 0.52\,R_\odot$). The second example (Chapter 6) dealt with the optically thin and optically thick radio sources pulsating in antiphase due to the sausage oscillations. The spectral index of the emitting electrons $\delta \approx 4.4$ was estimated, and the magnetic field value in the flaring loop $B \approx 190$ G was determined using formulas for the gyrosynchrotron radiation. This section is devoted to the diagnostics of the parameters of solar and stellar flaring loop plasma using both nonleaky and leaky magnetohydrodynamic (MHD) mode approaches as well as to the diagnostics of electric currents in stellar atmosphere.

8.1 Modulation of Flaring Emission by MHD Oscillations

Several reasons exist for periodic modulations of electromagnetic emission by an oscillating coronal loop containing thermal background plasma and accelerated particles. The example of modulation of the radio emission propagated through the coronal MHD resonator owing to variations of the optical thickness τ_ν driven by periodical variations of the plasma density was considered in Chapter 3. Radio and X-ray emission from high-energy electrons in a loop will also be modulated due to periodic variations of background plasma parameters (magnetic field, gas pressure, and density) as well as because the periodic changes of the loss-cone volume. If the coronal magnetic loop is unstable against various loss-cone instabilities, then both radiation from energetic particles and flux of precipitating high-energy particles toward the footpoints would be modulated. In the last case, this brings the modulation of optical, hard X-ray, and γ-ray emission.

Coronal Seismology: Waves and Oscillations in Stellar Coronae, First Edition.
A. V. Stepanov, V. V. Zaitsev, and V. M. Nakariakov.

8.1.1
Modulation of Gyrosynchrotron Emission

The quite suitable band for the study of the variations of parameters of stellar flaring plasma by MHD waves and oscillations is the microwave band, as modern microwave observations have excellent time, spectral, and spatial resolution. Consider once more the modulation of the microwave emission produced by mildly relativistic electrons by MHD oscillations via the gyrosynchrotron mechanism. Following the simplified Eqs. (6.12) and (6.13) [1], the dependence of the intensity of gyrosynchrotron emission on the magnetic field strength B for the power-law distribution function of electrons (Eq. (6.8a)) in optically thin ($\tau_\nu \ll 1$) and optically thick ($\tau_\nu \gg 1$) sources is

$$I_\nu \propto \begin{cases} B^{0.9\delta-0.22}, \tau_\nu \ll 1 \\ B^{-0.52-0.08\delta}, \tau_\nu \gg 1 \end{cases} \tag{8.1}$$

In the optically thin regime, the emission is very sensitive to the magnetic field, for example, $I_\nu \propto B^{3.4}$ for $\delta = 4$. Thus, the modulation depth of the gyrosynchrotron emission is several times deeper than the relative amplitude of the magnetic field perturbation causing the modulation. For any plausible value of δ, the increase in the magnetic field increases the radio intensity at $\tau_\nu \ll 1$ and decreases it at $\tau_\nu \gg 1$. In particular, a sausage oscillations would produce antiphase oscillations in optically thin ($\nu > \nu_{\rm peak}$) and optically thick ($\nu < \nu_{\rm peak}$) parts of gyrosynchrotron spectrum (see also Section 6.1.1). Here, $\nu_{\rm peak}$ is determined by Eq. (6.14) and usually $\nu_{\rm peak} \approx 10$ GHz for solar flaring loops.

Similarly, modulation of the intensity of gyrosynchrotron emission occurs due to the variation of the angle ϑ between the magnetic field and the line of sight, for example, by a kink mode or torsion waves. In general case, the modulation depth of the intensity can be defined as $M = |\Delta I_\nu(\vartheta)|/I_\nu(\vartheta)$, where $\Delta I_\nu(\vartheta) = I_\nu(\vartheta + \Delta\vartheta) - I_\nu(\vartheta)$. Expanding the $\Delta I_\nu(\vartheta)$ in a series of a small parameter $\Delta\vartheta$ and restricting ourselves to the first-order approximation, we obtain from Eq. (8.2) the modulation depths for the optically thin and optically thick cases [2]:

$$M_1 = (-0.43 + 0.65\delta)\cot\vartheta\,\Delta\vartheta,\, M_2 = (0.34 + 0.07\delta)\cot\vartheta\,\Delta\vartheta \tag{8.2}$$

In addition, in the optically thin regime, the degree of circular polarization is [1]

$$r_{\rm c} = 1.26 \times 10^{0.035\delta} 10^{-0.071\cos\vartheta} \kappa^{0.78-0.55\cos\vartheta}$$

where $\kappa = \nu/\nu_{\rm c} = 10–100$, $\nu_{\rm c} = eB/(2\pi mc)$. Then, we find that the modulation depth is

$$M_c = \frac{|\Delta r_c|}{r} = (0.545\ln\kappa - 0.163)\sin\vartheta\,\Delta\vartheta \tag{8.3}$$

Figure 8.1 shows the dependence of modulation depths on ϑ for optically thin and optically thick sources at $\delta = 3, 4, 5$. One can see that $M_1 > M_2$ and the modulation depth of the optically thin source can achieve sufficiently large values, $M_1 > 0.1$ at $\vartheta < 60^\circ$ and it decreases with an increase of ϑ. However, Figure 8.1b shows that the modulation depth of the polarization $M_{\rm c}$ increases with an increase of ϑ, but its value is smaller than M_1.

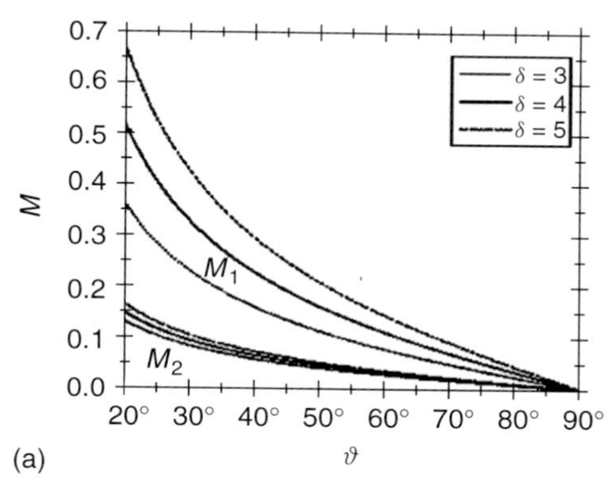

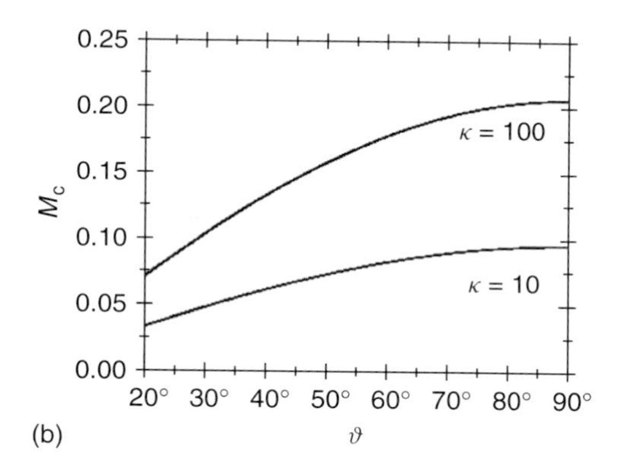

Figure 8.1 (a) Modulation depths for the optically thin (M_1) and optically thick (M_2) sources of gyrosynchrotron emission versus the angle ϑ between the direction of the magnetic field and the line of sight at different values of the electron spectral index δ. (b) The same dependence but for the degree of circular polarization M_c at different values of $\kappa = \nu/\nu_c$. The magnetic field direction is changed with the amplitude $\Delta\vartheta = 5^\circ$ [2].

8.1.2
Modulation of Plasma Radiation

Coronal loops can be sources of intense plasma radiation from solar and stellar flares [3–5]. Plasma radiation means a two-step process. At the first step, Langmuir waves are generated due to the beam or loss-cone instabilities. After that the nonlinear conversion of plasma waves into electromagnetic emission occurs due to Rayleigh and Raman scattering processes, which result in radio emission at the fundamental tone ($\approx \omega_p$) and at the second harmonic ($\approx 2\omega_p$), respectively. Therewith, the intensity of the second harmonic emission is one to two orders usually larger than the fundamental tone emission under solar and stellar conditions. Large coronal loops are the source of type IV solar radio emission at meter–decimeter wavelengths, whereas microwave plasma emission originates from compact and dense flaring loops, as described in Chapter 6.

Nonlinear transformation of plasma waves into electromagnetic ones as the result of Raman scattering is described by Eqs. (6.31) and (6.36)–(6.38). For the loss-cone instability of Langmuir waves, two cases should be considered: (i) weak source of accelerated particles when the loss cone is empty. This case corresponds to the weak diffusion regime $\tau_D > \sigma\tau_0 = \sigma l/v$. (ii) Sufficiently strong source of energetic particles when loss cone is almost full, which corresponds to the intermediate ($\tau_0 < \tau_D < \sigma\tau_0$) or strong ($\tau_D < \tau_0 = l/2v$) diffusion regimes.

The first case can be realized, for example, in large coronal solar arches ($n_0 \approx 10^8$ cm^{-3}, $n_1/n_0 \approx 10^{-6}$, $T_0 \approx 100$ eV) – the sources of pulsed type IV radio emission, with typical period of a few seconds, which is larger compared to $\tau_0 = l/2v \leqslant 0.3$ s. When the variations in the magnitude of the magnetic field are slow in comparison with the Larmor period, the magnetic moment $m_e v_\perp^2/2B$

of energetic collisionless particles is conversed, while their distribution function varies by the law:

$$f_1\left(v_{||}, v_{\perp}, t\right) = f_1\left[v_{||}, v_{\perp}/\psi(t)\right], \psi^2(t) = B(t)/B(0) \tag{8.4}$$

The size of loss cone and, hence, the instability growth rate of plasma waves (Eq. (6.16)) vary in the process. As a result, the depth of modulation of the radio emission intensity[1)] [6] for the optically thin source $\tau_{\text{nl2}} = \int \mu_{\text{nl2}} \text{d}l < 1$ is [7]

$$\delta I_{\omega}/I_{\omega} = (7/6)\delta B/B, \delta B/B \ll 1 \tag{8.5}$$

The second case can take a place in the flaring loops ($n_0 \approx 10^{11}$ cm^{-3}, $n_1/n_0 \approx 10^{-4}$–10^{-5}, and $T_0 \approx 1$keV). The typical time of wave–particle interaction is usually much less than the mean time required for a particle to travel the loop length $\tau_D < \tau_0 = l/2v$. Hence, the distribution of nonthermal electrons has a sufficiently large isotropic part everywhere except for the regions near loop footpoints. It was shown in Ref. [7] that depending on the plasma turbulence level $w = W/n_0\kappa_B T$, for example, on the optical thickness $\tau_{\text{nl2}} = \int \mu_{\text{nl2}} \text{d}l$ of the decay process of electromagnetic waves into two plasma waves $t(2\,\omega_{\text{p}}) \rightarrow l + l'$, the intensity can be expressed as

$$\delta I_{\omega}/I_{\omega} = -(1/2)\delta B/B \text{ for } \tau_{\text{nl2}} < 1, \text{and } \delta I_{\omega}/I_{\omega} = (1/12)\delta B/B \text{ for } \tau_{\text{nl2}} > 1 \tag{8.6}$$

where $\tau_{\text{nl2}} > 1$ at $w > w^* \approx \dfrac{7.5}{(2\pi)^2}\dfrac{c}{\omega_{\text{p}}}\dfrac{m_{\text{e}}c^2}{\kappa_{\text{B}}T}\dfrac{1}{NL_n}$, $L_n = n_0/|\nabla n_0|$, N-refractive index.

Hence, one can expect that the modulation of plasma radio emission from the optical thick and optical thin sources will be in antiphase and the modulation depth in the case of $\tau_{\text{nl2}} < 1$ is larger when compared with the case of $\tau_{\text{nl2}} > 1$.

8.1.3
Modulation of the Electron Precipitation Rate

Quite strong modulation depth of flaring white light and hard X-ray emission can be produced by the periodic variations of the nonthermal electron precipitation rate by loop MHD oscillations [8]. In Chapter 6, it was shown that flaring loops are, in fact, the magnetic mirror traps for charged particles, accelerated by flares. The electrons with sufficiently large pitch angle θ bounce between the magnetic mirrors created by the converging magnetic flux tube at the loop legs. The range of the pitch angles of the trapped particles (Eq. (6.5)) is determined by the ratio on the magnetic fields at the loop top, B_{top}, and near the footpoint, B_{fp},

$$\cos^2\theta < \frac{B_{\text{fp}} - B_{\text{top}}}{B_{\text{fp}}} = 1 - \frac{1}{\sigma}. \tag{8.7}$$

1) The intensity of radio emission at the frequency ω is connected with the brightness temperature by the formula $I_{\omega} = \frac{N^2\omega^2\kappa_B T_b}{(2\pi)^3 c^2}$, and the radio emission flux density is $F_{\omega} = I_{\omega}\Omega$, where $\Omega = S/d^2$ is the source solid angle [7].

Magnetoacoustic oscillations, in particular sausage modes, periodically change the strength of the magnetic field at the loop top, varying the mirror ratio σ given by Eq. (8.7). Hence, the critical pitch angle will be periodically modulated, leading to the periodic escape of the particles from the trap downward the footpoints. Reaching the lower layers of the atmosphere, the electrons interact with dense plasma and cause hard X-ray and white light emission.

For example, the intensity of hard X-rays is proportional to the flux of fast electrons leaving the trap, $J_X \propto S_e \approx n_1 l/\sigma \tau_D(\sigma)$ [8]. Here, $\tau_D(\sigma)$ represents the timescale for diffusion process involved. It can be particle–particle collisions or wave–particle interaction. From this relation, one can see that to a first approximation the relative changes in the hard X-ray intensity will be proportional to the changes in the mirror ratio σ due to the pulsations of the trap magnetic field: $\delta J_X/J_X \sim \delta\sigma/\sigma \sim \delta B/B$.

8.2 Global Sausage Mode and Diagnostics of the Solar Event of 12 January 2000

This event gives a very clear example of global sausage mode oscillations, demonstrating, in particular, the applicability of the estimation for the global sausage mode period $P_{GSM} = 2l/c_p$, where c_p is the phase velocity of the sausage mode corresponding to the wave number $k_{||} = \pi/l$, $c_{Ai} < c_p < c_{Ae}$ [9]. Figure 8.2 shows the characteristics of quasi-periodic pulsations (QPOs) in an off-limb flaring loop observed on 12 January 2000 at microwaves, X-rays, and optical band. There are two clear spectral peaks at 8–11 and 14–17 s. The longer period component is more intensive at the loop apex than in the regions close to the footpoints. The shorter period ($\sim$9 s) may be associated with sausage mode of higher spatial harmonics.

The length of the flaring loop was estimated as $l \approx 2.5 \times 10^9$ cm, and its width at half intensity at 34 GHz as $2a \approx 6 \times 10^8$ cm using the Nobeyama Radio Heliograph (NoRH) measurements. These estimations were confirmed by *Yohkoh/SXT* images. Microwave diagnostics (in terms of *gyrosynchrotron mechanism*) together with soft X-ray Geostationary Operational Enviromental Satellites (*GOES*), and magnetograph (Solar and Heliospheric Observatory) data show that the flaring loop to be filled with a dense plasma with the number density $n_0 = 10^{11}$ cm^{-3} and the magnetic field changing from $B_0 \approx 100$ G near the loop top up to $B_0 \approx 200$ G near the footpoints [10]. From the global sausage mode period $P_{GSM} = 2\,l/c_p$, one can obtain the phase speed $c_p \approx 3.2 \times 10^8$ cm s^{-1}. This value is close to and less than the cutoff value $c_p(k_c) = c_{Ae}$. This allows us to estimate the Alfvén speed outside the loop as $c_{Ae} > 3.2 \times 10^8$ cm s^{-1}. Moreover, from Eq. (3.28), we obtain the upper limit on the Alfvén velocity inside the loop: $c_{A0} < 5.1 \times 10^7$ cm s^{-1}. Assuming that the plasma β is small and the magnetic field inside and outside the loop has almost equal strength, we can obtain the estimation of the density contrast ratio $n_0/n_e \approx 40$. The properties of pulsations indicate the possibility of the simultaneous existence of two modes of oscillations in the loop: the global one with the period of $P_1 = 14$–17 s and nodes at the footpoints and the second harmonics

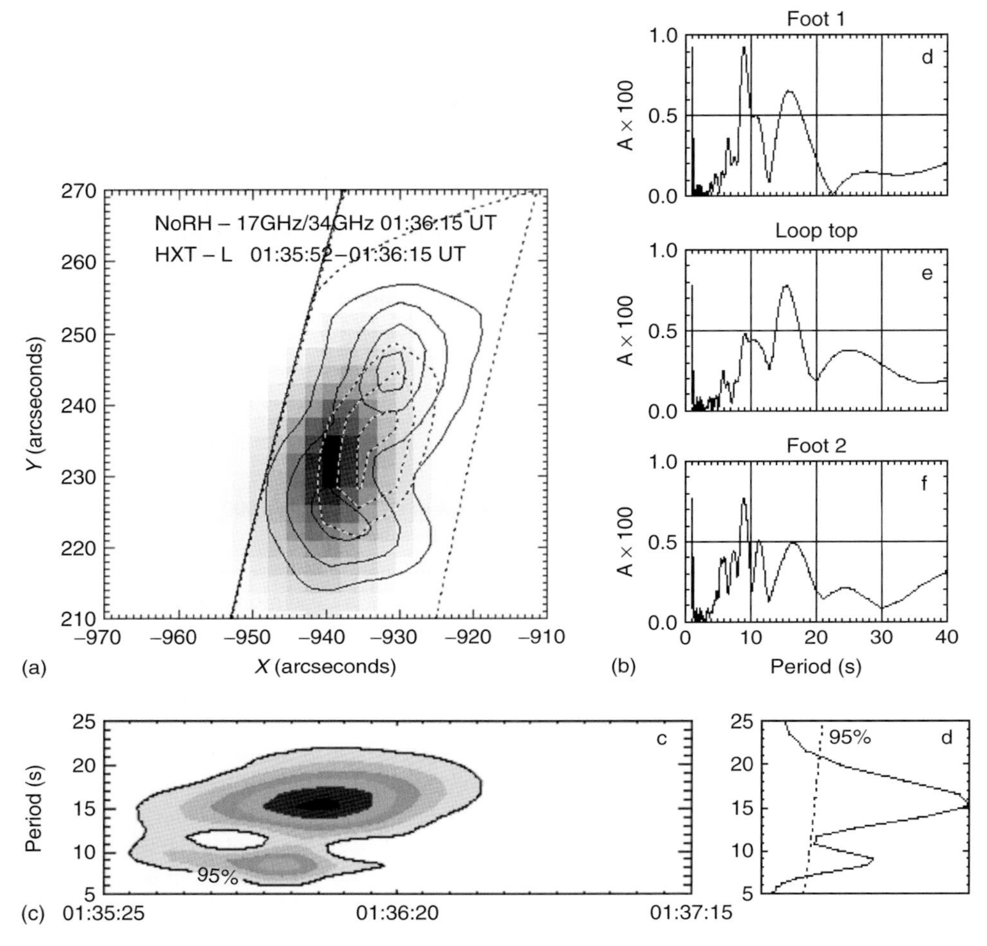

Figure 8.2 (a) Radio images made in the flare maximum at 01:36:15 UT at 34 GHz (filled pixels) and 17 GHz (dotted contour curves). The black contour shows the positioning of the hard X-ray sources, obtained in the channel L (13.9–22.7 keV) of Yohkoh/HXT (Hard X-ray Telescope). (b) Wavelet power spectrum (using Morlet wavelet) of 34 GHz flux variations observed by NoRH. (c) Fourier power spectra of radio flux variations from different parts of the flaring loop. The spectra show that there are both 8–11 and 14–17 s spectral components everywhere in the loop; the shorter period component dominates at the loop legs, while the longer period component – at the apex.

with $P_2 = 8–11$ s [9]. Note that the interpretation of the observed periodicity in terms of *kink global mode* should be excluded because the period of this mode is >75 s, which is very far from the observed period.

There are also two nice examples of the global sausage modes of flaring loops in the solar QPP events on 8 May 1998 and 3 July 2002 presented in Ref. [10].

8.3 Dissipative Processes in Coronal Loop for MHD Modes

It is evident that the damping of nonleaky (trapped) MHD modes of oscillating loops is determined by the dissipative processes in the loop itself. Nevertheless, this is also true for leaky modes of thin ($l \gg a$) coronal loops. Indeed, in many events, there is a jump of specific impedance for MHD waves at the interface between the loop and surrounding plasma. For example, in the case of normal incidence, the intensity ratio of the reflected and incident waves is

$$\frac{I_{\text{refl}}}{I_{\text{inc}}} = \frac{(Z_{\text{i}} - Z_{\text{e}})^2}{(Z_{\text{i}} + Z_{\text{e}})^2} \tag{8.8}$$

In particular, for sausage mode, the specific impedance is $Z = \rho\sqrt{c_A^2 + c_s^2}$, where $\rho = m_i n$ is the plasma density. If the density ratio inside (i) and outside (e) a loop $\geq 10^2$, but the velocity $\sqrt{c_A^2 + c_s^2}$ of sausage waves are roughly the same, the waves are reflected almost completely from the interface ($I_{\text{refl}} \approx I_{\text{inc}}$). In other words, the loop would be a rather good resonator for sausage oscillations, so that one may neglect the acoustic losses compared with the dissipative losses.

Sausage oscillations of a flare loops give the main effect to the modulation of flaring emission in various wave bands, from radio to optical emission because this kind of oscillations are accompanied by the variations of the magnetic field, plasma pressure, and loss-cone volume. Sausage modes undergo different types of dissipations in stellar coronae. The total decrement includes Joule dissipation υ_J, electron thermal conductivity ν_{cond}, ion viscosity ν_{v}, and radiation losses ν_{rad}, and can be written as [11–13] $\nu_\Sigma = \nu_{\text{J}} + \nu_{\text{c}} + \nu_{\text{v}} + \nu_{\text{r}}$, where

$$\nu_{\text{J}} = \frac{1}{2}\frac{m_{\text{e}}}{m_{\text{i}}}\frac{\omega^2}{\omega_{\text{i}}^2}\nu_{\text{ei}} \tag{8.9}$$

$$\nu_{\text{cond}} = \frac{1}{3}\frac{m_{\text{e}}}{m_{\text{i}}}\beta^2\frac{\omega^2}{\nu_{\text{ei}}}\sin^2\vartheta\cos^2\vartheta \tag{8.10}$$

$$\nu_{\text{v}} = \frac{1}{12}\sqrt{\frac{m_{\text{i}}}{2m_{\text{e}}}}\left(\frac{V_{\text{Ti}}}{c_{\text{A}}}\right)^2\frac{\omega^2}{\nu_{\text{ei}}}\sin^2\vartheta \tag{8.11}$$

$$\nu_{\text{rad}} = \frac{2\pi}{3}\frac{n_0^2\phi(T_{\text{e}})}{B^2}\sin^2\vartheta \tag{8.12}$$

Here, ϑ is the angle between the wave vector k and the magnetic field B, $\beta = 8\pi n_0 \kappa_{\text{B}} T/B^2$, and $\phi(T_{\text{e}}) = 5 \times 10^{-20}/\sqrt{T_{\text{e}}}$ is the loss function at $10^6 < T_{\text{e}} < 10^7$ K [13]. The effective electron–ion collisions frequency is determined by Eq. (6.2), which can be represented in the following form:

$$\nu_{\text{ei}} \approx \frac{60 n_0}{T^{3/2}}\ \text{s}^{-1} \tag{8.13}$$

Here, it was suggested that $T_e = T_i = T$K. Note that the approximation used in Eqs. (8.9)–(8.12) is true for the wave length $\lambda = 2\pi/k_{||} > l_{fp} = V_{Te}/\nu_{ei}$ and for the pulsation period $P > 1/\nu_{ei}$. For example, with $n_0 = 10^{10}$ cm^{-3} and $T = 3 \times 10^6$ K, above mentioned inequalities satisfied quite easily for $\lambda \geq 10^8$ cm and pulsation period $P \geq 1$ s. Estimations made in Ref. [14] have shown that under solar and red dwarf coronae conditions, the major contribution to the internal dissipation of sausage oscillations gives the ion viscosity and the electron thermal conductivity. Thus, the total damping decrement is

$$\nu_\Sigma = \nu_v + \nu_{cond} = \frac{1}{12\sqrt{2}}\sqrt{\frac{m_i}{m_e}}\frac{\omega^2}{\nu_{ei}}\beta \sin^2\vartheta \left(1 + \sqrt{\frac{32 m_i}{m_e}}\beta \cos^2\vartheta\right). \quad (8.14)$$

Hereafter to construct the diagnostic model for flare loop parameters, we will use the Q-factor of pulsation $Q = \omega/\nu_\Sigma$ together with the pulsation period and modulation depth.

8.4 The Stellar Flare Plasma Diagnostics from Multiwavelength Observations Stellar Flares

Haisch [15] suggested quite popular method for stellar plasma diagnostics based on quasi-static radiative and conductive cooling during the earliest phases of flare decay ($\nu_{rad} \approx \nu_{cond} \approx 1/\tau_d$). This method is known as the Haisch's simplified approach (HSA) [15, 16]. Given an estimate of two measured quantities, the emission measure (EM) in the flare and the decay timescale of the flare τ_d, HSA leads to the following expression for the temperature, density, and loop length:

$$T(\mathrm{K}) = 4 \times 10^{-5}(\mathrm{EM})^{0.25}\tau_d^{-0.25}, n(\mathrm{cm}^{-3}) = 10^9(\mathrm{EM})^{0.125}\tau_d^{-0.125},$$
$$l(\mathrm{cm}) = 5 \times 10^{-6}(\mathrm{EM})^{0.25}\tau_d^{0.75}$$

The attractiveness of these formulae is obvious: even for a star where no imaging is possible we can using measured EM and τ_d, extract information about parameters of a flare loop. Additionally keeping in mind that plasma $\beta = 8\pi n \kappa_B T/B^2 < 1$, a minimum magnetic field can be estimated.

Meanwhile, HSA has serious disadvantages. For example, it suggests the zero thermal flux at the footpoints. Because this approximation and of the large number of unknown parameters, the results of the diagnostics obtained by different authors are substantially different. Coronal seismology opens new possibilities for flare plasma diagnostics. Indeed, in order to obtain the plasma temperature, magnetic field, and plasma density in a flaring loop, the MHD-oscillation approach uses, as a rule, three equations for the pulsation period, Q-factor, and modulation depth. Moreover, one can get additional information on magnetic field and plasma density from radio and soft X-ray data. Below, there are several examples of this diagnostic method.

8.4.1
Pulsations in the Optical Range (U,B) in an EV Lac Flare

Zhilyaev *et al.* [17] presented simultaneous U–B observations of the EV Lac flare of 11 September 1998 at the Terskol Peak (Russia), Stephanion (Greece), Crimean Observatory (Ukraine), and Belogradchik (Bulgaria). EV Lac is located 5.05 pc apart and has the radius $\approx 0.36 R_\odot$. The pulse period was $P = 13$ s, pulsation quality $Q = 50$, and modulation depth $M = 0.2$. Stepanov *et al.* [14] suggested that these pulsations are due to sausage oscillations of the flaring loop (Figure 8.3). The pulsed U–B emission is the bremsstrahlung from the loop footpoints, which are bombarded by modulated fluxes of precipitating energetic electrons.

Using Eq. (3.38) for the period of the radiative sausage mode with $\eta_j = \eta_0 \approx 2.4$, Eq. (3.53) for the modulation depth $M = \Delta J/J = 4\pi k_B T/B^2 = \beta/2$, and the Q-factor $Q = \omega/\nu_\Sigma$ with $\nu_\Sigma = \nu_{cond} + \nu_v$ from Eq. (8.14), one can find the expression for the plasma temperature, density, and loop magnetic field:

$$
\begin{aligned}
T &\approx 1.2 \times 10^{-8} \frac{\tilde{r}^2 \beta}{P^2 \chi} \mathrm{K} \\
n &\approx 3.5 \times 10^{-13} \frac{\tilde{r}^3 \kappa \beta^{5/2} Q \sin^2 \vartheta}{P^4 \chi^{3/2}} \mathrm{cm}^{-1} \\
B &\approx 3.8 \times 10^{-18} \frac{Q^{1/2} \tilde{r}^{5/2} \kappa^{1/2} \beta^{5/4} \sin \vartheta}{P^3 \chi^{5/4}} \mathrm{G}
\end{aligned}
\tag{8.15}
$$

Here $\tilde{r} = 2\pi a/j_0$, $\kappa = 243\beta \cos^2 \vartheta + 1$, and $\chi = 10\beta/3 + 2$. Supposing, similar to the solar case, $l/a = 10$, that is, $\vartheta = 76^\circ$, we obtain from Eq. (8.15) the physical parameters of the plasma in the flaring loop in EV Lac: $T \approx 3.7 \times 10^7$ K, $n \approx 3 \times 10^{11}$ cm^{-3}, and $B \approx 300$ G.

Here, the solar–stellar analogy was used and suggested that the optical emission source is localized at the loop footpoints. If we assume, however, that the optical emission from flares could be determined by the bremsstrahlung of hot plasma in

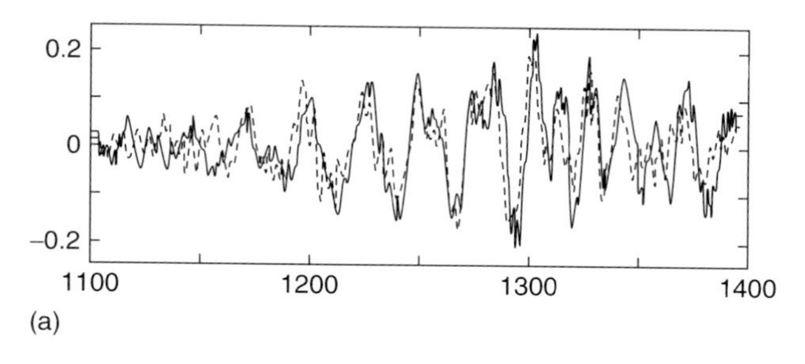

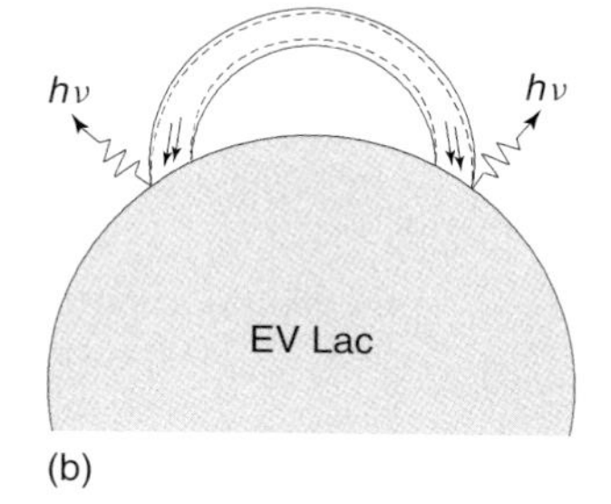

Figure 8.3 (a) Oscillations of the U (solid line) and B (dotted line) emission with the period of 13 s separated from the light curve of EV Lac flare on 11 September 1988. (b) Schematic view of the model for pulsations of optical flare emission.

the coronal part of a flare loop [18], we can overestimate the plasma beta parameter. Indeed, the bremsstrahlung fluxes for optically thin and optically thick sources, respectively, can be represented as [14]

$$J_1 \propto \frac{n^2}{\sqrt{T}} V, \quad J_2 \propto TV \tag{8.16}$$

where $V \approx \pi a^2 l$ is the emitting volume. Taking into account the adiabatic equation $n^{3/2}/T = \text{const}$, we can obtain instead of Eq. (8.16)

$$J_1 \propto n^{5/4} V, \quad J_2 \propto n^{3/2} V \tag{8.16a}$$

The excitation of sausage mode causes the emitting volume to change due to variation in the loop radius $\delta V/V = 2\delta a/a$. In turn, from the conservation of the magnetic flux $Ba^2 = \text{const}$, it follows that $\delta B/B = -2\delta a/a$, so we can find the relation

$$\frac{\delta V}{V} = -\frac{\delta B}{B} \tag{8.17}$$

The following relation holds for the sausage modes [11]:

$$\frac{\delta n}{n} = \frac{\delta B}{B} \sin \vartheta \tag{8.18}$$

and we obtain from Eqs. (8.16a)–(8.18)

$$\frac{\delta J_1}{J_1} = (1.25 \sin \vartheta - 1)\frac{\delta B}{B}, \frac{\delta J_2}{J_2} = (1.5 \sin \vartheta - 1)\frac{\delta B}{B} \tag{8.19}$$

Taking into account that $\beta = 2|\delta B/B|$ and the modulation depth $M = \beta/2$ from Eq. (8.19), we have

$$M_1 = (1.25 \sin \vartheta - 1)\beta, \quad M_2 = (1.5 \sin \vartheta - 1)\beta \tag{8.20}$$

Substituting into Eq. (8.15) $P = 13$ s, $Q = 50$ and using Eq. (8.20), we obtain $T \approx 9 \times 10^7$ K, $n \approx 10^{13}$ cm^{-3}, and $B \approx 2000$ for an optically thin source and $T \approx 6 \times 10^7$ K, $n \approx 1.4 \times 10^{12}$ cm^{-3}, and $B \approx 850$ G for an optically thick source; whence, we find the plasma parameters $\beta_1 \approx 0.9$ and $\beta_2 \approx 0.5$. Coronal loops with such large plasma beta are very unstable against flute perturbations. The growth time of flute instability under coronal loop condition is [19] $\tau_{fl} \sim \sqrt{al}/c_s \sim 30 - 100$ s. Because pulse period $P = 13$ s and Q-factor $Q = 50$, we conclude that the hypothesis about location of the optical emission source in the coronal part of the loop is in poor agreement with observations of EV Lac flare.

8.4.2
Quasi-Periodic Oscillations (QPOs) from EQ Peg

EQ Peg (Gl 896AB) is a visual binary system with a separation of 5″. Both components are M-type flare stars with visual magnitudes of 10.3 and 12.4, respectively. Flare activity of EQ Peg has been observed over a large band, from X-rays to radio wavelengths. The flare frequency in white light is ~0.8 flares h^{-1}

for flares with energy in excess of 10^{30} erg. There is evidence that EQ PegB is a more active flare star.

Mathioudakis *et al.* [20] presented the observations of the EQ PegB flare on 4 November 2003, obtained with the triple-beam CCD instrument *ULTRACAM* with the 4.2 m William Herschel Telescope on La Palma. It was suggested that the 10 s optical oscillations were caused by nonleaky sausage modes, and the parameters of the flare plasma were obtained using Haisch's scaling laws: $T \approx 5 \times 10^7$ K, $n \approx 4 \times 10^{12}$ cm^{-3}, $B \approx 1100$ G, and the loop length $l \approx 1.8 \times 10^9$ cm $\approx 0.07R_*$. Here $R_* = 2.7 \times 10^{10}$ cm is the radius of EQ PegB.

Under such parameters, the plasma $\beta \approx 1.2$, and the oscillations should be destroyed very rapidly due to the ballooning instability. On the other hand, Eq. (8.15) for sausage mode with $P = 10$ s, $\Delta = 0.1$, and $Q = 30$ yields $T \approx 6 \times 10^7$ K, $n \approx 2.7 \times 10^{11}$ cm^{-3}, $B \approx 540$ G, and $\beta \approx 0.2$ [21].

From wavelet analysis performed in [20], it follows that the pulsation period increased from 8 to 12 s during the impulse phase of the flare in the time interval $\Delta t \approx 45$ s (Figure 8.4).

This fact can be understood in terms of the chromosphere evaporation, which increases the density of the plasma in the loop [21]. The period of sausage oscillations (Eq. (3.39)) for $\beta \ll 1$ and $B \approx$ const is $P \sim n^{1/2}$. Hence, the ratio of plasma density at the final stage of the pulse train to the density at the initial phase is $n_2/n_1 \approx (P_2/P_1)^2 \approx 2.3$. The velocity of evaporated matter is close to the sound velocity $c_s = (10\ k_B T/3\ M)^{1/2}$; therefore, hot evaporated plasma fills the entire loop, if its length $l \leq 2\ c_s\ \Delta t$. For $T = 6 \times 10^7$ K and $\Delta t \approx 45$ s, we obtained the estimate

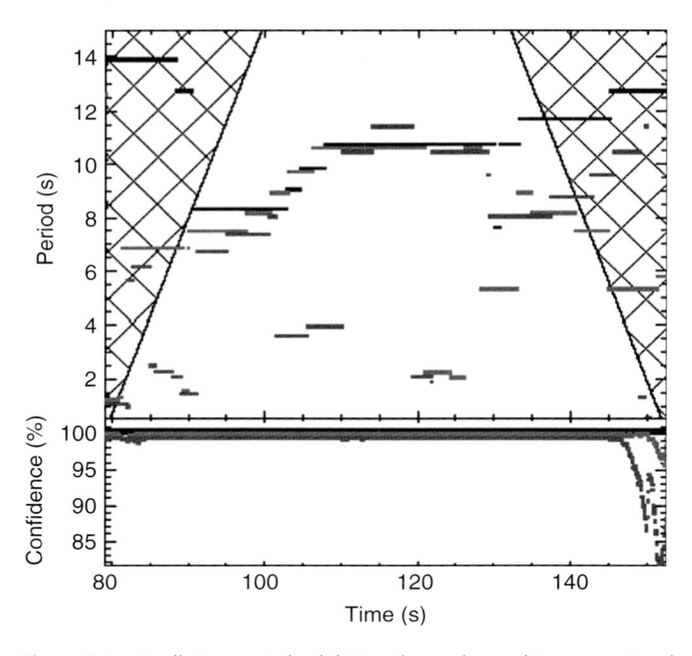

Figure 8.4 Oscillation period exhibiting the peak wavelet power at each point in time [20].

for the length of the loop $l \leq 1.2 \times 10^{10}$ cm $\approx 0.4R_*$. This means that active dMe stars have more extended, hotter, and denser coronae compared to the Sun's.

8.4.3
Soft X-Ray Pulsations from a Flare of AT Mic

AT Mic (Gj 799A/B) is an M-type binary dwarf, with both stars of the same spectral type (dM4.5e + dM4.5e). The stellar radius is $\approx 3.3 \times 10^{10}$ cm $\approx 0.47 R_\odot$ at the distance is 10.2 pc and AT Mic mass is $m_* = 0.4\, m_\odot$.

Mitra-Kraev *et al.* [22] presented the first soft X-ray (0.2–12 keV) observations with XMM-*Newton* of the flare on 16 October 2000, on the red dwarf AT Mic, which revealed oscillations with the period $P \approx 750$ s and the exponential damping time $t_d \approx 2000$ s (Figure 8.5).

Using the X-ray spectra and the multitemperature model, the flare plasma temperature $T \approx 2.4 \times 10^7$ K and the density $n \approx 4 \times 10^{10}$ cm^{-3} were obtained by Raassen *et al.* [23] for this event. To estimate of magnetic field strength, the centrifugal force model of Zaitsev and Stepanov [24] was applied. The magnetic loop is stretched by a centrifugal force, which is caused by the upwardly evaporated plasma. The flux tube starts to oscillate up and down. Consequently, the particle density inside the loop also oscillates, causing oscillations in the thermal radiation. Following this scenario, the relative amplitude of the soft X-ray oscillations are determined by the additional energy from filling the loop with hot plasma:

$$\frac{\Delta J}{J} = \frac{4\pi n \kappa_B T}{B^2} \tag{8.21}$$

With $\Delta J/J = 0.15$, $n = 4 \times 10^{10}$ cm^{-3}, and $T = 2.4 \times 10^7$ K the magnetic field $B = 105$ G. Mitra-Kraev *et al.* [22] interpreted this oscillations in terms of slow magnetoacoustic mode oscillating at the fundamental frequency. Using the relations from Ref. [25], they found a flare loop length $l = 2.5 \times 10^{10}$ cm. This value is consistent with estimating the loop length from radiative cooling as well as from pressure balance considerations [26].

In order to determine the number density of soft X-ray emitting plasma independently, the information on the damping rate of slow magnetoacoustic waves can be used. Estimates performed in Ref. [27] show that under such conditions the dissipation of slow magnetosonic oscillations is governed by the electron thermal conduction (Eq. (8.10)). In this case, the damping rate is

$$\frac{1}{t_d} \approx 76 \frac{T^{3/2}}{P^2 n}\ \mathrm{s}^{-1} \tag{8.22}$$

At $T = 2.4 \times 10^7$ K, $P = 750$ s, and $t_d \approx 2000$ s, Eq. (8.22) yields the plasma density $n \approx 3.2 \times 10^{10}$ cm^{-3}, which is in agreement with the result presented in Ref. [23] for an optically thin X-ray loop.

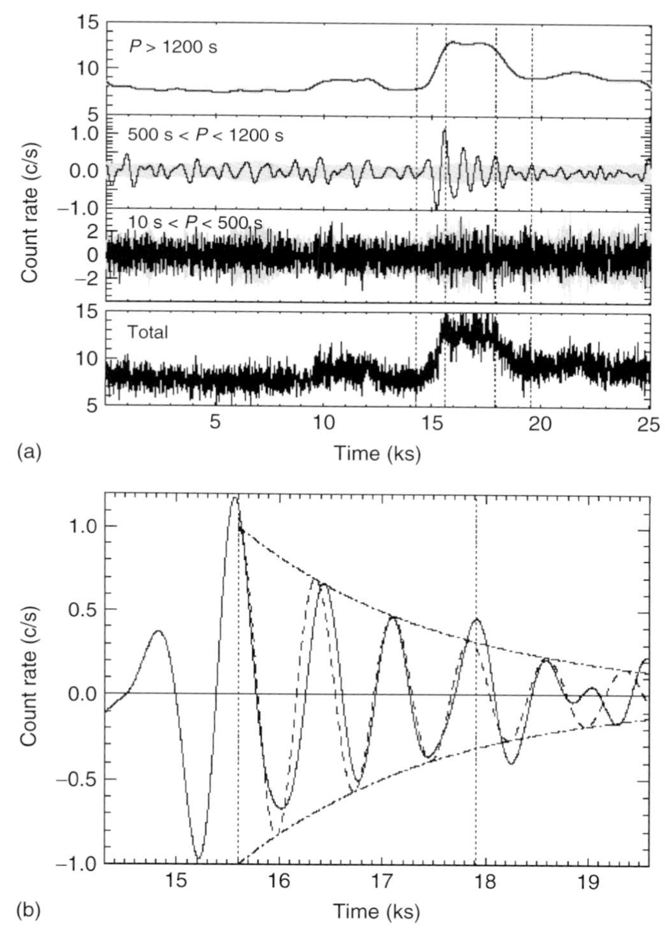

Figure 8.5 (a) The reconstructed light curve divided into three frequency bands together with the original data (lowest panel on the left) and the flare part of the $P = 500–1200$ s. (b) The solid line shows the reconstructed light curve from the data, whereas the dashed line is a damped sine curve with an oscillation period $P = 750$ s and an exponential decay time of $t_d \approx 2000$ s. The relative peak-to-peak amplitude is initially 15% [22].

8.5 Diagnostics of Electric Currents in Stellar Atmospheres

There are several diagnostic methods of the electric currents in stellar atmospheres. Electric currents in the vicinities of sunspots ($\geq 10^{11}$ A) were first determined by Severny [28] based on magnetograph measurements. An approach was proposed by Spangler [29] to measure electric currents in solar and stellar coronae based on differential Faraday rotation of radio emission. The efficiency of the method was illustrated by applying it to observations of radio source 3C 228, carried out with the radio telescope Very Large Array, showing that the current value in the corona of 3C 228 varied from 2.3×10^8 to 2.5×10^9 A in various events.

In Chapter 2, the RLC approach for describing of physical processes in coronal current-carrying loops is presented. The examples of determination of the electric currents ($1.7 \times 10^8 - 2.3 \times 10^{10}$ A) in two inductively interacting solar loops are given. Relaying on this method, it is possible to detect the accumulation of electric current energy before flares, current dissipation during flare, and postflare current increase in active region.

8.5.1
Observational Evidence for Energy Accumulation and Dissipation in Coronal Magnetic Loop

From Eqs. (2.50), (2.64), and (2.67), it follows that the loop as an equivalent electric circuit has a period of eigen oscillations, which is inversely proportional to the current amplitude:

$$P = \frac{1}{\nu_{\mathrm{RLC}}} = \frac{2\pi}{c}\sqrt{LC(I)} \approx \frac{10 S_{17}}{I_{11}} \mathrm{s} \tag{8.23}$$

where S_{17} is the cross-sectional area of the coronal part of the loop in units of 10^{17} cm^2 and I_{11} is the current in the loop in units of 10^{11} A. This dependence arises from the circuit capacitance being depended on Alfvén velocity within the loop, that is, the magnetic field whose magnitude, in turn, is a function of current in the loop. Indeed, since the capacitance is determined by the coronal part of the loop, namely

$$C(I) = \frac{c^4 \rho_2 S^2}{2\pi l_2 I^2} \tag{8.24}$$

and the Biot–Savart law links the current and magnetic field value as $I \propto B\sqrt{S}$, Eq. (8.24) is equivalent to the expression for the capacitance of a capacitor, $C = \varepsilon_A S/l_2$, with l_2 being the distance between its plates, S their areas, and $\varepsilon_A = c^2/c_A^2$ the permittivity of the medium with respect to Alfvén waves.

RLC oscillations of a coronal magnetic loop modulate its microwave emission of both thermal and nonthermal origin. The analysis of solar flares observed in 1989–1993 with the radio telescope in Metsähovi (Finland) at 22 and 37 GHz aimed at the determination of electric current magnitude has been performed in Ref. [30]. Table 8.1 lists the characteristics pertaining to a set of bursts with pulsations of flare loops on the Sun of typical size $S_{17} = 1$ and $l_2 = 5 \times 10^9$ cm. Spectral analysis revealed the modulation of radiation with period P ranging from 0.7 to 17 s, which yields the magnitudes of electric currents $I \approx 6 \times 10^{10} - 1.4 \times 10^{12}$ A. Table 8.1 also lists total energies of electric currents stored in respective circuits: $LI^2/2 \approx 10^{30} - 5 \times 10^{32}$ erg.

For same events, it was possible to compare the energy stored in the magnetic loops with the flare energy. For instance, for the flare on 22 June 1989, Eq. (8.23) gives $I = 2 \times 10^{11}$ A and $LI^2/2 \approx 10^{31}$ erg. On the other hand, an estimate of

Table 8.1 Bursts of solar emission in millimeter wave range characterized by pulsations with a high *Q*-factor and parameters of the RLC circuit. (1 SFU = 10^{-22} W m^{-2} Hz^{-1}.)

Date	Burst in millimeter band (UT)	F_{max}(SFU)	P (s)	$I(10^{11}$ A)	$LI^2/2(10^{31}$ erg)
22 June 1989	14:47–14:59	<150	5.2	2.0	1.0
19 May 1990	13:15–13:40	10	0.7	14.2	50.0
01 September 1990	7:06–7:30	27	1.1	9.1	20.9
24 March 1991	14:11–14:17	<700	10.0	1.0	0.25
16 February 1992	12:36–13:20	≈ 2000	5.0	2.0	1.0
08 July 1992	9:48–10:10	≈ 2500	3.3	3.0	2.3
08 July 1992	10:15–11:00	15	16.7	0.6	0.08
27 June 1993	11:22–12:00	40	3.5	2.8	2.0

thermal energy in the evaporating chromospheric plasma for this flare, derived from microwave and soft X-ray data emission gives $E_{th} = (1.0–4.5) \times 10^{29}$ erg [31]. The thermal energy of hot plasma, as is known, makes up a marked part of flare energy. Thus, a conclusion can be drawn that in flaring <5% of electric current energy stored in the magnetic loop is released. This seems plausible for cases when the flare magnetic structure is not destroyed in the process of energy release.

Since a flare is accompanied by dissipation of the current, the frequency of RLC oscillations should decrease as the flare develops. In contrast, if the loop current increases by the action of photospheric electro-motive force (EMF), the frequency of RLC oscillations increases with time. The search for linearly frequency-modulated (LFM) signals (i.e., signals with a frequency obeying the expression $\omega = \omega_0 + Kt$, where K is a constant and t is the time) with positive and negative frequency drifts in the spectrum of low-frequency modulation of microwave emission from solar flares was carried out in Ref. [32]. To perform it, the Wigner–Ville transform [33–35] was applied to the analysis of microwave emission from solar flares at 37 GHz observed in Metsähovi.

Figure 8.6 shows an example of such an analysis. A time profile (Figure 8.6a) of the microwave burst on 24 March 1991 (14:05 UT) in a solar active region S25W03 is shown. The maximum of radio emission flux in the impulsive phase amounted to 700 Solar Flux Units (1 SFU = 10^{-22} W m^{-2} Hz^{-1}) with a total burst duration of about 30 min. Figure 8.6b shows the dynamic spectrum of low-frequency modulation of radio emission flux. It embraces all phases of the burst and consists of three components.

1) A narrow-band signal whose frequency grows linearly from ~0.4 to ~0.45 Hz. The frequency drift rate is $d\nu/dt \approx 1.5 \times 10^{-3}$ Hz min^{-1}. This type of modulation is observed at the preflare stage.
2) A high *Q*-signal at frequencies of 0.5–0.1 Hz with a negative drift rate $d\nu/dt \approx 1.3 \times 10^{-2}$ Hz min^{-1}. This modulation appears during the impulsive phase of the burst and disappears when the intensity of the radiation decreases to its

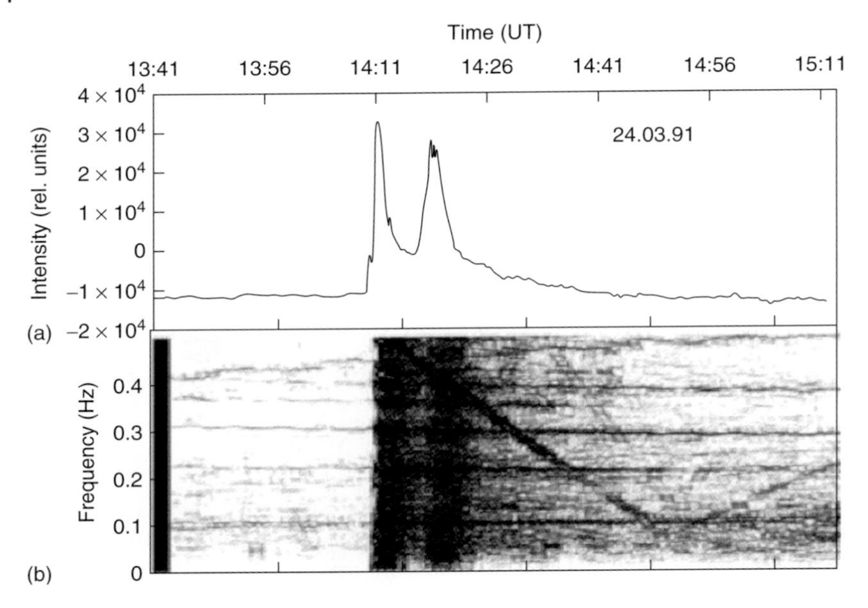

Figure 8.6 (a) Time profile of solar flare burst on 24 March 1991 at 37 GHz (Metsähovi). (b) The dynamic spectrum of low-frequency modulation of the radio emission flux obtained using the Wigner–Ville transform [32].

preflare level (14:50 UT). This stage corresponds to the dissipation of electric current energy in the flare.

3) A signal with a positive frequency drift $d\nu/dt \approx 8 \times 10^{-3}$ Hz min^{-1} that follows the signal with a negative frequency drift. It corresponds to the beginning of the phase of electric current energy accumulation in the flare loop.

Taken as a whole, the dynamic spectrum represents an LFM high Q-signal ($Q \approx 10^3$) with the frequency drifting from 0.4 to 0.1 Hz. The data analysis performed in Ref. [32] for the four radio bursts yielded values of currents in flaring loops falling in the range $(0.75\text{–}9.0) \times 10^{11}$ A and stored electric current energies lying within the range $1.4 \times 10^{30}\text{–}2 \times 10^{32}$ erg. For the particular event of 24 March 1991 (Figure 8.6), the electric current decreased from 9×10^{11} A at the beginning of the burst to 10^{11} A at its final stage, that is, by almost one order of magnitude. The energy release rate reached 10^{28} erg s^{-1}. After the flare, the frequency drift of the LFM signal became positive. This corresponds to the renewal of the energy accumulation process in the coronal magnetic loop.

The examples given here can be considered as fairly important experimental evidence of the presence of dissipation and accumulation of the electric current energy in coronal flaring loops. Estimates of electric current, stored energy, and the rate of its dissipation satisfactorily agree with respective values characteristic for flare processes on the Sun and obtained from other data. This demonstrates the efficiency of applying "circuit" flare models based on the analogy between a flare and an electric circuit with a large inductance, in which the energy is stored in the form of electric current and is released at some instant of time following a drastic increase in the resistance.

8.5.2
Pulsating Microwave Emission from AD Leo

Radio emission from the active red dwarf AD Leo (Gl 388, dM3.5e) located 4.85 pc apart has been repeatedly observed with large radio telescopes in the range of 1.3–5.0 GHz. Observations have revealed the existence of QPPs with periods in the range 1–10 s at the continuum background [36–38]. The modulation depth in pulsations can reach 50% and the brightness temperature of radio emission is in the range of 10^{10}–10^{13} K, which hints at coherent mechanism of radio emission. The pulsations of AD Leo bursts exhibit a frequency drift of 100–400 MHz s^{-1} and posses a high degree of polarization, 50–100%. The presence of frequency drift and the high degree of polarization testify in favor of a periodical regime of particle acceleration in the stellar atmosphere, as well as the plasma mechanism of emission in the inhomogeneous atmosphere of the star.

Detailed study of pulsed radio emission from the flare of AD Leo on 19 May 1997 (18:45 UT), observed with 100 m radio telescope in Effelsberg (Germany) in the frequency range 4.6–5.1 GHz, was performed in Refs. [37, 39]. Figure 8.7a displays the time profile of radio emission at 4.85 GHz near the instant of the burst with the total duration of impulsive phase of about 50 s. A preliminary analysis [37] revealed a component with a period of 2 s (0.5 Hz) in the pulsation spectrum, which is apparently different from the natural oscillation periods of the telescope.

Figure 8.7b,c presents the spectrum of low-frequency pulsations in the range of 0–9 Hz, obtained through the Wigner–Ville transform. Apart from the noisy component, one can see nearly equidistant narrow-band signals drifting over the frequency and exhibiting frequency splitting. The pulsation spectrum indicates that the emission source affected by two types of modulations: (i) a periodic

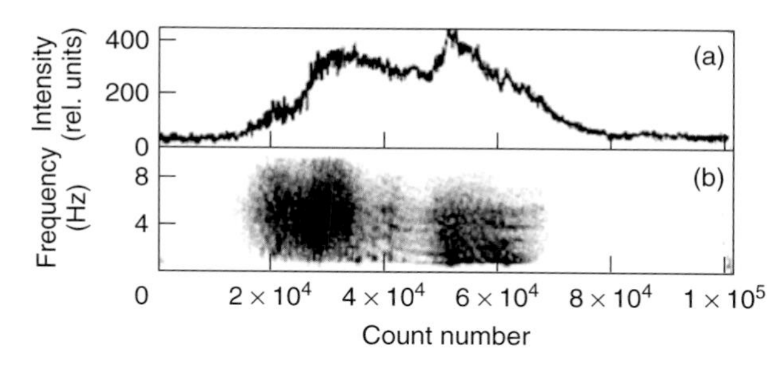

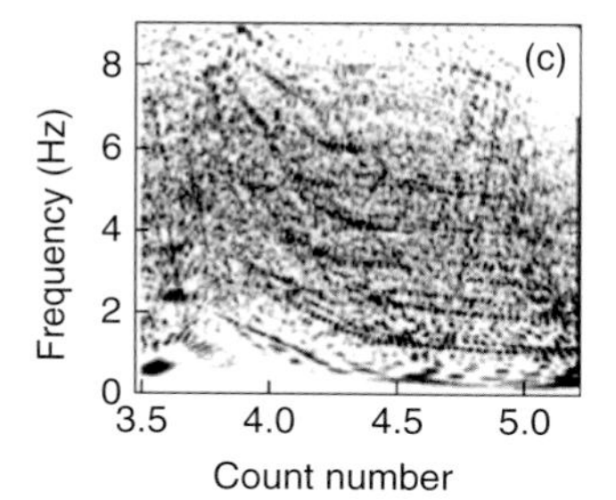

Figure 8.7 (a) The light curve of the flare of AD Leo on 19 May 1997 at 4.85 GHz with the impulse phase duration of about 50 s. (b) The dynamic spectrum of pulsations obtained with Wigner–Ville transform. (c) A zoomed fragment of the spectrum.

sequence of short pulses with repetition frequency $\nu_1 \approx 2$Hz and (ii) narrow-band ($\Delta\nu/\nu \leqslant 0.1$) harmonic whose frequency gradually decreases ($\nu_2 \approx 2.0 \rightarrow 0.5$ Hz), which is called here an *linearly frequency-modulated* signal. The obtained frequencies are apparently different from the natural oscillation frequencies of the telescope.

The preliminary analysis of the pulsations [37] demonstrated that the main source of the radio flare was a coronal magnetic loop with the electron number density $n \approx 2.3 \times 10^{11}$ cm^{-3}, temperature $T \approx 3 \times 10^7$ K, and magnetic field $B \approx 730$–810 G. On the basis of this data, it was concluded that 2 Hz modulation is due to sausage oscillations of a loop with the frequency $\nu_1 \approx c_A/a$. For $\nu_1 = 2$ Hz and the estimated Alfvén velocity $c_A = 7 \times 10^8 \left(\omega_c/\omega_p\right) \approx 3.5 \times 10^8$ cm s^{-1} $\left(\omega_c/\omega_p \approx 0.5\right)$, we obtain the loop radius $a \approx 1.8 \times 10^8$ cm, which is compatible to the radius of flare loops on the Sun.

As concern the gradually decreasing modulation, it, most likely, is caused by eigenmodes of the magnetic loop as an RLC oscillator with a frequency obeying Eq. (2.65). For our purpose here, we rewrite Eq. (2.65) as [39]

$$\nu_{\mathrm{RLC}} \approx \frac{1}{2\pi\sqrt{2\pi\Lambda}} \left(1 + \frac{c^2 a^2 {B_z}^2}{4{I_z}^2}\right)^{1/2} \frac{I_z}{ca^2\sqrt{nm_i}}, \Lambda = \ln\frac{4l}{\pi a} - \frac{7}{4} \tag{8.25}$$

This frequency depends on the strength of current flowing along the loop and decreases as the current dissipates during the flare, which explains the observed negative frequency drift of the modulating signal. When the current satisfies the inequality $I_z \ll caB_z/2$, the frequency ν_{RLC} depends weakly on the current. Thus, at the certain time, the frequency drift rate $\mathrm{d}\nu_{\mathrm{RLC}}/\mathrm{d}t$ must experience a break and decrease sharply. This occurs when the condition $I_z(t) < caB_z/2$ is satisfied. The dynamic spectra in Figure 8.7c exhibit a similar time dependence of the frequency drift rate. Using the value of $\nu_2 = \nu_{\mathrm{RLC}} = 0.5$ Hz, corresponding to the instant of "slope break" in the spectrum in Figure 8.7c, and the relationship $I = caB_z/2$, from Eq. (2.67) and the formula $\nu_2 = c_A/\left(a\sqrt{2\pi\Lambda}\right)$, one can estimate the loop length $l \approx 4 \times 10^{10}$ cm, which proves to be of the same order as the radius of the star (3.5×10^{10} cm). It should be noted that the very long baseline interferometry (VLBI) observations of red dwarfs [40] supported the existence of large coronal loop whose size is on the order or larger than stellar radii.

Adopting further a maximum value of $\nu_2^{\max} \approx 2$ Hz (the lower track in Figure 8.7c), an opportunity appears to estimate the magnitude of electric current corresponding to the maximum in the first phase of the radio burst, $I \approx 4.5 \times 10^{12}$ A. The current energy stored in the flare is $W = LI^2/2 \approx 5.5 \times 10^{26}$ J, and the energy release rate $\mathrm{d}W/\mathrm{d}t \approx 10^{25}$ W. The latter values are by three to four orders of magnitude higher than these in a typical solar flare. This is due to the large magnetic field on red dwarf surfaces and the augmented activity of photospheric convection.

Note that a decade before the event of 19 May 1997, a similar event in AD Leo was observed by Bastian *et al.* [36] using the 300 m Arecibo radio telescope at the wavelength 21 cm. In accordance to the RLC model, in the flare on 4 May 1987, which revealed high Q-quasi-periodic radio pulsations with the period ≈ 0.7 s, the current was estimated as $I \approx 10^{13}$ A.

References

1. Dulk, G.A. (1985) *Ann. Rev. Astron. Astrophys.*, **23**, 16.
2. Tsap, Y.T., Kopylova, Y.G., Stepanov, A.V., Melnikov, V.F., and Shibasaki, K. (2008) *Sol. Phys.*, **253**, 161.
3. Kaplan, S.A. and Tsytovich, V.N. (1973) *Plasma Astrophysics*, Pergamon Press, Oxford.
4. Zaitsev, V.V. and Stepanov, A.V. (1983) *Sol. Phys.*, **88**, 297.
5. Stepanov, A.V., Kliem, B., Krüger, A. *et al.* (1999) *Astrophys. J.*, **524**, 961.
6. Zheleznyakov, V.V. (1996) *Radiation in Astrophysical Plasmas*, vol. **204**, Kluwer Academic Publishers, 462 p.
7. Rozenraukh, Y.M. and Stepanov, A.V. (1988) *Sov. Astron.*, **32**, 153.
8. Zaitsev, V.V. and Stepanov, A.V. (1982) *Sov. Astron. Lett.*, **8**, 132.
9. Nakariakov, V.M., Melnikov, V.F., and Reznikova, V.E. (2003) *Astron. Astrophys.*, **412**, L7.
10. Nakariakov, V.M. and Melnikov, V.F. (2009) *Space Sci. Rev.*, **149**, 119.
11. Braginskii, S.I. (1965) *Rev. Plasma Phys.*, **1**, 205.
12. Hollweg, J.V. (1985) *J. Geophys. Res.*, **90**, 7620.
13. Priest, E. (1982) *Solar Magnetohydrodynamics*, D. Reidel Publishing Company, Dordrecht, Holland.
14. Stepanov, A.V., Kopylova, Y.G., Tsap, Y.T., and Koupriyanova, E.G. (2005) *Astron. Lett.*, **31**, 612.
15. Haisch, B.M. (1983) in *Activity in Red-Dwarf Stars* (eds X-ray Observations of Stellar Flares P.B. Birne and M. Rodono) D. Reidel Publishing Company, p. 255.
16. Mullan, D.J., Mathioudakis, M., Bloomfield, D.S., and Christian, D.J. (2006) *Astrophys. J. (Suppl)*, **164**, 173.
17. Zhilyaev, B.E., Romanyuk, Y.O., Verlyuk, I.A. *et al.* (2000) *Astron. Astrophys.*, **364**, 641.
18. Mullan, D.J., Herr, R.B., and Bhattacharyya, S. (1992) *Astrophys. J.*, **391**, 256.
19. Shibasaki, K. (2001) *Astrophys. J.*, **557**, 326.
20. Mathioudakis, M., Bloomfield, D.S., Jess, D.B. *et al.* (2006) *Astron. Astrophys.*, **456**, 323.
21. Tsap, Y.T., Stepanov, A.V., Kopylova, Y.G., and Zhilyaev, B.E. (2010) *Astron. Lett.*, **36**, 49.
22. Mitra-Kraev, U., Harra, L.K., Williams, D.R., and Kraev, E. (2005) *Astron. Astrophys.*, **436**, 1041.
23. Raassen, A.J.J., Mewe, R., Audard, M., and Güdel, M. (2003) *Astron. Astrophys.*, **411**, 509.
24. Zaitsev, V.V. and Stepanov, A.V. (1989) *Sov. Astron. Lett.*, **15**, 66.
25. Roberts, B., Edwin, P.M., and Benz, A.O. (1984) *Astrophys. J.*, **279**, 857.
26. Shibata, K. and Yokohama, T. (2002) *Astrophys. J.*, **577**, 422.
27. Stepanov, A.V., Tsap, Y.T., and Kopylova, Y.G. (2006) *Astron. Lett.*, **32**, 569.
28. Severny, A.B. (1964) *Space Sci. Rev.*, **3**, 451.
29. Spangler, S.R. (2007) *Astrophys. J.*, **670**, 841.
30. Zaitsev, V.V., Stepanov, A.V., Urpo, S., and Pohjolainen, S. (1998) *Astron. Astrophys.*, **337**, 887.
31. Urpo, S., Zaitsev, V.V., and Stepanov, A.V. (1994) *Sol. Phys.*, **154**, 317.
32. Zaitsev, V.V., Kislyakov, A.G., Urpo, S., Stepanov, A.V., and Shkelev, I.E. (2003) *Astron. Rep.*, **47**, 873.
33. Wigner, E. (1932) *Phys. Rev.*, **40**, 749.
34. Ville, J. (1948) *Cables Transm.*, **2A**, 61.
35. Cohen, L. (1989) *Proc. IEEE*, **77**, 941.
36. Bastian, T.S., Bookbinder, J.A., Dulk, G.A., and Davis, M. (1990) *Astrophys. J.*, **353**, 265.
37. Stepanov, A.V., Kliem, B., Zaitsev, V.V. *et al.* (2001) *Astron. Astrophys.*, **374**, 1072.
38. Abada-Simon, M., Lecacheux, A., Aubier, M., and Bookbinder, J.A. (1997) *Astron. Astrophys.*, **321**, 841.
39. Zaitsev, V.V., Kislyakov, A.G., Stepanov, A.V., Kliem, B., and Fürst, E. (2004) *Astron. Lett.*, **30**, 319.
40. Benz, A., Conway, J., and Güdel, M. (1998) *Astron. Astrophys.*, **331**, 596.

9
Heating Mechanisms in Stellar Coronae

Determination of the energy source responsible for the heating of highly magnetized and rarefied plasmas is vital for the understanding of heating in the solar and stellar coronae. Optical radiation of the photosphere is unable to heat the corona to the temperature of the order of several million degrees of Kelvin, as the temperature of, for example, the solar photosphere, is about 6000°. On the other hand, radiation losses in the corona are very high, and, in order to compensate for these losses and maintain the corona in the quasi-stationary state, intense sources of heat should exist in it. Currently, the solution of the problem of heating in the coronae has not been finally reached, although some perspective approaches to it are briefly considered in this chapter. In the distribution of the temperature of the solar atmosphere along the height [1], a very narrow transition region between the chromosphere and corona exists, within which the temperature sharply increases from $\sim 10000^\circ$ in the chromosphere to several million degrees in the corona. In this case, the concept of the coronal temperature is slightly conditional, since the corona is extremely inhomogeneous: as a rule, it consists of numerous magnetic loops, the temperature and density in which may noticeably differ from corresponding values in the surrounding plasma. In addition to that, it is within coronal magnetic loops that energy release and other active processes frequently occur; therefore, the problem of the heating of coronal plasma should also be considered in the context of the heating of coronal magnetic loops. For the sake of simplicity, we will thereafter restrict the consideration of the heating of coronal plasma with the example of heating in coronal magnetic loops, leaving the energy exchange between loops and adjacent plasma, as well as heating of open coronal magnetic structures, beyond the scope.

We consider quasi-steady-state magnetic loops, which exist on time intervals of several hours. They do not include the so-called flare loops, which may be sporadically heated to the temperature of several tens of millions degrees, owing to the flare energy release. Among the quasi-steady-state loops, the most thoroughly studied are the so-called warm loops and hot X-ray loops.

In "warm" solar magnetic loops, the temperature varies within 1.5–2 million degrees. These loops have the characteristic length $l \approx (1.5-7) \times 10^{10}$ cm and are concentrated above the sunspots' penumbra. The heating source in "warm" loops is inhomogeneous along the height and displays its maximum power in the proximity

Coronal Seismology: Waves and Oscillations in Stellar Coronae, First Edition.
A. V. Stepanov, V. V. Zaitsev, and V. M. Nakariakov.

of the footpoints of the loops [2]. It was found that the temperature in these loops is almost homogeneously distributed along several height scales. Owing to minor temperature gradients, radiation losses cannot be compensated by electron thermal conductivity; therefore, the heating source should be balanced with radiative losses along the entire length of the loop, which provides the quasi-steady state of the loops during at least several hours, in accordance with observations. Since radiative losses in loops are proportional to the square of electron concentration, the heating power should decrease exponentially with the height in the corona, while the corresponding height scale should be equal to approximately a half of that for the plasma concentration. In addition to that, observations by Aschwanden *et al.* [2] revealed a square dependence of the heating power on plasma pressure within a magnetic tube, and also sporadic increases of the emission measure of extreme ultraviolet (EUV) radiation in the footpoints of the loops. The latter fact may provide evidence for injection of dense chromospheric plasma into coronal magnetic loops.

Quasi-steady-state X-ray loops in the solar corona were studied with the use of *Yokhoh* soft X-ray telescope [3]. These loops display the temperature in the range of 3–7 million degrees, relatively small lengths $l \approx (0.2-3) \times 10^{10}$ cm, and may be remote from the sunspots. For the characteristic volume $V = 10^{28}$ cm^3, plasma number density $n = 3 \times 10^9$ cm^{-3}, and temperature $T = 5 \times 10^6$ K, the optical radiation energy losses from a volume of a hot X-ray loop

$$W_{\mathrm{R}} = n^2 \chi(T) V \tag{9.1}$$

where $\chi(T)$ is the function of radiative losses [4], are approximately equal to 10^{25} erg s^{-1}. Which heating mechanism may compensate for these very high losses and maintain the loop in the quasi-steady-state mode? Several heating models are generally discussed: by magnetohydrodynamic (MHD)-waves, owing to the Ohmic dissipation of electric currents, and by microflares.

9.1 Wave Heating

One of the possible heating mechanism is based on waves, which are excited, for example, due to photospheric convection, penetrate into the corona, and heat it in the course of their dissipation. One of the most energetically capacious mechanisms of wave generation in coronal magnetic loops may be photospheric convection. This model was applied to the Sun by Ionson [5] and to red dwarfs by Mullan and Cheng [6]. The most suitable wave modes that may be excited by photospheric convection are torsional Alfvén waves, kink magnetoacoustic waves, and slow magnetoacoustic waves. Alfvén and kink waves, however, are less likely to play this role than are slow magnetoacoustic waves, for two reasons. First, for typical parameters of hot X-ray loops, the periods of Alfvén oscillations of coronal magnetic loops as MHD resonators are, as a rule, within the interval 10–50 s [3], which differs substantially from characteristic timescales for oscillations of the velocity of photospheric plasma (150–400 s [4]). Therefore, the problem of "resonant" excitation of Alfvén oscillations in coronal magnetic loops, that is, the problem of the transfer of large fluxes of energy from photospheric oscillations to the corona, emerges. Second, torsional Alfvén waves poorly decay in the corona. The rates of their attenuation due to Ohmic losses and viscosity, respectively, are [7]

$$\gamma_A^{\text{joul}} \approx \frac{\pi^2 c^2 \nu_{\text{ei}}}{\omega_{\text{pe}}^2 l^2} \approx 0.5 \times 10^{-16}\,\text{s}^{-1}$$

$$\gamma_A^{\text{visc}} \approx \pi^2 \left(\frac{m_e}{m_i}\right)^{1/2} \frac{V_{\text{Ti}}^2 \nu_{\text{ei}}}{\omega_{\text{Bi}}^2 l^2} \approx 0.6 \times 10^{-18}\,\text{s}^{-1} \tag{9.2}$$

where ν_{ei} is the effective frequency of electron–ion collisions, $\omega_{\text{pe}}, \omega_{\text{Bi}}$ the plasma frequency of electrons and gyrofrequency of ions, respectively, V_{Ti} the thermal velocity of ions, l the length of the loops, m_{ei} the mass of electrons and ions, and c the speed of light. Therefore, the corresponding rate of heating,

$$q_A^{\text{joul}} \approx \gamma_A^{\text{joul}} B_{\approx}^2/8\pi \approx 2 \times 10^{-13} \ll n^2 \chi(T) \approx 6 \times 10^{-4}\,\text{erg}\,\text{cm}^{-3}\,\text{s}^{-1} \tag{9.3}$$

which may be estimated as the product of the decrement and the energy density of Alfvén waves, appears to be many orders of magnitude smaller than radiative losses, even when the amplitude of the magnetic field in the wave is of an order of magnitude of nonperturbed field in a coronal magnetic loop. The dissipation rate may be increased, however, in the case of resonant absorption of kink waves in coronal magnetic loops [5].

Slow magnetoacoustic waves in coronal magnetic loops display periods from 3 to 20 min depending on the length of the loops and the temperature. In the corona, they decay due to electron thermal conductivity with the decrement

$$\gamma_T = 0.12 \left(\frac{m_i}{m_e}\right) \omega^2 \tau_e \approx 1.5 \times 10^{-3}\,\text{s} \tag{9.4}$$

where ω is the frequency of acoustic oscillations, $\tau_e = \nu_{\text{ei}}^{-1}$ [7]. The rate of decay is so high that the rate of heating caused by acoustic waves

$$q_s \approx \gamma_T W_s \approx (6 \div 20) \times 10^{-3} \frac{W_s}{n k_B T} \geq n^2 \chi(T) \approx 6 \times 10^{-4}\,\text{erg}\,\text{cm}^{-3}\,\text{s}^{-1} \tag{9.5}$$

exceeds optical radiation losses in coronal plasma, provided the energy density of acoustic oscillations exceeds the density of thermal energy of the plasma by $W_s/nk_B T \approx 3$–10%. The question arises, in what way slow magnetoacoustic waves in coronal magnetic loops could be excited. It appears that the excitation may occur due to parametric resonance with 5 min oscillations of the photospheric convection velocity. The mechanism for this resonance may be as follows [8]: the 5 min oscillations of the velocity of the photospheric matter modulate the electric current in a coronal magnetic loop, due to "catching" of the convective motion of plasma by the magnetic field in the footpoints of the loops. The modulation of the current results in that of the sound speed in a coronal magnetic loop. Therefore, all eigenfrequencies of slow magnetoacoustic oscillations of loops appear to be modulated with the period 5 min, and parametric excitation of acoustic eigenmodes in a coronal magnetic loop by 5 min oscillations of the velocity of photospheric plasma (p-modes) becomes possible. The occurrence of parametric resonance in active regions was confirmed with observations of microwave radiation at the frequency 37 GHz [9]. In the study [9], quasi-periodic modulation of microwave radiation of solar flares at 37 GHz was considered. Approximately 90% of the observed microwave bursts reveal low-frequency modulation with the period 5 min. Approximately in 70% of the events, simultaneously with 5 min oscillations, oscillations with the period 10 min were observed. In 30% of the cases, simultaneous modulation of microwave radiation by three low-frequency signals with the periods 3.3, 5, and 10 min was detected. The reason for the "double" and "triple" modulation may be related to parametric excitation of acoustic oscillations with the given periods in a coronal magnetic loop; in this case, the periods correspond to the pumping frequency (5 min), subharmonics (10 min), and the first upper frequency of parametric resonance (3.3 min). Later on, the presence of parametric resonance in active regions was confirmed with observations of microwave emission at 11 GHz [10]. Since 5 min photospheric oscillations cannot penetrate to the corona directly, parametric resonance may serve as an efficient channel of transfer of energy of photospheric oscillations to the upper layers of the solar atmosphere. This fact is essential for the understanding of heating mechanisms of coronal plasma.

9.1.1 Parametric Excitation of Acoustic Oscillations

The excitation of acoustic oscillations in current-carrying coronal magnetic loops by parametric resonance with 5 min oscillations of the photospheric convection rate was considered in the study [9]. According to this study, if there are converging flows of plasma in footpoints of a loop, caused by photospheric convection, they generate electric current along the axis of the loop, which results from electromotive force produced by "catching" between the radial component of the photospheric convection rate and the azimuthal component of the magnetic field in the loop. Suppose that the photospheric convection rate oscillates around its mean value (for example, due to 5 min photospheric oscillations), following a $|V_{\mathrm{r}}| = V_0 + V_{\approx} \sin \omega t$ dependence with $|V_{\approx}| \ll V_0$. Then forced oscillations of the current flowing along the loops $I_z = I_0 + I_{\approx}$ will emerge, with the relative amplitude

$$\frac{I_{\approx}^m}{I_0} \approx \frac{h V_{\approx}}{\omega L r_0} \tag{9.6}$$

where h is the height interval in which e.m.f. is acting, r_0 the radius of the magnetic tube in the footpoints, and L the inductance of the loop as an equivalent electric circuit [9]. The height interval h usually extends approximately from lower layers of the solar photosphere to the transition region between the photosphere and chromosphere and is of an order of $h = 500–1000$ km.

Since variations of electric current I_z (and hence the magnetic field B_φ) are slow compared to the period of fast magnetoacoustic (sausage) oscillations of the tube, they may be considered adiabatic. Then from the condition of the equilibrium of the tube along the radial variable, it follows that the pressure in the tube will also vary periodically with the amplitude:

$$p_{\approx} = \frac{4}{3}\frac{I_0 I_{\approx}}{\pi c^2 r^2} \tag{9.7}$$

where c is the light speed and r the radius of the tube in its coronal part, considered to be constant. As a result, the speed of sound will be modulated with the period of 5 min oscillations, and the equation for acoustic oscillations takes the following form:

$$\frac{\partial^2 V_z}{\partial t^2} + \omega_0^2(1 + q\cos\omega t)V_z = 0 \tag{9.8}$$

where

$$\omega_0^2 = k_{||}^2 c_{s0}^2, k_{||} = \frac{s\pi}{l}, \qquad s = 1, 2, 3, \ldots \tag{9.9}$$

$$q = \frac{4}{3}\frac{\gamma - 1}{\gamma}\frac{I_0 I_{\approx}}{\pi c^2 r^2 p_0} \tag{9.10}$$

In Eqs. (9.8)–(9.10), $c_{s0} = (\gamma k_B T_0/m_i)^{1/2}$ is the speed of sound, $p_0 = 2nk_B T_0$ the pressure, T_0 – unperturbed temperature, and $\gamma = c_p/c_v$ the heat capacity ratio. Equation (9.8) is a Mathieu equation [11], which describes parametric instability. The latter arises in narrow zones close to the frequency

$$\omega_n = \frac{n\omega}{2}, n = 1, 2, 3, \ldots \tag{9.11}$$

This means that if, for example, a coronal magnetic loop is affected by 5 min photospheric oscillations, then acoustic oscillations with the periods 10, 5, 3.3 min, and so on may be excited in the loop. The excitation, however, will occur only when the eigenfrequency of acoustic oscillations in the loop ω_0 falls on the first zone of instability, that is, close to $\omega/2$. The width of the zone is of the order of q:

$$-\frac{q\omega_0}{2} < \frac{\omega}{2} - \omega_0 < \frac{q\omega_0}{2} \tag{9.12}$$

This implies that the coronal magnetic loop should have an appropriate length, so that 5 min photospheric oscillations could excite acoustic oscillations in it. In this case, the energy of 5 min photospheric oscillations, which in ordinary conditions are reflected from the temperature minimum, will penetrate high to the corona and may serve as the source of plasma heating in the coronal loop.

9.1.2 The Energy of Acoustic Oscillations

In order to estimate a possible role that slow magnetic sound excited in a coronal magnetic loop plays in the plasma heating, we should, first, determine the energy of acoustic oscillations generated as a result of parametric instability, and, second, make sure that dissipation of these oscillations is efficient enough to provide the necessary heating rate, which should exceed losses on thermal conductivity and optical radiation of the plasma. The rate amplitude in the acoustic oscillations $v_{\approx}$is related to the pressure amplitude $p_{\approx}$ as $v_{\approx} = p_{\approx}/\rho_0 c_{s0}$, where ρ_0 is the density of unperturbed plasma. Then, taking into account the relation (Eq. (9.7)), we shall obtain the following formula for the average energy density of acoustic oscillations:

$$W_s = \frac{\rho_0 v_{\approx}^2}{2} = \frac{8}{9}\frac{1}{\rho_0 c_{s0}^2}\left(\frac{I_0^2}{\pi c^2 r^2}\right)^2\left(\frac{I_{\approx}}{I_0}\right)^2 \tag{9.13}$$

It follows from this formula that W_s depends on the amplitude of the oscillations of electric current modulated by 5 min oscillations of the photospheric convection rate. The amplitude $I_{\approx}$ may be determined from low-frequency modulations of microwave radiation of a coronal magnetic loop [9]. In microwave radiation of flares, narrow-band modulation with the frequency of the order of fractions of a Hertz is often observed, which could be due to intrinsic oscillations of loops as an equivalent electric circuit (see Chapter 2). In this case, the frequency of the oscillations depends on the electric current in the circuit through the self-consistent magnetic field, while the large inductivity of the loops provides a high-quality factor of the oscillations. The eigenfrequency of the equivalent electric circuit is proportional to the current [12]:

$$\nu_{\mathrm{RLC}} \approx \frac{I}{2\pi c r^2 \sqrt{2\pi \Lambda n_0 m_i}}, \quad \Lambda = \ln\frac{4l}{\pi r} - \frac{7}{4} \tag{9.14}$$

where l is the length of the loop. Therefore, when 5 min photospheric oscillations modulate the electric current in the loop, the eigenfrequency of the equivalent circuit will also be modulated with the period 5 min, and the relative depth of the frequency modulation will coincide with the relative depth of the modulation of the electric current. Figure 9.1 presents an example of such modulation [9].

Figure 9.1 presents two solar microwave bursts observed on 28 August 1990: the time profile for the intensity at the frequency 37 GHz and the dynamic spectrum of low-frequency modulation of the radio intensity obtained by Wigner–Ville method. The latter transform displays appreciably higher frequency–time resolution; here, the modulation by 5 min oscillations is seen more clearly. Figure 9.1 shows that the relative depth of the frequency modulation ν_{LRC} is of the order of one to several percents. Similar values are obtained for a number of events [13]. Therefore, for further estimates, we can take

$$\frac{I_{\approx}}{I_0} \approx \frac{\Delta\nu_{\mathrm{LRC}}}{\nu_{\mathrm{LRC}}} \approx 10^{-2} \tag{9.15}$$

It follows from Eqs. (9.13) and (9.15) that for sufficiently large electric current in a coronal magnetic loop, the energy density for excited acoustic oscillations may be comparable with the density of the thermal energy of plasma.

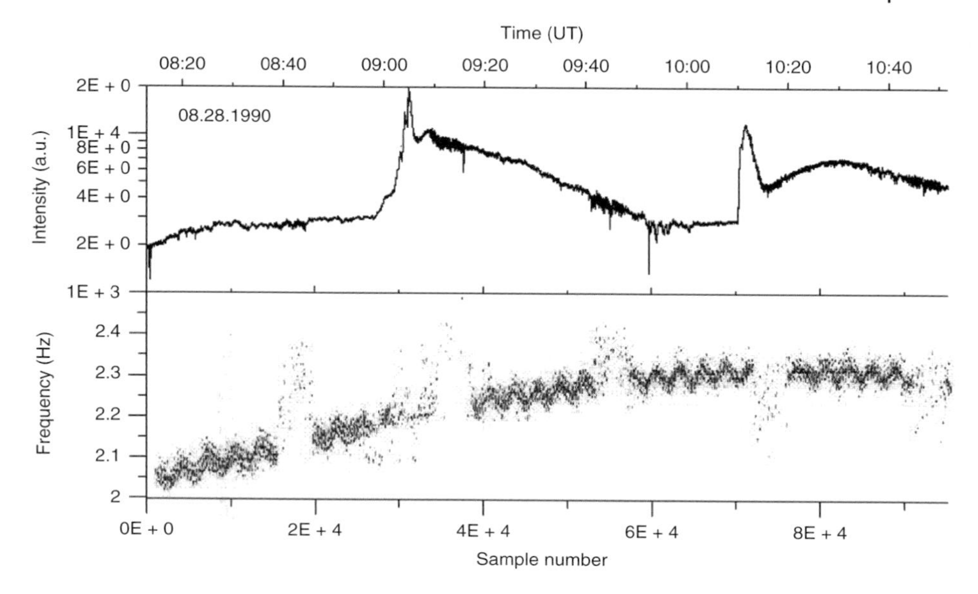

Figure 9.1 Modulation of flare microwave emission at 37 GHz by 5 min photospheric oscillations [9].

9.1.3
Acoustic Wave Heating Function

Dissipation effects, such as conductivity, viscosity, and thermal conductivity, result in transformation of the energy of acoustic oscillations into heat. If the damping rate for the energy of acoustic oscillations γ_s is substantially smaller than frequency, that is, $\gamma_s \ll \omega$, then the energy that transforms into heat owing to dissipation of acoustic oscillations, for a unit of time and in a unit volume or the "heating function," will be

$$H = \gamma_s W_s \tag{9.16}$$

In the case of acoustic waves, the decrement γ_s may be presented in the following form [7]:

$$\gamma_s = \gamma_{joul} + \gamma_{visc} + \gamma_T \tag{9.17}$$

The summands related to conductivity, viscosity, and thermal conductivity are, respectively,

$$\gamma_{joul} = \frac{c}{4\pi\sigma}\frac{c_s^2}{c_A^2}k, \gamma_{visc} = 0.18\omega^2\tau_i, \gamma_T = 0.12\frac{m_i}{m_e}\omega^2\tau_e \tag{9.18}$$

where σ is the conductivity, c_A Alfvén speed, $\tau_{i,e}$ characteristic times of ion–ion and electron–ion collisions, k wave number respectively. The estimate of the terms on the right-hand side of Eq. (9.17) shows that the main contribution to the dissipation of the ion sound is brought by electron thermal conductivity with the decrement

$\gamma_T \approx 7 \times 10^{-2}\omega$ (for $n \approx 10^9\,\text{cm}^{-3}$, $T \approx 10^6\,\text{K}$, $\omega \approx 0.02\,\text{s}^{-1}$). Therefore, taking into account Eqs. (9.13) and (9.18), the following formula for the heating function may be written:

$$H = b\frac{T^{1/2}}{n^2}\left(\frac{I_0^2}{c^2 r^2}\right)^2 \omega^2\,\text{erg}\,\text{cm}^{-3}\,\text{s}^{-1}, \qquad b = 1.5 \times 10^{15}\left(\frac{hV_{\approx}}{\omega L r_0}\right)^2 \quad (9.19)$$

It is apparent that the necessary condition for heating is that the heating function should exceed radiative losses, that is, $H \geq H_R = n^2\chi(T)$. The maximum value $\chi(T) = 10^{-21.2}\,\text{erg}\,\text{cm}^3\,\text{s}^{-1}$ is reached for $T \approx 10^6\,\text{K}$ [14]. Assuming $n \approx 10^9\,\text{cm}^{-3}$, $T \approx 10^6\,\text{K}$, $\omega \approx 2 \times 10^{-2}\,\text{s}^{-1}$ (which corresponds to the period of oscillations 5 min), we obtain the heating condition $B_\varphi \approx (I_0/cr) \geq 10\,\text{G}$. Here, B_φ represents the nonpotential part of the magnetic field related to the electric current along a coronal magnetic loop and corresponds to very weak twist of the magnetic field in the observed loops (10^2–10^3 G). The obtained estimate for B_φ corresponds to the electric current through the cross section of a loop $I_0 \geq 10^{10}\,\text{A}$.

Thereby, the parametric resonance becomes an efficient source of plasma heating, provided the electric current in the loop exceeds some critical value. The second necessary condition is that the length l of the loop should make it possible for the parametric resonance to appear:

$$l \approx \frac{2\pi c_{s0}}{\omega}\,\text{cm} \quad (9.20)$$

Characteristic frequencies of 5 min photospheric velocity oscillations are within the interval

$$1.5 \times 10^{-2}\,\text{s}^{-1} < \omega < 6 \times 10^{-2}\,\text{s}^{-1} \quad (9.21)$$

with the spectral maximum at the frequency $\omega \approx 2 \times 10^{-2}\,\text{s}^{-1}$. Therefore, the resonance will appear for coronal magnetic loops with the lengths

$$2 \times 10^9 \leq l \leq 2 \times 10^{10}\,\text{cm} \quad (9.22)$$

Note that "warm" magnetic loops with the temperature $\sim 1.5 \times 10^6$ K observed with transition region and coronal explorer mission [2] display the lengths $l \approx (1.5 - 7) \times 10^{10}$ cm, which are generally beyond the interval (Eq. (9.22)), whereas "hot" X-ray loops with the temperature $\sim(3$–$6) \times 10^6$ K observed with Yohkoh [3] have the lengths $l \approx (0.2$–$3) \times 10^{10}$ cm, which is consistent with the interval given by Eq. (9.22). This may mean that parametric resonance plays an important role in the origin of hot X-ray loops.

9.1.4 Thermal Balance in a Coronal Magnetic Loop

Consider now heating of a magnetic loop by acoustic waves excited in the case of parametric resonance between intrinsic acoustic oscillations of the loop and 5 min photospheric oscillations (p-modes). The thermal balance equation for a steady-state tube will be in the following form [4]:

$$\frac{\partial}{\partial s}\kappa_e T^{\frac{5}{2}}\frac{\partial T}{\partial s} = n^2\chi(T) - H \quad (9.23)$$

where $\kappa_e = 0.92 \times 10^{-6}$ erg cm^{-1} K$^{-7/2}$ is the thermal conductivity and s is the coordinate along loop axis. Here, it was also supposed in accordance with observations [15] that the loop cross-sectional area is constant in the corona.

If the loop height is smaller than the characteristic scale of pressure variations in the corona ($5 \times 10^3 T \approx (0.75 - 2.5) \times 10^{10}$ cm), the pressure inside the loop may be considered constant and the plasma number density may be expressed through pressure and temperature. In addition to that, for certainty, we will use as the function of radiative losses one of possible approximations in the temperature interval 10^5 K $< T < 10^7$ K [4]: $\chi(T) = \chi_0 T^{-1/2}$ erg cm^3 s^{-1}, where $\chi_0 = 10^{-19}$.

We will count the s-coordinate along the loop from the chromosphere temperature minimum. Besides, let us use the symmetry condition at the top of the loop. Then, the boundary conditions for Eq. (9.23) are

$$\frac{\mathrm{d}T}{\mathrm{d}s} = 0 \ (s = 0, T = T_0), \ \frac{\mathrm{d}T}{\mathrm{d}s} = 0 \left(s = \frac{l}{2}, T = T_1\right) \tag{9.24}$$

Integrating Eq. (9.23) along the length of the loop starting from the top, we obtain

$$\frac{1}{2}\kappa_e T^5 \left(\frac{\mathrm{d}T}{\mathrm{d}s}\right)^2 = \frac{\chi_0 p^2}{4k_B^2}(T - T_1) - \frac{4bk_B^2\omega^2}{6p^2}\left(\frac{I_0^2}{c^2r^2}\right)^2 (T^6 - T_1^6) \tag{9.25}$$

Using the first boundary condition from Eq. (9.24) and assuming $T_0 \ll T_1$, from Eq. (9.25), we find the relation between pressure, current, and temperature at the top of the loop:

$$\frac{\chi_0 p^2}{4k_B^2} = \frac{4bk_B^2\omega^2}{6p^2}\left(\frac{I_0^2}{c^2r^2}\right)^2 T_1^5 \tag{9.26}$$

With the use of Eq. (9.26), Eq. (9.25) may be transformed as

$$T^2\frac{\mathrm{d}T}{\mathrm{d}s} = \left(\frac{\chi_0 p^2}{2k_B^2\kappa_e}\right)^{1/2}\left(1 - \frac{T^5}{T_1^5}\right)^{1/2} \tag{9.27}$$

Integrating this equation from the footpoints to the top of the loop, we will obtain the known relation between the pressure, loop length, and temperature [4, 14]:

$$lp = A\left(\frac{8k_B^2\kappa_e}{\chi_0}\right)^{1/2} T_1^3, \ A = \int_0^1 x^2(1 - x^5)^{-1/2}\mathrm{d}x \approx 0.55 \tag{9.28}$$

From Eqs. (9.26) and (9.28), we can eliminate the pressure and find the dependence of the temperature at the top of the loop on the length:

$$T_1 = \left(\frac{b\chi_0\omega^2}{24\kappa_e^2 A^4}\right)^{1/7}(B_\varphi l)^{4/7} = 9.5\left(\frac{hV_\approx}{Lr_0}\right)^{2/7}(B_\varphi l)^{4/7} \approx 7\left(\frac{h}{r_0}\right)^{2/7} V_\approx^{2/7} l^{2/7} B_\varphi^{4/7} \tag{9.29}$$

where $B_\varphi \approx (I_0/cr)$ is the mean value of the azimuthal component of the electric current in the loop. In the transition to the latter formula, we expressed the inductance of the loop through its length [16]. It follows from Eq. (9.29) that the temperature in the loop increases with the loop length, the electric current along the loop, and the rate amplitude in photospheric oscillations. In the final result, the maximum temperature does not depend on the frequency of photospheric oscillations, since the factor b in Eq. (9.19) decreases as a square of the frequency. For $l = 10^{10}$ cm (the characteristic length of hot X-ray loops), from Eq. (9.29) we obtain an estimate for a typical temperature in hot X-ray loops observed with *Yohkoh*, $T_1 = (4-6) \times 10^6$ K, if we assume the azimuthal component of magnetic field $B_\varphi = 5{-}10$ G, which corresponds to the electric current $5 \times 10^9{-}10^{10}$ A through the cross section of the coronal magnetic loop. These values exceed the currents in "warm" magnetic loops, but they are by the factor of a few smaller than currents in flare loops [17].

9.1.5
Hot X-Ray Loops in the Solar Corona

As it was already noted before, the parameters of numerous magnetic loops that form the structure of the solar corona may differ substantially. For example, warm magnetic loops display the temperature $(1.5{-}2) \times 10^6$ K and a large length $(1.5{-}7) \times 10^{10}$ cm. Mechanisms for plasma heating in these loops are still unclear. Single flare loops have small lengths $5 \times 10^8 - 5 \times 10^9$ cm and high temperature, which may in some cases reach 4×10^7 K. The heating source here results from flare energy release, owing to which the loops become filled with a chromosphere plasma, suddenly heated to high temperature. Apparently, the so-called quasi-steady-state hot X-ray loops observed with Yohkoh [3] form a separate group. They display lengths $2 \times 10^9 \leq l \leq 2 \times 10^{10}$ cm and temperature $T = (4{-}6) \times 10^6$ K, intermediate between warm and simple flare loops. A specific feature of hot X-ray loops is that a doubled frequency of their acoustic oscillations $2 \times (\pi c_{S0}/l)$ belongs to the interval given by Eq. (9.21), $1.5 \times 10^{-2}\,\text{s}^{-1} < \omega < 6 \times 10^{-2}\,\text{s}^{-1}$, which corresponds to 5 min oscillations of the photospheric convection rate. In this case, as it was shown in Section 9.1.1, parametric excitation of acoustic oscillations in a coronal magnetic loop may occur, as a result of which the energy of 5 min photospheric oscillations penetrate into the corona and may heat the coronal plasma. As discussed in Section 9.1.2 that for sufficiently large electric current in a coronal magnetic loop, the density of the energy of the excited acoustic oscillations may be comparable with the density of the thermal plasma energy. Compared to, for example, Alfvén waves, acoustic oscillations display very efficient damping. Under the solar corona conditions, this damping is basically due to electron thermal conductivity, which results in the formation of a heating function exceeding maximum losses on optical radiation of the plasma for the current $I_0 \geq 10^{10}$ A. In hot X-ray loops, critical currents decrease, since for the temperature $T > 10^6$ K losses on optical radiation decrease with the increase of temperature (up to that of the order of 10^7 K). The analysis of the thermal balance equation with the obtained heating function indicates that the temperature in a loop increases from footpoints to the top. The observed temperature in hot X-ray loops $T = (4{-}6) \times 10^6$ K appears when the electric current in the loop reaches $5 \times 10^9 - 10^{10}$ A, which exceeds currents in "warm" magnetic loops, but is by the factor of a few smaller than currents in flare loops.

Quasi-steady-state X-ray loops in the solar corona are different from similar nonsteady-state structures both in the pressure and temperature. It is not excluded that a specific heating mechanism operates there. The data analysis yields the dependence of the temperature on the loop length $T \sim l^{0.27 \div 0.52}$ [3]. We saw that in this case the heating mechanism may be provided by dissipation of acoustic oscillations excited by photospheric 5 min oscillations in the presence of parametric resonance with intrinsic acoustic oscillations in a coronal magnetic loop. This mechanism provides a similar temperature-length dependence: $T \sim l^{0.286}$ (Eq. (9.29)).

9.1.6
Magnetic Loops in Late-Type Stars

Currently, it is commonly accepted that coronae of active stars, such as the solar corona, are filled with magnetic loops. In particular, evidence for that is provided by observations of pulsations of optical and X-ray radiation of active stars, which are considered to be related to oscillations of magnetic loops in stellar coronae [18–23]. The pulsations were observed both during flares and out of them. For example, Mullan and Jonson [18] observed oscillations in soft X-ray radiation of several dMe stars, which were not associated with any flares and displayed periods P from several tens to several hundreds of seconds. For the same stars, characteristic lengths of coronal magnetic loops and temperature were determined [23]:

$$\text{ADLeo: } R = 0.35 R_{\odot}, l = (0.4{-}4) \times 10^{10}\,\text{cm}, \; T = (2{-}3) \times 10^{7}\,\text{K}, P = 150{-}190\,\text{s}$$

$$\text{Proxima Cen: } R = 0.145 R_{\odot}, \; l = (0.5{-}9) \times 10^{9}\,\text{cm}, \; T = (1{-}3) \times 10^{7}\,\text{K}, \; P = 66{-}68\,\text{s}$$

$$\text{UVCet: } R = 0.15 R_{\odot}, \; l = (0.7{-}2) \times 10^{9}\text{cm}, \; T = (2{-}3) \times 10^{7}\text{K}, \; P = 56\;\text{s}$$

It is easy to check that the observed pulsation periods for each star are situated inside the corresponding periods of slow magnetoacoustic oscillations, which are specified by the scatter of l and T. Since these pulsations were not related to any flares, they are likely to have been excited by photospheric convection in the presence of parametric resonance between oscillations of photospheric plasma velocity and intrinsic acoustic oscillations of magnetic loops, as for dMe-type stars, the timescale of photospheric convection $\tau_c \approx 60{-}180$ s is close to the observed pulsation periods. The increase in the plasma temperature in coronal magnetic loops in the case of late-type stars may be due to the increase in electric current in the loops and the increase in the amplitude of the photospheric velocity oscillations. In red dwarfs, coronal magnetic fields may reach several kilogausses. The photospheric convection rate also increases, since with the decrease in the radius and temperature, the role of convection in the energy transfer from the center to the surface of the star increases. Estimates show that, for example, for AD

Leo the temperature of loops $T = (2-3) \times 10^7$ K may be obtained if in Eq. (9.29) we assume $l \approx 2 \times 10^{10}$ cm, $V_{\approx} \approx 10^6$ cm s^{-1}, and $B_\varphi \approx 500-1000$ G.

Thereby, slow magnetoacoustic waves may provide efficient heating in coronal magnetic loops, if the periods of these waves satisfy the condition of "resonance" with characteristic periods in oscillations of the photospheric convection rate. The energy of the excited acoustic oscillation may be compared to the thermal energy of the plasma. The heating mechanism is of a threshold type, that is, the heating function exceeds losses on optical radiation if the electric current in the magnetic loop exceeds some threshold value of the order of 10^9 A, under the solar coronal conditions. Suitable "resonance" lengths needed for parametric excitation of sonic oscillations are displayed by quasi-steady-state hot X-ray loops with the temperature $T = (4-6) \times 10^6$ K. In these loops, the observed temperature rises with the current $5 \times 10^9 - 10^{10}$ A. The considered heating mechanism yields the temperature $T = (2-3) \times 10^7$ K, characteristic for magnetic loops in the coronae of late-type stars, if we assume higher photospheric convection rate and the electric current in the magnetic loop.

9.2 Ohmic Dissipation of Electric Currents

Ohmic dissipation of the electric currents flowing along a loop was also considered as a possible heating mechanism [24, 25]. The rate of plasma heating due to electric currents dissipation one can obtain from the generalized Ohm's law (see Chapter 2, Eq. (2.3)):

$$q_J(r) = \left(\vec{E} + \frac{1}{c}\vec{V} \times \vec{B}\right)\vec{j} = \frac{j^2}{\sigma} + \frac{F^2}{(2-F)c^2 n m_i \nu'_{ia}} \left(\vec{j} \times \vec{B}\right)^2 \tag{9.30}$$

The first term on the right-hand side of Eq. (9.30) describes the current dissipation in a magnetic flux tube due to the Spitzer (classical) conductivity caused by electron–ion and electron–atom collisions. The second term in the right part of Eq. (9.30) appears only in the case when the neutral atoms exist in the plasma. This term describes the current dissipation due to ion–atom collisions. Here, ν'_{ia} is the effective frequency of ion–atom collisions, F the relative density of neutrals, and magnetic field $\boldsymbol{B}$ includes both an ambient field $\boldsymbol{B}_0$ and the magnetic field caused by the electric current. Let us consider a simple case when the current flows along the ambient magnetic field $\boldsymbol{B}_0$ and is distributed uniformly in the flux tube of radius r_0 with the axis coincided with $\boldsymbol{B}_0$. Taking into account that $\left(\vec{j} \times \vec{B}\right) = -jB_\varphi$ where $B_\varphi = 2\pi jr/c$, $r \leq r_0$ is azimuthal component of the magnetic field driven by the current, and the total current through the cross section of the flux tube is $I = \pi r_0^2 j$, we obtain for the Joule heating rate:

$$q_J = \frac{j^2}{\sigma} + \frac{F^2 B_\varphi^2 j^2}{(2-F)c^2 n m_i \nu'_{ia}} \tag{9.31}$$

If there are no neutral atoms in the plasma ($F = 0$), then to fulfill the necessary condition for heating in the case of classical conductivity σ

$$q_{J_{||}} = \frac{j_{||}^2}{\sigma} \geq n^2 \chi(T) \tag{9.32}$$

too strong currents are required ($I \geq 10^{14}$–10^{15} A), which exceed the observed values [26] by four to five orders of magnitude. The required value of the current can be decreased if the regime of anomalous resistivity takes a place, for example, owing to the Buneman instability when electron–ion relative velocity exceeds the electron thermal velocity. Anomalous resistivity, however, occurs under the condition of an extremely large filling factor, $\sim 10^{-6}$, of the current in the flux tube: the thickness of the filaments should be of the order of several meters, which is rather unlikely. In Eq. (9.32), $j_{||}$ is the component of the density of the electric current parallel to the magnetic field.

On the other hand, if there are some neutral atoms in the plasma, then Joule dissipation can be essentially increased. In that case, the second term in the right part of Eq. (9.31) can exceed the first one. For example, the dissipation due to ion–atom collisions caused by the so-called Cowling conductivity exceeds the dissipation due to classical Spitzer conductivity by more than four orders of magnitude for the magnetic loops of the thickness 2×10^8 cm, plasma density $n = 10^9$ cm^{-3}, temperature $T = 10^6$ K, and the azimuthal component of the magnetic field $B_\phi = 60$ G even for very few neutral atoms, $F \sim 10^{-6}$–10^{-7}. The current dissipation increases in the course of ion–atom collisions due to two reasons. First, the ampere force in Eq. (9.30) can provide the acceleration of the ions to the velocity that exceeds ion–electron relative velocity in the electric current and thereby supply the relative large energy to the ions. Second, substantial part of acquired kinetic energy of ions transfers in the course of ion–atom collisions to the atoms, whereas the energy transfer during electron–ion collisions is slower because the large ion–electron mass ratio. A decisive role of ion–atom collisions in the electric current dissipation was drawn into attention in [27].

As discussed in Chapter 2, very large electric currents, up to 10^{11} A, can exist in coronal magnetic loops. The sources of "superhot" plasma can appear inside a loop due to such currents. Let us discuss this point in more detail. For example, if the effectiveness of ion–atom collisions is determined by the recharge cross section, then in the temperature range $10^5 \leq T \leq 10^7$ K the effective collision frequency is $\nu'_{\mathrm{ia}} \approx 10^{-11} F n T^{1/2}$. The relative mass of the neutrals can be represented as [28, 29]

$$F(T) = \frac{\xi(T)}{T} \tag{9.33}$$

where the function $\xi(T) \approx 0.15$ has almost no temperature dependence for $T \geq 10^5$ K. Taking into account this circumstances and omitting the first term on the right-hand side of Eq. (9.31), we obtain the formula for the specific heating rate:

$$q \approx 2.6 \times 10^{-9} \frac{I^4}{n^2 r_0^6 T^{3/2}} \,\mathrm{erg\ cm^{-3}\ s^{-1}} \tag{9.34}$$

The heating rate decreases when the temperature increases. When the temperature is high enough, the heating will compensate the radiative losses. This radiative losses for $T \geq 5 \times 10^6$ K can be approximated by the function $q_R \approx 3 \times 10^{-27} n^2 T^{1/2}$ [30]. Then from the equality $q = q_R$, one can estimate the plasma temperature:

$$T_h \approx 10^9 \frac{I^2}{n^2 r_0^3} \tag{9.35}$$

The cross-sectional area in the coronal part of a loop is quite large, so the electric current density is not enough for the plasma heating up to high temperature. In the chromospheres, however, the loop radius decreases due to an increase in the ambient pressure, the current density grows, and at some height above the photosphere quite effective plasma heating is possible. Indeed, for the current $I = 3 \times 10^{10}$ A $\approx 10^{20}$ cgs and the loop radius 2×10^7 cm, the plasma temperature can reach 10^7 K in the loop domains with plasma density $n \approx 3 \times 10^{10}$ cm^{-3}, which corresponds approximately to the transition region from the chromosphere to the corona. Apparently this heating process cannot be a steady state. "Superhot"

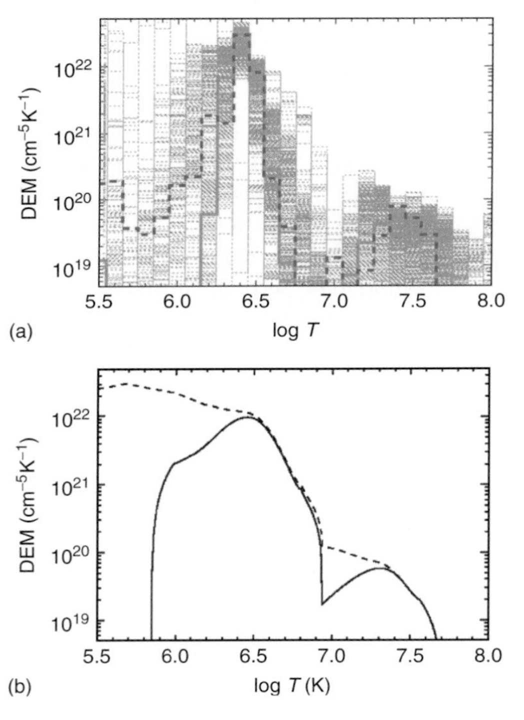

Figure 9.2 Differential measure of emission as a function of temperature for the flare-free active region AR 10955 (S09W30), 13 May 2007, 1800 UT (a). Modeling of heating in the above region by microflares with the average energy flux through the footpoints of loops $5.6 \times 10^6 - 3.0 \times 10^7$ erg cm^{-2} s^{-1} (b) [32].

plasma that is injected into the coronal part of a loop coexists with own loop plasma, thereby being the heating source of its. The energy exchange between both plasma components occurs, for example, owing to the particle collisions and thermal conductivity.

Data obtained with Hinode and Reuven Ramaty High-Energy Solar Spectroscopic Imager space observatories [31, 32] indicate that in flare-free active regions, the differential measure of emission of soft X-ray radiation displays, along with the component with the temperature 2–3 million degrees, a less intense, but sufficiently distinct, component with the temperature about 20 million degrees (Figure 9.2). Existence of such high-temperature component testifies to the important role of electric current dissipation in the heating of coronal loop plasma.

9.3 Heating by Microflares

The "superhot" component in the differential emission measure can also provide evidence for microflares, which transform the energy of the magnetic field to the thermal energy of plasma [31, 32]. The microflares may result, for example, from tearing instability [33] or magnetic reconnection processes [34, 35]. They may also originate due to variations of the magnetic field in the photospheric footpoints of the loops, which, in turn, stem from stochastic variations of the photospheric convection rate. As a result, inside the loop, current sheets may be formed and electric fields may be induced, which, in turn, will heat the plasma and accelerate charged particles. Such an event may be interpreted as a nano- or microflare, depending on the energy release rate.

Below, we consider the possibility for microflare generation owing to the flute instability at the footpoints of coronal magnetic loops. Note that for the average energy of a microflare $\sim 10^{26}$ erg, they should occur in a loop rather frequently, once in every 100 s, to compensate for radiative losses. Studies of low-frequency modulation of the microwave radiation of coronal magnetic loops [36] showed that directly before the flare energy release, the electric current in the loop increases substantially. We analyzed microwave radiation of two flares detected at Nobeyama radio observatory on 2 November 1992 and 30 March 2001 at the frequencies 35, 17, 9, 3.75, 2, and 1 GHz. In both cases, the radiation went from coronal magnetic loops. The Wigner–Ville analysis of the low-frequency modulation showed (Figure 9.3) that at the preflare stage, quasi-steady-state oscillations with the frequency $\nu \approx 0.005$ Hz (the corresponding period $P \approx 200$ s) exist in the loop, which cannot be explained by intrinsic oscillations in the loop as an MHD resonator.

Indeed, fast magnetoacoustic, Alfvén, kink, and slow magnetoacoustic oscillations of the measured parameters of the loop displayed periods appreciably different from the observed modulation period. It was suggested that in this case the modulation results from oscillations in a coronal magnetic loop as an equivalent electric circuit. Then, the frequency of the oscillations depends on the electric

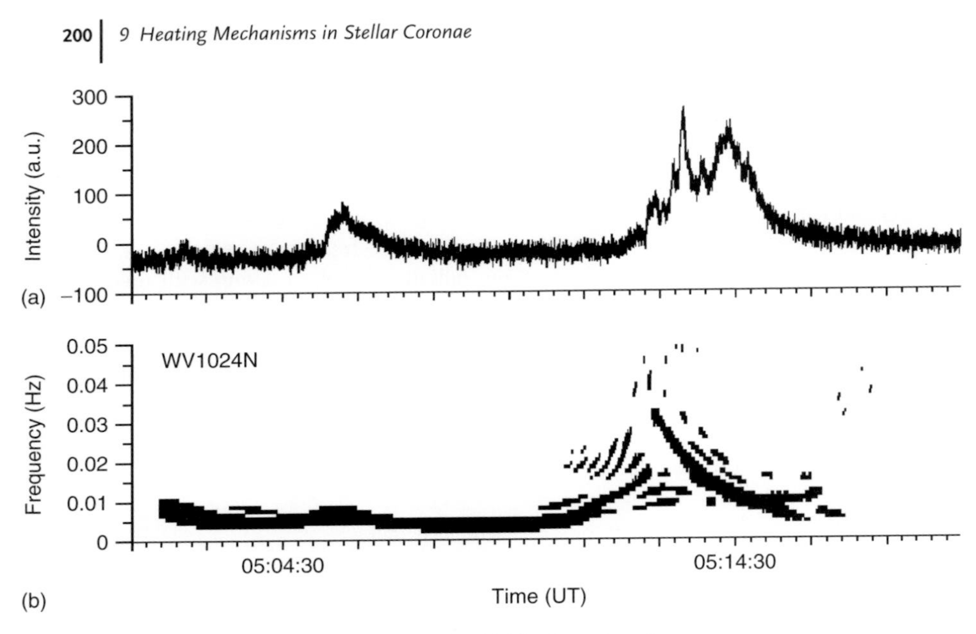

Figure 9.3 The 30 March 2001 event. (a) The intensity of microwave radiation of a coronal magnetic loop at 17 GHz, according to Nobeyama spectropolarimetric data. (b) The low-frequency modulation spectrum of the microwave radiation obtained with the Wigner–Ville transformation.

current flowing along the magnetic loop and generated as a result of the interaction between convective flows of photospheric plasma and the magnetic field in the loop footpoints (Chapter 2). The estimates indicate that the modulation of the microwave radiation with the frequency $\nu \approx 0.005$ Hz at the preflare phase appears when the current in the coronal magnetic loop reaches $I_0 \approx 10^{10}$ A. Approximately 50 s before the flare, the modulation frequency increases relatively rapidly from 0.005 to 0.035 Hz, that is, by a factor of 7. Consequently, the electric current in the coronal magnetic loop increases by the same factor. The characteristic time of the current increase is about 50–100 s; at the final stage, it decreases to 20 s. After the maximum has been reached, the modulation frequency and, hence, the electric current decreases to the preflare values for the time of the order of the duration of the flare ($t_f \approx 200$ s). A similar pattern was also seen in two flares in 2 November 1992 event.

Thereby, the data show that the flare process is preceded by a substantial increase in the electric current in the coronal magnetic loop. This may result from developing the balloon mode of the flute instability in the footpoints of the coronal magnetic loop, which, in this case, triggers the flare. Penetration of a plasma "tongue" into the current channel in the loop causes supplementary contraction of the magnetic field and the increase in the current, which, in turn, increases dissipation [36]. Besides, the magnetic field variation is accompanied by generation of induction electric field, which results in particle acceleration and supplementary heating of plasma. If these variations of the current occur sufficiently frequently and if their amplitude is not too large, these processes may be treated as microflares, and they may result in the heating of coronal plasma. Figure 9.4 presents an example of this, for the event on 2 November 1992.

Here, Figure 9.4a presents the intensity of microwave radiation from a coronal magnetic loop at the frequency 2 GHz, according to Nobeyama radio data. Figure 9.4b show the variations of the frequency of the oscillations in the loop as an equivalent electric circuit. These variations correspond to minor increases in the current in the loop, with the amplitude $(1-3) \times 10^{10}$ A. The duration and pulse–duration ratio of such increases are approximately equal, around 100 s. With these parameters, the average energy release rate in the "microflares" caused by the current dissipation is $(10^{24}-10^{25})$ erg s^{-1}.

The flute instability at the footpoints of coronal magnetic loops may be explained as follows [27]. Owing to the decrease of the pressure of the ambient plasma, the loop radius in its chromospheric footpoint increases with the height; as a result, the curvature of the magnetic field appears, directed toward the loop center. The radius of the curvature is of the order of the height of the inhomogeneous atmosphere at the loop footpoints. The curvature causes centrifugal force directed inside the tube; thereby, conditions for the flute instability are formed. The instability criterion in this case is

$$2\frac{n}{n+n_{\mathrm{a}}} - \cos\theta > 0 \tag{9.36}$$

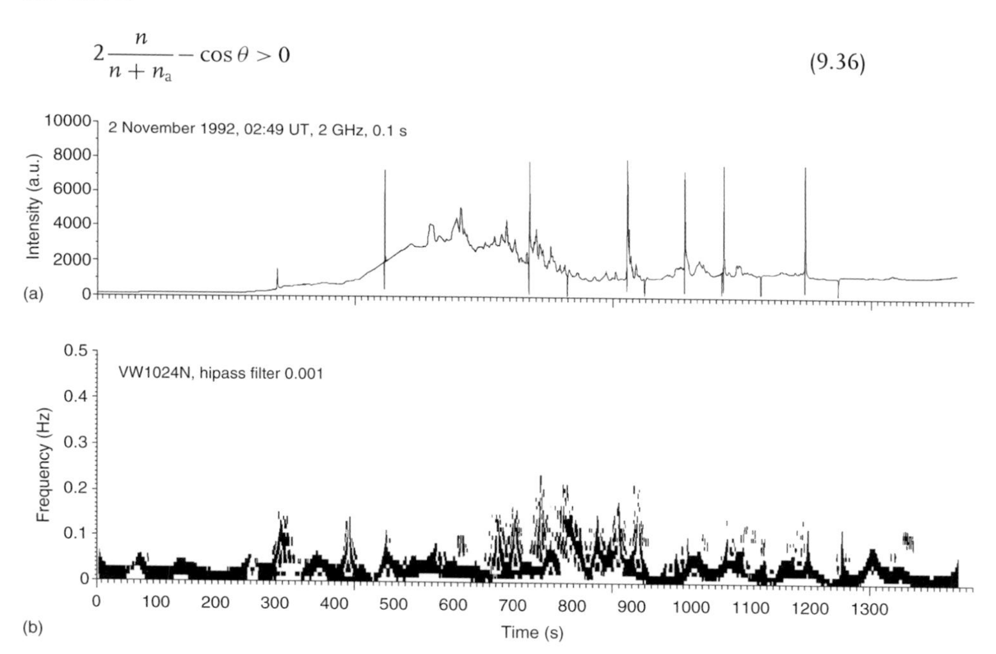

Figure 9.4 (a) The intensity of microwave radiation from a coronal magnetic loop at the frequency 2 GHz, according to Nobeyama data for 2 November 1992 event. (b) Dynamic spectra for the low-frequency modulation of the intensity, obtained with Wigner–Ville transformations.

where n and n_a are the number densities for electrons and neutrals, respectively, and θ is the angle between the curvature radius and the vertical line [27]. It follows from Eq. (9.36) that in order for the flute instability to develop, the ionization rate should not be very small, which means that the loop footpoints should be heated previously. The characteristic growth time for the ballooning mode of the flute instability is

$$t_b \approx 2 \times 10^{-3} \sqrt{H(\text{cm})}\ \text{s} \tag{9.37}$$

Here, $H = \kappa_B T/\mu m_i g$ is the scale height of the inhomogeneous atmosphere, where in the solar atmosphere $\mu \approx 1.4$ for H:He = 10:1. For the plasma temperature in the vicinity of the loop footpoints $T \approx 10^5$ K, the growth time for the ballooning instability is about $t_b \approx 35$ s, which is comparable with the observed growth time for the electric current in the loop before the impulse phase of a flare. The flute instability drives penetration of additional plasma from the vicinity of the footpoints into the loop. As a result, as it follows from the induction equation

$$\frac{\partial \vec{B}}{\partial t} = \nabla \times \left[\vec{V} \times \vec{B}\right] \tag{9.38}$$

that the electric current strengthens. For example, while before the development of the flute instability the azimuthal component of the magnetic field depends on the loop radius linearly, that is, $B_{\varphi 0}(r) = B_{\varphi 0} r/r_1$, where r_1 is the loop radius in the instability domain, during the plasma tongue penetration with the velocity

$$V_r(r,t) = \frac{-V_r(t) r}{r_1} \tag{9.39}$$

the longitudinal electric current in the loop will grow as follows:

$$I(t) = I_0 \exp\left[2 \int_0^t \frac{V_r(t')}{r_1}\right] dt' \tag{9.40}$$

The exponent index in Eq. (9.40) depends on the current. The increase in the current results in the heating of the plasma and the increase in the internal pressure in the loop, and the plasma tongue eventually decelerates. In other words, the exponential term is determined by the nonlinear stage of the instability. An approach to the study of the nonlinear stage of the flute instability is presented in [37]. The observed increase in the electric current by the factor of a few before a microflare corresponds to the exponential term in Eq. (9.40) of the order of a unity.

The microflare mechanism of the heating in a coronal magnetic loop in the presence of the flute instability can be understood as follows. The flute instability results in the increase in the current in the loop. When the current reaches $\sim 3 \times 10^{10}$ A, which is roughly an order of magnitude smaller than the current emerging in regular solar flares, hot ($\sim 10^7$ K) plasma is formed because of the dissipation of the current in the transition region. Then, the hot plasma evaporates into the coronal part of the loop and heats the plasma in the loop through particle collisions and thermal conductivity. The emission measure of the hot component is $\text{EM} = n_h^2 l \approx 10^{26} \div 10^{27}\ \text{cm}^{-5}$ [31, 32], which for the characteristic length of a loop $l \approx 5 \times 10^9$ cm yields the number density of the hot plasma in the loop $n_h \approx (1-4) \times 10^8$ cm^{-3}. The evaporation rate is roughly equal to the ion thermal velocity $V_{Ti} = (\kappa_B T_h/m_i)^{1/2}$. The decrease in the plasma density in the heating region of the loop is compensated by the cold plasma coming from the lower chromosphere. For the time of the order of a microflare duration t_f, the energy $W \approx 2n\kappa_B T_h \pi r_0^2 V_{Ti} t_f$ enters the coronal part of the loop. Here, n and r_0 are the density and the loop radius in the domain of the hot plasma origin, respectively. This energy is distributed over the coronal part of the loop with the volume $\pi r^2 l$, where r is loop radius in the corona. Then, the average heating function has the following form:

$$H \approx 2n\kappa_B T_h \frac{r_0^2 V_{Ti}(T_h) t_f}{r^2 l t_{ff}} \tag{9.41}$$

where t_{ff} is the characteristic time interval between microflares. Similar to the case of heating by magnetoacoustic waves (Section 9.1), we restrict consideration with compact magnetic loops; therefore, the pressure inside a loop may be considered to be constant. Then, using the same problem formulation as that in Section 9.1.4, we obtain the dependences of the maximum plasma temperature at the top of the loop on the heating function, loop length, and pressure:

$$lp = A_1 \left(\frac{8k_B\kappa_e}{\chi_0}\right)^{1/2} T_1^3 \tag{9.42}$$

$$T_1 = \left(\frac{1}{7{A_1}^2\kappa_e}\right)^{2/7} H^{2/7} l^{4/7} \approx 10^6 \left(\frac{I(A)}{3 \times 10^{10}}\right)^{6/7} \left(\frac{t_f}{t_{ff}}\right)^{2/7} \text{K} \tag{9.43}$$

where $A_1 = \int_0^1 (1-x^{-5/2})x^2 dx \approx 0.72$. The latter equality in Eq. (9.43) corresponds to the loop length $l = 5 \times 10^9$ cm and the loop thickness in the coronal part $2r = 4 \times 10^8$ cm. From Eq. (9.43), it follows that microflares driven by the flute instability at loop footpoints may lead to heating of the loop if the characteristic time interval between the microflares t_{ff} does not differ substantially from the microflare duration t_f, in spite of the fact that in Eq. (9.43) these two parameters are rather insignificant factors.

References

1. Aschwanden, M.J. (2005) *Physics of Solar Corona*, Springer, Praxis Publishing, Chichester.
2. Aschwanden, M.J., Newmark, J.S., Delaboudiniere, J.-P. *et al.* (1999) *Astrophys. J.*, **515**, 842.
3. Kano, R. and Tsuneta, S. (1995) *Astrophys. J.*, **454**, 934.
4. Priest, E. (1982) *Solar Magnetohydrodynamics*, D. Reidel Publishing Company, Dordrecht, Holland.
5. Ionson, J.A. (1982) *Astrophys. J.*, **254**, 318.
6. Mullan, D.J. and Cheng, Q.Q. (1994) *Astrophys. J.*, **420**, 392.
7. Braginskii, S.I. (1965) *Rev. Plasma Phys.*, **1**, 205.

8. Zaitsev, V.V. and Kislyakova, K.G. (2010) *Astron. Rep.*, **54**, 367.
9. Zaitsev, V.V. and Kislyakov, A.G. (2006) *Astron. Rep.*, **50**, 823.
10. Kislyakova, K.G., Zaitsev, V.V., Urpo, S., and Riehokainen, A. (2011) *Astron. Rep.*, **55**, 575.
11. Landau, L.D. and Lifshitz, E.M. (1976) *Mechanics*, Pergamon Press, Oxford.
12. Zaitsev, V.V., Stepanov, A.V., Urpo, S., and Pohjolainen, S. (1998) *Astron. Astrophys.*, **337**, 887.
13. Zaitsev, V.V., Kislyakov, A.G., and Urpo, S. (2003) *Radiophys. Quantum Electron.*, **46**, 999.
14. Rosner, R., Tucker, W.H., and Vaiana, G.S. (1978) *Astrophys. J.*, **230**, 643.
15. Klimchuk, A., Lemen, J.R., Feldman, U., Tsuneta, S., and Uchida, Y. (1992) *Publ. Astron. Soc. Jpn.*, **44**, L181.
16. Landau, L.D. and Lifshitz, E.M. (1984) *Electrodynamics of Continuous Media*, Pergamon Press, Oxford.
17. Zaitsev, V.V. and Kruglov, A.A. (2009) *Radiophys. Quantum Electron.*, **52**, 323.
18. Mullan, D.J. and Johnson, M. (1995) *Astrophys. J.*, **444**, 350.
19. Rodono, M. (1974) *Astron. Astrophys.*, **32**, 337.
20. Andrews, A.D. (1974) *Astron. Astrophys.*, **239**, 235.
21. Mullan, D.J., Herr, R.B., and Bhattacharyya, S. (1992) *Astrophys. J.*, **391**, 265.
22. Mitra-Kraev, U., Harra, L.K., Williams, D.R., and Kraev, E. (2005) *Astron. Astrophys.*, **436**, 1041.
23. Mullan, D.J., Mathioudakis, M., Bloomfield, D.S., and Christian, D.J. (2006) *Astrophys. J. Suppl.*, **164**, 173.
24. Holman, G.D. (1985) *Astrophys. J.*, **293**, 584.
25. van Balegooijen, A.A. (1986) *Astrophys. J.*, **311**, 1001.
26. Hagyard, M.J. (1988) *Sol. Phys.*, **115**, 107.
27. Zaitsev, V.V. and Shibasaki, K. (2005) *Astron. Rep.*, **49**, 1009.
28. Verner, D.A. and Ferland, G.J. (1996) *Astrophys. J. Suppl.*, **103**, 467.
29. McWhirter, R.W.P. (1965) Spectral intensities, in *Plasma Diagnostic Techniques* (eds R.H. Huddlestone and S.I. Leonard), Academic Press, New York 201.
30. McWhirter, R.W.P., Thonemann, P.C., and Wilson, R. (1975) *Astron. Astrophys.*, **40**, 63.
31. Schmelz, J.T., Saar, S.H., DeLuca, E.E., Golub, L. *et al.* (2009) *Astrophys. J.*, **693**, L131.
32. Schmelz, J.T., Kashyap, V.L., Saar, S.H., Dennis, B.R. *et al.* (2009) *Astrophys. J.*, **704**, 863.
33. Galeev, A.A., Rosner, R., Serio, S., and Vaiana, G.S. (1979) *Astrophys. J.*, **243**, 301.
34. Parker, E.N. (1988) *Astrophys. J.*, **330**, 474.
35. Vlahos, L., Isliker, H., and Lepreti, F. (2004) *Astrophys. J.*, **608**, 540.
36. Zaitsev, V.V., Kislyakova, K.G., Altyntsev, A.T., and Meshalkina, N. (2011) *Radiophys. Quantum Electron.*, **51**, 219.
37. Zaitsev, V.V., Urpo, S., and Stepanov, A.V. (2000) *Astron. Astrophys.*, **357**, 1105.

10
Loops and QPOs in Neutron Stars and Accretion Disk Coronae

Solar–stellar analogy is used also for the interpretation of flaring processes in quite exotic astrophysical objects, neutron stars, and accretion disks. In this chapter, the origin of Quasi Periodical Oscillations (QPOs) from the single-neutron stars – magnetars – is considered and some peculiarities of accretion disk coronae are discussed. There is quite a lot astrophysical literature devoted to these objects [1–3]. Nevertheless, the gap between theory and observations is still exists.

10.1
The Origin of Fast QPOs from Magnetars and Diagnostics of Magnetar Corona

Quasi-periodical pulsations (QPOs) were observed also from the magnetars – a class of neutron stars with external magnetic fields in excess of 10^{14} G and internal fields that could be as high as 10^{16} G [4]. Their activity is powered by the decay of the ultrastrong magnetic fields and lasts about 10^4 years. They are observed as either soft gamma repeaters (SGRs) or anomalous X-ray pulsars.

The three giant flares of neutron stars (Table 10.1) occurred on 5 March 1979 (SGR 0526-66), 27 August 1998 (SGR 1900 + 14), and 27 December 2004 (SGR 1806-20), with an energy release of $10^{44}-10^{46}$erg, were accompanied by high-frequency (from tens to thousands hertz) quasi-periodic oscillations (QPOs) of the X-ray emission [5–7].

Such oscillations were observed not only in tails of the flares, 100–300 s after the main pulse with a duration ≤ 1 s, but also at the growth phase of the main pulse [5, 8, 9]. The greatest variety of the oscillations with frequencies from 18 to 2384 Hz was recorded by the Rossi X-Ray Timing Explorer (RXTE) and Reuven Ramaty High-Energy Solar Spectroscopic Imager (RHESSI) space observatories on the "ringing tail" of the flare from SGR 1806-20 [10]. The amplitude of the high-frequency oscillations is modulated by the magnetar's rotation with a spin period of the order of several seconds (Figure 10.1).

The model of high-frequency pulsations must explain not only their period and excitation mechanism but also their high-quality factor, $Q \geq 10^4-10^5$. For example, during the flare of SGR 1806-20 on 27 December 2004, oscillations with a frequency 1840 Hz were observed for 50 s, while those with a frequency 625 Hz

Coronal Seismology: Waves and Oscillations in Stellar Coronae, First Edition.
A. V. Stepanov, V. V. Zaitsev, and V. M. Nakariakov.

Table 10.1 Three Giant Flares from SGRs.

	Energy in initial spike (erg)	Energy in pulsating tail (erg)
5 March 1979 SGR 0526-66	1.6×10^{44}	4×10^{44}
27 August 1998 SGR 1900+14	$>7 \times 10^{43}$	5×10^{43}
27 December 2004 SGR 1806-20	2×10^{46}	10^{44}

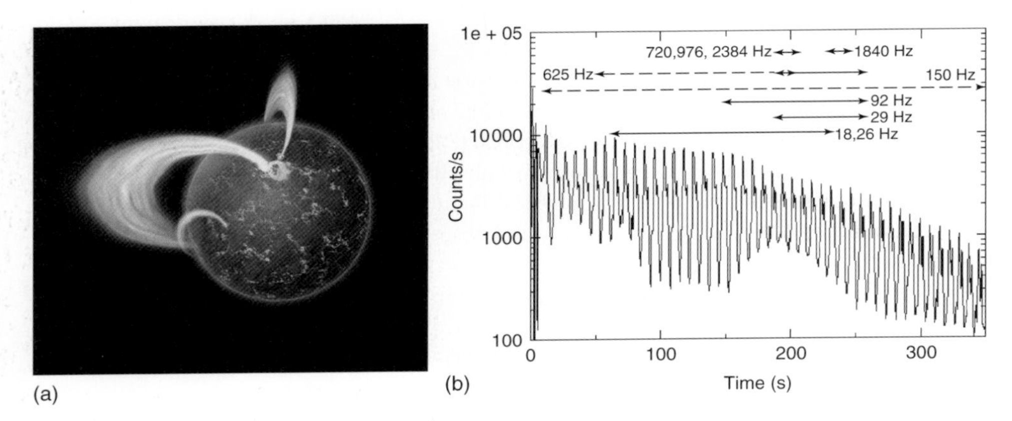

Figure 10.1 (a) Artist impression of a neutron star with coronal loops. The frequencies of fast QPOs detected by RXTE and RHESSI in the "ringing tail" of SGR 1806-20. (b) The spin period is 7.56 s [10].

lasted for 200 s [7]. A significant fraction of the existing models cannot explain the complete set of the observed characteristics for the oscillations with frequencies from 20 to 2400 Hz. The approach presented here is based on the representation of the oscillation source – a hot electron–positron plasma in a *trapped fireball* – as a system of magnetic loops with an electric current [11], and by analogy with a flare loop, an equivalent RLC circuit [12, 13]. As an illustration of the efficiency of the proposed model, the diagnostics of the corona of the SGR 1806-20 is presented.

10.1.1
A Brief Overview of the Existing Models

According to present views, the giant flares of SGRs – single-neutron stars with the radius ~10–30 km, a mass $\sim 1.5\ M_{\mathrm{Sun}}$ (Figure 10.1), and a magnetic field $B \sim 10^{14}$–10^{15} G – result from sudden disturbances in the metal stellar crust: tectonic plate motions [14], starquakes triggering a catastrophic rearrangement of electric currents and fields in the magnetosphere that is accompanied by an enormous energy release, 10^{44}–10^{46}erg. The forming fireball of hot (~1 MeV) electron–positron plasma and high-energy photons [15] is the source of the main pulse of the flare, which lasts for several fractions of a second. The hot plasma fireball leaves the stellar surface at the speed of light. The remnant – the trapped fireball, that is, plasma trapped by the magnetic field frozen into the stellar crust, decays within several minutes and provides the flare tail, on which high-frequency quasi-periodic pulsations, along with spin second-scale oscillations, are prominent.

The models of high-frequency QPOs based on global seismic vibrations of a magnetar are most popular [16]. The popularity of these models is attributed to a rapid development of asteroseismology, which opens new possibilities in studying the interior structure and crust of neutron stars. Early models of neutron stars implied the excitation of torsional vibrations of a crust with a shear during starquakes. The crustal motions cause a modulation of superstrong ($10^{14}-10^{15}$ G) magnetic fields and electric currents in the magnetosphere of a neutron star, which produce variations in the X-ray flux [10]. The oscillations with frequencies from 30 (the $n = 0$, $l = 2$ mode) to 1840 Hz ($n = 3$) can be explained in this way. However, the observed pulsations with the frequency $\leq$20 Hz cannot be explained by the torsional vibrations of a crust with a shear [10]. The cause of the high Q-factor of oscillations is not investigated in seismic models either. Moreover, Levin [17] drew attention to the rapid, on a timescale of the order of 10 oscillations, decay of torsional crustal modes due to the transfer of their energy into Alfvén waves, which effectively decay in the inner layers of the magnetar. Therefore, Levin suggests that either QPO should be of magnetospheric origin or the magnetic field of the core should display a specific configuration before a flare. At present, the role of the fine crustal structure and peculiarities of the magnetic field configuration in neutron stars are discussed in seismic QPO models [18].

Beloborodov and Thompson [11] proposed a possible alternative to the mechanism of global seismic vibrations: the QPO source is located in the magnetosphere. They drew attention to the fact that nonlinear oscillations of the production of electron–positron pairs emerge during the formation of the magnetar corona consisting of a set of magnetic loops (Figure 10.2). The electric field generated in an archlike magnetic flux tube owing to its twisting and the coronal plasma regulates each other self-consistently. More specifically, the electric field must be strong enough for electron–positron pairs to be produced, but the field decreases because of screening when the corona is saturated with plasma. The process is then repeated. A numerical experiment showed that the electric current oscillations in a coronal loop can have a fairly high frequency ~10 kHz [11].

Ma *et al.* [19] proposed an interpretation of the QPOs based on magnetohydrodynamic (MHD) oscillations of coronal magnetic loops in the magnetar corona. These authors believe that oscillations of magnetic loops are excited by the turbulence at the loop footpoints in the metal stellar crust. Standing slow magnetoacoustic modes of coronal magnetic loops modulating the SGR emission are invoked to explain the observed QPO frequencies in SGR 19001 + 14 and SGR 1806-20 (18, 26, 30, 92, and 150 Hz). However, it was noted in Ref. [19] that the proposed mechanism, nevertheless, does not explain the high-frequency oscillations, with frequencies of 625 and 1840 Hz.

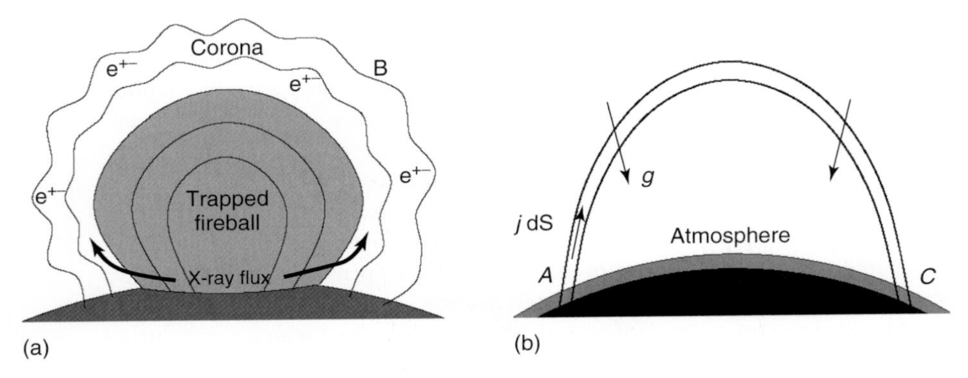

Figure 10.2 (a) Schematic views of trapped fireball [15] and (b) a current-carrying coronal loop frozen into the neutron star surface. The electric current is initiated by starquakes, which lead to a twisting of one or two loop footpoints (A and/or C). The current flows along the loop and is closed in the metal crust of the magnetar [11].

10.1.2 An Equivalent Electric RLC Circuit

The model for the high-Q high-frequency oscillations in magnetars proposed in Ref. [20] is based on the ideas of coronal seismology, which studies oscillations and waves in stellar coronae. Coronal seismology has turned out to be very efficient in diagnosing the coronal parameters and flare plasma not only on the Sun but also on latetype stars [13, 21]. At present, two approaches are developed in coronal seismology. The first approach, which is presented in Chapter 3, investigates MHD oscillations and waves in the fundamental structures of stellar coronae–coronal magnetic loops. Flare energy release takes place in such structures. The second approach is based on the idea of Alfvén and Carlqvist [12] about a flare loop as an equivalent electric circuit. The basic principles of this approach can be found in Chapter 2.

The corona of a magnetar and the trapped fireball with which the ringing tail is associated can be represented as a set of current-carrying magnetic loops (Figure 10.2) with various sizes whose eigenfrequencies and Q-factors are given by the well-known relations:

$$\nu = \left(2\pi\sqrt{LC}\right)^{-1}, \quad Q = \frac{1}{R}\sqrt{\frac{L}{C}} \tag{10.1}$$

where R and C are the resistance and capacitance of the coronal loop and L is the inductance determined by the loop geometry (see Eq. (2.50)):

$$L = 2l\left(\ln\frac{4l}{\pi a} - \frac{7}{4}\right) \tag{10.2}$$

Given the energy $E = LI^2/2$ released in the flare tail, we can determine the electric current I in the "trapped fireball" coronal loops and, hence, the coronal plasma density and the φ-component of the magnetic field. The observed power of the energy release $W = RI^2$ allows us to find the resistance R of the current-carrying coronal loop and to estimate its capacitance $C = \varepsilon_A S/l$, that is, also the Q-factor from the oscillation frequency. Let us illustrate the efficiency of the proposed model using the most powerful known SGR event with developed QPOs as an example.

10.1.3 The Flare of SGR 1806-20 on 27 December 2004

The energy released during the flare was of the order of 5×10^{46}erg and the stored magnetic energy $\sim 10^{47}$ erg [9]. The energy of the ringing tail of the flare was $\sim 10^{44}$ erg [7]. Since several frequency oscillations were observed in the flare tail, we will assume that the set of current-carrying coronal loops are responsible for the oscillations. For an "average" loop, we will assume that the energy stored in it to be $\approx 2 \times 10^{43}$ erg. Taking the loop length and radius to be $l = 3 \times 10^6$ cm and $a = 3 \times 10^5$ cm, respectively, we find its inductance $L \approx 5 \times 10^6$ cm $= 5 \times 10^{-3}$ H from Eq. (10.2). Assuming that a considerable fraction of the energy stored in the "average" loop was released ($E \approx 2 \times 10^{43}$ erg $= 2 \times 10^{36}$ J), we determine the current $I = (2E/L)^{1/2} \approx 2.8 \times 10^{19}$A. Further we estimate the φ component of the magnetic field in the loop from the electric current to be $B_\varphi \approx I/ca \approx 10^{13}$ G. The number density of electron–positron pairs n in the source of the flare tail can be determined from the electric current $I = encS$ and the cross section of the coronal loop S with the radius $a = 3 \times 10^5$ cm. For $I = 2.8 \times 10^{19}$A, $n = 2 \times 10^{16}$ cm^{-3}, that is, the Langmuir frequency $\omega_p = 8 \times 10^{12}$ s^{-1} ($\nu_p = \omega_p/2\pi = 1.3 \times 10^{12}$ Hz).

The power of the energy release in the flare tail with the duration $\approx$200 s is $\approx 10^{41}$ erg s^{-1} [7], that is, $W = RI^2 \approx 2 \times 10^{40}$ erg s^{-1} $= 2 \times 10^{33}$ W for the average loop. The resistance of a loop in the ringing tail of the flare from SGR 1806-20 is $R = W/I^2 \approx 2 \times 10^{-6}$ Ω. The origin of such a resistance of the coronal loop can be associated with an anomalous conductivity that arises in the course of excitation of small-scale plasma waves. The effective (turbulent) resistance can be represented as

$$R = R_{\text{eff}} = \frac{l}{\sigma_{\text{eff}} S} = \frac{l}{S} \frac{4\pi \nu_{\text{eff}}}{\omega_p^2} \tag{10.3}$$

where the anomalous (turbulent) conductivity is $\sigma_{\text{eff}} = e^2 n / m\nu_{\text{eff}}$, $\nu_{\text{eff}} = (W_p/nk_B T)\omega_p \approx 10^{-1}\,\omega_p$. Thus, the level of small-scale plasma turbulence $W_p/nk_B T$ in the QPO source must be fairly high. One of the possible causes of such a level of plasma waves can be the instability of fairly dense ($n_b/n \sim 0.1$) beams of high-energy electrons, accelerated in electric fields of the magnetar corona [11]. Here, n_b is the density of the beam of accelerated electrons. The second possibility for the appearance of an anomalous resistance in the magnetar corona is associated with the instability of ion-sound waves [11] that arises when the relative velocity of ions and electrons exceeds the speed of sound, $v_d > c_s = (5k_B T/3m_i)^{1/2}$. Note that, according to the present views, apart from electron–positron pairs the magnetar corona contains about 10% ions. At the temperature of the cooling trapped fireball $T = 3 \times 10^9$ K, $c_s \approx 6 \times 10^8$ cm s^{-1} $\ll c$.

The minimum ($\nu_1 = 18$ Hz) and maximum ($\nu_2 = 2384$ Hz) frequencies of the ringing tail of SGR 1806-20 allow the capacitance of the current-carrying magnetic loops, equivalent RLC circuits, to be estimated from Eq. (10.1) for the frequency. As a result, we obtain $C_1 \approx 1.5 \times 10^{-2}$F and $C_2 \approx 8 \times 10^{-7}$F. On the other hand, the capacitance of a coronal loop can be approximately represented as $C \approx \varepsilon_A S/l$, where $\varepsilon_A = c^2/c_A^2$ is the dielectric permittivity of the medium for Alfvén waves [13]. It follows from the dispersion relation for Alfvén waves

$$\frac{\omega}{k} = c\left(l + \frac{4\pi\rho c^2}{B^2}\right)^{-1/2} \tag{10.4}$$

that the Alfvén velocity in the magnetar corona is approximately equal to c. Therefore, $\varepsilon_A \approx 1$ and for adopted cross-sectional area of the loop $S = \pi a^2 \approx 3 \times 10^{11}$ cm^2 and length $l = 3 \times 10^6$ cm, we obtain $C \approx 10^5$ cm $= 10^{-7}$F, which is several times lower than the value of C_2 calculated from Eq. (10.1). Note that the sizes of coronal loops in the trapped fireball can differ by several orders of magnitude. It is easy to see that with increasing S as l decreases ("thick" loop), the coincidence between of the capacitance with C_2 and C_1 can be achieved. Applying the second relation from Eq. (10.1), we find the Q-factors for the minimum and maximum frequencies, $Q_1 \approx 3 \times 10^5$ and $Q_2 \approx 10^7$, which exceed the observed Q-factors of QPOs by 1 or 2 orders of magnitude. This discrepancy can be attributed to both insufficient sensitivity of the recording equipment or to "cooling" of the trapped fireball.

10.1.4
Excitation of High-Frequency Current Oscillations of the Current in Coronal Loop

For small deviations of the electric current $|\tilde{I}| \ll I$, the equation describing the electric current oscillations in a loop can be represented as [22]

$$L\frac{\partial^2 \tilde{I}}{\partial t^2} + \alpha\left(I^2 - I_{max}^2\right)\frac{\partial \tilde{I}}{\partial t} + \frac{\tilde{I}}{C} = 0 \tag{10.5}$$

In Eq. (10.5), we take into account the fact that, since the anomalous resistance $R_{eff} \sim \nu_{eff}$ is proportional to the power of energy release, $W \sim I^2$, the resistance can be represented from dimensional relations as $R_{eff} \sim \alpha I^2$, where α is some coefficient. We see from Eq. (10.5) that the oscillations will be excited at currents lower than the maximum one in the giant pulse of the flare, $I < I_{max}$, that is, not only at the descending stage of the flare but also at its ascending stage. Recall that oscillations with a frequency 43 Hz in SGR 0526-66 [5] and 50 Hz in SGR 1806-20 [8, 9] were also observed at the ascending stage of the impulse phase.

Note another possible mechanism for the generation of oscillations in coronal magnetic loops – the excitation through a parametric resonance (see Chapter 3 and [23]). The electric current oscillations owing to magnetar crust disturbances with a pumping frequency ν through a parametric interaction with a coronal loop can trigger oscillations in the loop both at the frequency ν, at the subharmonic $\nu/2$, and at the first upper frequency of the parametric resonance $3\nu/2$. A similar phenomenon is observed in the optical and microwave emission from solar flares [23]. The variations in coronal loop parameters can be described by the following equation:

$$\frac{d^2y}{dt^2} + \nu_0^2(1 + q\cos \nu t)y = 0 \tag{10.6}$$

Here, ν_0 is the frequency of the eigenoscillations of the coronal loop. The parameter q defines the width of the zone near the parametric resonance frequencies $\nu_n = n\nu/2, n = 1, 2, 3, \ldots$ [24], namely $(-q\nu_0/2 < \nu/2 - \nu_0 < q\nu_0/2)$. The excitation takes place when the frequency of loop eigenmodes ν_0 falls into the first instability zone, that is, it is close to $\nu/2$. This means that for a coronal loop to be excited parametrically, it must have suitable sizes, density, temperature, and magnetic field. It may well be that the QPOs in the 27 August 1998 flare on SGR 1901 + 14 27 were excited through a parametric resonance: $\nu = 53$ Hz, $\nu/2 = 26.5$ Hz (a frequency of 28 Hz was observed), and $3\nu/2 = 79.5$ Hz (a frequency of 84 Hz was observed). Note that we obtain the observed frequencies ($\nu/2 = 28$ and $3\nu/2 = 84$ Hz) with a high accuracy for ν equal to 56 Hz rather than 53 Hz. Inductive interaction of current-carrying coronal loops can also increase the number of observed QPO frequencies [22].

Thus, RLC model can explain the entire set of observed oscillation frequencies, from 20 to 2400 Hz, the excitation of the oscillation both in the flare tail and at the beginning of the main pulse, and, what is especially important, the high Q-factor of the oscillations $Q > 10^4$. The high-frequency current variations in coronal loops cause periodic magnetic field variations, leading to a modulation of the magnetar emission. Although the proposed in Ref. [20] approach – the representation of coronal loops as equivalent RLC circuits – is largely phenomenological, it, nevertheless, allows the parameters of the neutron star magnetosphere to be estimated independently.

Currently, there exist various methods for determining the magnetic field of magnetars. The field strength found from the spindown of magnetars is $B \approx 10^{14}$–10^{15} G. The field strength obtained from the proton cyclotron absorption is $B \approx 10^{15}$ G. From the model of QPOs as torsion Alfvén oscillations, the magnetic field of SGR 1806-20 within the interval $(3–7) \times 10^{15}$ G was determined [25]. Having studied the luminosity variations of the oscillations with frequencies of 625 and 1840 Hz in the ringing tail of SGR 1806-20, Vietri *et al.* [7] concluded that the magnetic field strength in the QPO source (in the magnetar corona) is $B \approx 6.6 \times 10^{13}$ G and the dipole approximation on the stellar surface yields $B \approx 2 \times 10^{15}$ G. The estimation of the minimum magnetic field strength in the QPO source in the flare of SGR 1806-20 on 27 December 2004, performed in Ref. [20] obtained from the electric current $I = (2E/L)^{1/2} \approx 3 \times 10^{19}$A gives $B_{\min} \approx 10^{13}$ G. Note that obtained magnetic field strength in flare tail is less as compared with the reference field

$$B_{\rm QED} = \frac{m_e^2 c^3}{e\hbar} = 4.4 \times 10^{13}\ \mathrm{G} \tag{10.7}$$

at which the nonrelativistic Landau energy $\hbar eB/m_e c$ becomes comparable to the electron rest energy $m_e c^2$. Hence, the nonquantum approach can be used for the description of flare tail plasma. On the basis of the RLC model, the current value makes it possible to estimate also the electron density in the magnetosphere of SGR 1806-20: $n \approx 2 \times 10^{16}$ cm^{-3}.

One should emphasize that when determining the electric current and the φ-component of the magnetic field in the ringing tail the energy relation $E = LI^2/2$ was used. In this case, the energy of the ringing tail of the flare ($\sim 10^{44}$ erg) is lower than the total flare energy by more than two orders of magnitude. If the energy release in a single magnetic loop is assumed to be responsible for the main pulse of the flare from SGR 1806-20 (5×10^{46} erg), then the current $I \approx 10^{21} A$ and the φ-component of the magnetic field reaches $B_\varphi \approx 4 \times 10^{14}$ G, consistent with the present-day models of magnetars.

10.2
Coronae of Accretion Disks

Solar paradigm is applied quite often for the description of physical processes in the coronae of accretion disks. There are various kinds of accretion disks: around magnetized neutron stars and white dwarfs, around stellar mass black holes, and in active galactic nuclei (AGN) [2, 26]. Protostellar disks and disks around Young stellar object are also the subjects of interest of astrophysicists.

Geometrically thin disk condition follows from vertical hydrostatic equilibrium $h/r \approx c_s/v_K \ll 1$, where $h(r)$ is the disk half-thickness at radius r, $c_s(r)$ is the sound velocity, and $v_K(r) = \Omega_K(r) r = (GM/r)^{1/2}$ is the Keplerian speed. Typical value for thin disk is $h/r = 10^{-1}$ to 10^{-3}. An illustration of thin accretion disk *around a magnetic white dwarf* is shown in Figure 10.3.

Density outside the accretion disk *around stellar mass black holes* decreases very steeply with distance $\sim\exp\left(-z^2/\Delta^2\right)$ for scale height Δ [27]. A disk corona extending over a substantial altitude, therefore, cannot be maintained by gas pressure. Like in pulsar magnetosphere, coronal gas can be produced and sustained by electric field if the magnetic fields are rooted in the accretion disk. Rotating magnetic fields produce electric field with a component parallel to the magnetic field. For the magnetic loop with footpoints separation D, the potential drop is $\Delta\Psi \approx \Omega_K(r) r D B(r)$ and can be enough to extract electrons from one footpoint and positive ions from other footpoint. Existence of high-energy electrons follows from the electron–positron annihilation lines ($\sim$0.5 MeV) observed from black hole candidates.

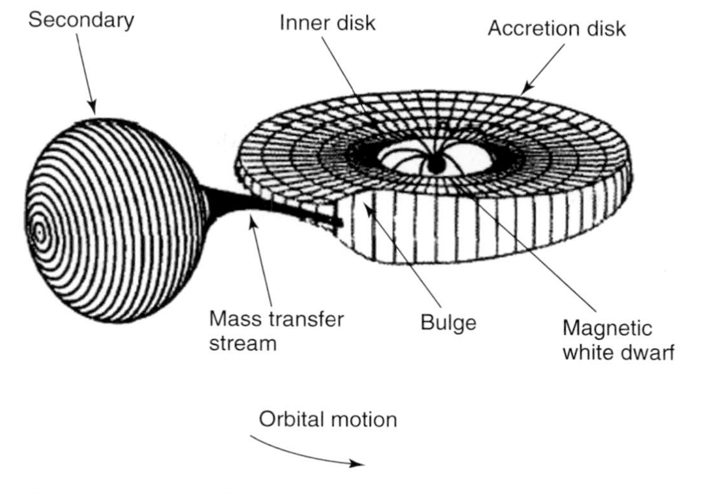

Figure 10.3 Sketch of thin accretion disk around a magnetic white dwarf in binary system [26].

Geometrically thin *accretion disk in AGN* orbiting a massive black hole ($M \approx 10^5 - 10^9$ M_{Sun}) and surrounded by a geometrically thick molecular dust torus. Disk emits X-ray, γ-ray (up to 1 GeV), and nonthermal radio emission owing to inverse Compton and synchrotron radiation.

10.2.1 Accretion Disk Corona of Cyg X-1

Cygnus X-1, the binary with $M \approx 10$ M_{Sun} is the best-studied black hole candidate. The radiation zone with $r \geq 10r_0$ is located in the inner part of disk where radiation pressure is large as compared to the gas pressure. Here, $r_0 = 2GM/c^2$ is gravitational radius. Corona with $T = 10^7 - 10^8$ K is located above the radiation zone [3]. Observations have shown that coronae of accretion disks are the sources of intense broadband emission, from radio to X-ray and γ-ray emission. For example, the energy spectrum of X-ray source Cyg X-1 reveals the presence of two components. The hard X-ray (>10 keV) component is always present; however, the soft X-ray (<10 keV) component has been detected only during the so-called high state when the 2–10 keV flux is relative strong. Linden-Bell [28] and Galeev *et al.* [29] were among the first to suggest that flares operate in and near accretion disks.

Present understanding of solar flares suggests that flare phenomena are common place in coronae of accretion disks. Galeev *et al.* [29] proposed a model for the fluctuating hard component of Cyg X-1 based on the amplification of magnetic field. Magnetic field $\sim 10^8$ G can be generated within a hot, $T \geq 10^6$ K and dense, $n \approx 10^{22}$ cm^{-3}, accretion disk [30] by the joint action of plasma thermal convection and differential rotation along Keplerian orbits. Magnetic flux tubes with such

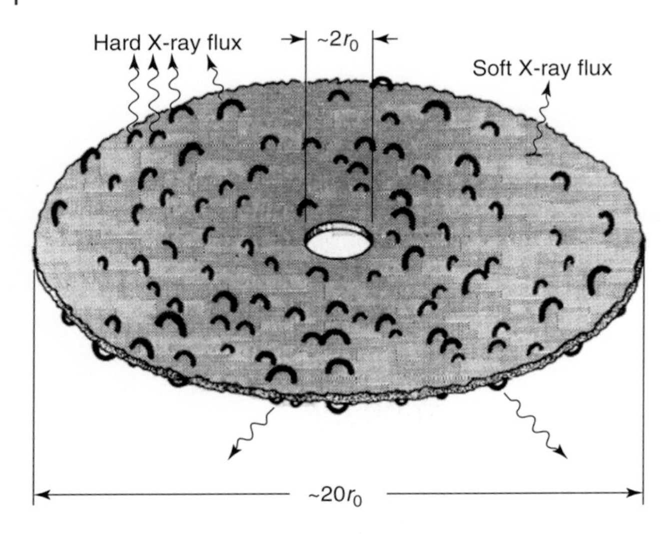

Figure 10.4 Schematic drawing of the inner accretion disk corona with $r_0 \approx 10^7$ cm in the case of Cyg X-1. The soft X-ray component derives from the relatively cool disk, while the hard X-ray emission originates from the ensemble of hot ($T \geq 5 \times 10^8$ K) plasma loops. The length of typical loop is of the order $\sim 10^6$ cm [29].

strong field contain less plasma than their ambient surroundings to mountain pressure balance; therefore, they are subject to buoyancy forces and will penetrate the accretion disk to form coronal loops with a very hot, $T \geq 10^8$ K, relatively low-density plasma as compared to the disk itself, analogous to the structure of solar corona (Figure 10.4). Coronal loops stored the energy and can provide an energy source for coronal plasma. Accretion disk coronae are heated by numerous reconnection events (flares) involving coronal magnetic loops. Therewith magnetic flaring activity is possible in a corona on both sides of the disk.

Modern models suggest, however, nonthermal corona because observations of Cyg X-1 have shown that the hard X-ray emission extending up to a few megaelectronvolts. Moreover, the protons are much hotter than electrons ($T_i \gg T_e$). For example, the model of spherical corona of Cygnus X-1 [31] suggests black hole mass $M = 10\ M_{Sun}$, corona radius $R_c = 5 \times 10^7$ cm, $n_i = n_e = 6 \times 10^{13}$ cm^{-3}, $T_i \approx 10^{12}$ K, $T_e \approx 10^9$ K, and $B = 6 \times 10^5$ G, with the disk characteristic temperature $k_B T = 0.1$keV. Two-temperature plasma of disk corona exists because electrons are cooled almost immediately by synchrotron losses in the magnetic fields of $\geq 10^5$ G.

There are proposals that acceleration of particles to the high-energy can be the loop–loop coalescence via reconnection process and/or by shock waves [31]. Magnetic pumping is also considered as the origin of coronal plasma heating and particle acceleration [32]. It is not excluded that disk coronae can be heated by current dissipation in numerous current-carrying loops.

10.2.2
QPOs in Accretion Disks

Our interest is dealing with the origin of quasi-periodic oscillations (QPOs) from accretion disks. Waves and oscillations in accretion disks are the subjects of *diskoseismology*. QPOs in the X-ray light curves of accreting neutron stars, black hole, and white dwarf binaries with frequencies ranging from a few 0.1 to 1300 Hz interpreted usually as being due to standing wave modes in accretion disks. Advanced theory of disk QPOs was done, for example, in Refs. [33, 34]. The origin of turbulence in the accretion disk is not trivial because Keplerian accretion disk is stable against small disturbances. The reason for turbulence excitation can be the magnetorotational instability (MRI) [35] that transforms the rotation energy into the energy of chaotic magnetic field and the energy of chaotic movement of the matter. This means the transition to the turbulent state. In particular, axisymmetric acoustic modes can be responsible for millisecond QPOs, which have been observed with RXTE in both neutron star and black hole binaries. Therewith, internal modes, called *g-modes* in the *diskoseismology*, can be destroyed by the turbulence [33].

In order to explain high-frequency QPOs in accretion disks, three kinds of resonances were considered by Petri [34]: a corotation resonance when the angular velocity of the disk equals the rotation speed of the star, a Linblad resonance due to a driven force, and a parametric resonance due to the time-varying epicyclic frequencies. In the last case, the rotation of the star induced a sinusoidal variation of the epicyclic frequency ν_0 leading to the Mathieu's equation (Eq. (10.6)). Here, y is the perturbation amplitude. The system becomes unstable if the excitation frequency $\nu = 2\nu_0/n$, where n is integer (see also Eqs. (9.8) and (9.11)).

In this context, it would be interesting to study a parametric resonance between disk turbulence driven by MRI and eigenmodes of coronal magnetic loops because exactly the disk corona consisting of set of magnetic loops (Figure 10.4) is responsible for X-ray emission. Our present understanding of solar flares is that flares result from distortion of coronal magnetic loops by subphotospheric flows. Similar explosions are expected in magnetic loops in accretion disk coronae when they become distorted by powerful fluid motion. Thus, coronal seismology can be quite promising tool for the accretion disk physics.

References

1. Shapiro, S.L. and Teukolsky, S.A. (1983) *Black Holes, White Dwarfs, and Neutron Stars: The Physics of Compact Objects*, Whiley-Interscience, New York.
2. Camenzind, M. (2007) *Compact Objects in Astrophysics: White Dwarfs, Neutron Stars and Black Holes*, Springer-Verlag, Berlin, Heidelberg.
3. Bisnovatyi-Kogan, G.S. (2010) *Stellar Physics, 2: Stellar Evolution and Stability*, 2nd edn, Springer-Verlag, Berlin, Heidelberg.
4. Thompson, C. and Duncan, R.C. (1995) *Mon. Not. R. Astron. Soc.*, **275**, 255.
5. Barat, S., Hayles, R.I., Hurley, K. *et al.* (1983) *Astron. Astrophys.*, **126**, 400.

6. Israel, G.L., Belloni, T., Stella, L. *et al.* (2005) *Astrophys. J.*, **628**, L53.
7. Vietri, M., Stella, L., and Israel, G.L. (2007) *Astrophys. J.*, **661**, 1089.
8. Palmer, D.M., Barthemly, S., Gehrels, M. *et al.* (2005) *Nature*, **434**, 1107.
9. Terasawa, T., Tanaka, T., Takei, Y. *et al.* (2005) *Nature*, **434**, 1110.
10. Strohmayer, T.E. and Watts, A.L. (2006) *Astrophys. J.*, **653**, 593.
11. Beloborodov, A.M. and Thompson, C. (2007) *Astrophys. J.*, **657**, 967.
12. Alfvén, H. and Carlqvist, P. (1967) *Sol. Phys.*, **1**, 220.
13. Zaitsev, V.V. and Stepanov, A.V. (2008) *Phys. Usp.*, **51**, 1123.
14. Ruderman, M. (1991) *Astrophys. J.*, **382**, 587.
15. Thompson, C. and Duncan, R.C. (2001) *Astrophys. J.*, **561**, 980.
16. Watts, A.L. and Strohmayer, T.E. (2007) *Adv. Space Res.*, **40**, 1446.
17. Levin, Y. (2007) *Mon. Not. R. Astron. Soc.*, **377**, 159.
18. Andersson, N., Glampedakis, K., and Samuelsson, L. (2009) *Mon. Not. R. Astron. Soc.*, **396**, 894.
19. Ma, B., Li, X.-D., and Chen, P.F. (2008) *Astrophys. J.*, **686**, 492.
20. Stepanov, A.V., Zaitsev, V.V., and Valtaoja, E. (2011) *Astron. Lett.*, **37**, 276.
21. Nakariakov, V.M. and Stepanov, A.V. (2007) *Lect. Notes Phys.*, **725**, 221.
22. Khodachenko, M.L., Zaitsev, V.V., Kislyakov, A.G., and Stepanov, A.V. (2009) *Space Sci. Rev.*, **149**, 83.
23. Zaitsev, V.V. and Kislyakov, A.G. (2006) *Astron. Rep.*, **83**, 921.
24. Landau, L.D. and Lifshitz, E.M. (1976) *Theoretical Physics*, Mechanics, Vol. **I**, 3rd edn. Pergamon Press, Oxford.
25. Sotani, H., Kokkotas, K.D., and Stergioulas, N. (2008) *Mon. Not. R. Astron. Soc.*, **385**, L5.
26. Kuijpers, J. (1995) *Lect. Notes Phys.*, **444**, 135.
27. Van Oss, R.F. (1994) *Space Sci. Rev.*, **68**, 309.
28. Linden-Bell, D. (1969) *Nature*, **223**, 690.
29. Galeev, A.A., Rosner, R., and Vaiana, G.S. (1979) *Astrophys. J.* **229**, 318.
30. Bisnovatyi-Kogan, G.S. and Blinnikov, S.I. (1977) *Astron. Astrophys.*, **59**, 111.
31. Romero, G.E., Vieyro, F.L., and Vila, G.S. (2010) *Astron. Astrophys.*, **519**, 109.
32. Belmont, R. and Tagger, M. (2005) *Chin. J. Astron. Astrophys.*, **5**, 43.
33. Arras, P., Blaes, O., and Turner, N.J. (2006) *Astrophys. J.*, **645**, L65.
34. Petri, J. (2006) *Astrophys. Space Sci.*, **302**, 117.
35. Velikhov, E.P. (1959) *Sov. Phys. JETP*, **36**, 995.

11
Conclusions

We have considered quasi-periodic pulsations and oscillations and related physical processes in fundamental magnetic structures of solar and stellar coronae – flux tubes and loops. Different representations have been invoked: an equivalent electric circuit, a resonator and wave guide for magnetohydrodynamic (MHD) oscillations and waves, and a trap with magnetic mirrors. With the aim of plasma parameter diagnostics in stellar coronae we have to study the modulation mechanisms of radio, X-ray, and optical emission driven by MHD waves and electric current oscillations. Therewith gyrosynchrotron, synchrotron, thermal bremsstrahlung, and plasma radio radiation mechanisms are important in this context.

The theory and the observational examples discussed in the book demonstrate the applicability of the coronal seismology as a quite effective diagnostic tool for stellar flaring plasma. Therewith solar paradigm is very useful in this context because it permits us to describe quasi-periodic oscillations (QPOs) in a variety of stellar objects, from red dwarfs to neutron stars and accretion disks, using solar terminology.

It should be noted that the authors' own interest in the coronal seismology problems is presented in the book. Nevertheless, we have discussed other ideas and approaches too.

Several important issues and topics are left out of the book, for example, the wave–flow interaction and its possible development in explosive and other negative energy wave instabilities; the possible role of fine, sub-resolution structuring of coronal loops, and arcades (there is a solid theoretical and observational indications of its presence); the effects of field-aligned nonuniformity of the equilibrium parameters of coronal loops, as well as their time variation, on properties of MHD oscillations; interaction of current-carrying loops that belong to different components of binary stellar systems (the possibility of this interaction that leads to particle acceleration and cause intensive radio emission has been discussed in Massi *et al.* [1]).

Coronal Seismology: Waves and Oscillations in Stellar Coronae, First Edition.
A. V. Stepanov, V. V. Zaitsev, and V. M. Nakariakov.

In addition, the approaches and models discussed in the book can also be useful in studying processes of energy release in a wider class of other astrophysical objects, such as close binary systems, neutron stars, and planetary systems similar to Jupiter-Io.

References

1. Massi, M., Ros, E., Menten, K.M. *et al.* (2008) *Astron. Astrophys.*, **480**, 489.

Index

Coronal Seismology: Waves and Oscillations in Stellar Coronae, First Edition.
A. V. Stepanov, V. V. Zaitsev, and V. M. Nakariakov.